# International Series in Operations Research & Management Science

Volume 325

More information about this series at https://link.springer.com/bookseries/6161

Wieslaw Kubiak

# A Book of Open Shop Scheduling

## Algorithms, Complexity and Applications

Springer

Wieslaw Kubiak
Memorial University of Newfoundland
St. John's, NL, Canada

ISSN 0884-8289 ISSN 2214-7934 (electronic)
International Series in Operations Research & Management Science
ISBN 978-3-030-91024-2 ISBN 978-3-030-91025-9 (eBook)
https://doi.org/10.1007/978-3-030-91025-9



This Springer imprint is published by the registered company Springer Nature Switzerland AG
The registered company address is: Gewerbestrasse 11, 6330 Cham, Switzerland

*To Inka, Ashley, and Michał*

# Preface

Open shop is often defined in the literature by what it is not—flow shop. This definition may have originated from the initial view of open shop as a manufacturing system, where operations of a job are done in any order, without preemptions on dedicated machines. The definition fits service operations as well. A popular example of open shop is that of a library and its readers. Each reader is a job, and each book is a machine. A reader wishes to read a collection of books, and reading a book is an operation that may take different time for different readers, in any order. A book can only be read by one reader at a time, and a reader can read at most one book at a time. Another example is class–teacher timetabling. Though non-preemptive open shop scheduling is a prominent part of open shop scheduling, it is preemptive open shop scheduling that makes modern open shop scheduling a universal approach to solving a broad range of real-life problems. Besides, the preemptive open shop scheduling is strongly linked to graph edge coloring and to doubly stochastic matrices, which makes it particularly attractive to study in scheduling theory.

Excellent chapters on open shop scheduling can be found in general books on scheduling by Pinedo [10], Błażewicz et al. [3], Tanaev et al. [13], and Brucker [4]. Naturally, those are limited in scope. In-depth reviews are given in chapters by Gonzalez [6], Prins [11], and Woeginger [14]. Recent literature reviews of open shop scheduling have been published by Anand and Panneerselvam [2] and Ahmadian et al. [1]. However, to the author's knowledge, there has not been any monograph focusing solely on open shop scheduling published thus far. This monograph is to fill in the void, to present a growing portfolio of applications of open shop scheduling, and to emphasize a great potential and need for novel theoretical results in the open shop scheduling field.

The intended audience of the book includes, but is not limited to, sophisticated practitioners, graduate students, and researchers in operations research, management science, computer science, and discrete mathematics.

Open shop scheduling has branched out into a growing number of models by adding new features, constraints, or objective functions over the years. Therefore, a selection of topics had to be made for this book intended for fewer than 300 pages.

Consequently, a balance had to be found between theory and applications, between traditional models captured by the Graham et al. [8] notation and new models that outgrew by far that notation, between graph theoretic approach to preemptive open shop scheduling and non-preemptive open shop scheduling, between existing results and open problems, and between the results that have been well-presented in the earlier books on scheduling and those that have been somewhat under-represented or more recent. As a result, the selection excludes some important open shop scheduling areas such as stochastic models, heuristics and meta-heuristics, models with machine availability, competing agents, transport delays, batch processing, and rejection, to name just few; see Ahmadian et al. [1] for a comprehensive list.

The plan of the monograph is as follows. Chapter 1 defines notation and terminology used in this book. It also reviews the key results on edge coloring of bipartite multigraphs (König's edge coloring theorem), on fractional chromatic index (Edmonds' theorem), on doubly stochastic matrices being a convex combination of permutation matrices (Birkhoff–von Neumann theorem), and the vector rearrangement theorem. Those results have proved to be the most insightful and influential for open shop scheduling. They also show an exciting diversity of perspectives that researchers have been taking at studying open shops scheduling.

Chapter 2 deals with makespan minimization for two-machine open shop problem. The problem is one of the first open shop scheduling problems studied in the literature. Gonzalez and Sahni [7] and Pinedo and Schrage [9] give linear-time algorithms for the problem. Both algorithms have been well-presented and analyzed in books on scheduling by Pinedo [10], Błażewicz et al. [3], Tanaev et al. [13], and Brucker [4] and reviews by Gonzalez [6] and Woeginger [14]. It is quite remarkable that the two-machine open shop problem with $n$ jobs can be reduced to a two-machine open shop with *three* jobs only, which is shown by de Werra [15]. The kind of reduction that results in instances of smaller sizes than the original instance is rare in scheduling theory and combinatorial optimization. The algorithm based on the reduction is presented in the chapter. Another quite remarkable observation is that preemptions do not make optimal schedules shorter for the two-machine open shop problem. The chapter shows that this highly desirable feature of optimal schedules holds for a more general two-machine open shop scheduling problem, where each operation requires a machine *and* a certain number of units, for instance, personnel, of a scarce additional resource to be processed. We refer to this open shop as two-machine open shop with additional renewable resources. The chapter presents a polynomial-time algorithm for makespan minimization for this open shop. The two-machine open shop with additional renewable resources seems to be the most general two-machine open shop currently known for which the makespan minimization can be done in polynomial time.

Chapter 3 considers general open shop scheduling with arbitrary number of machines $m$. The makespan minimization then leads to different complexity classes depending on whether preemptions are allowed or not. The chapter shows that the non-preemptive case with integer processing times is NP-hard in the strong sense even for optimal schedules not longer than 4. This consequently rules out the existence of Polynomial-Time Approximation Scheme (PTAS) for the problem unless $P = NP$. However 2-approximation solutions are provided by dense schedules for the non-preemptive open shop. The chapter also shows special cases of the non-preemptive open shop that can be solved in polynomial time by the application of the Vector Rearrangement Theorem, or by imposing additional constraints on short schedules. The makespan minimization with preemptions can be achieved in polynomial time by the algorithm of Gonzalez and Sahni [7]. The problem is one of the most versatile scheduling problems in scheduling theory and its applications. The chapter gives polynomial-time algorithm for the problem, which is based on Birkhoff–von Neumann theorem. The preemptions permit the problem to remain polynomial even if jobs are released at different release dates, and the objective becomes maximum lateness. The polynomial-time algorithm for the problem requires solving a linear program though. Open shop scheduling problems with other objective functions remain NP-hard even for two machines and remain so regardless of whether preemptions are allowed or not. This intractability changes for open shops with *all* operations being unit-time. Those open shop problems resemble problems with parallel identical processors. Therefore, polynomial-time algorithms for scheduling of the latter, if any, can be turned into polynomial-time algorithms for open shop scheduling with unit-time operations. The chapter shows how this can be done for a number of open shop scheduling problems. This resemblance no longer holds for open shops with 0–1 operations, i.e. where some operations are missing (their processing times equal 0). The chapter shows that total completion time then becomes NP-hard in the strong sense, though the makespan minimization is still polynomial.

Chapter 4 considers open shop scheduling with multiprocessor operations. A multiprocessor operation requires simultaneously all machines from a group of machines for execution. The multiprocessor operations are motivated by the University timetabling model. The model is a generalization of the well-known class–teacher timetabling model, where in addition to lectures given by a single teacher to a single class, there are some lectures given by a single teacher to a group of classes simultaneously. One looks for a minimum number of periods in which to complete all lectures without conflicts. The University timetabling problem is NP-hard in the strong sense even if the number of groups is three, but it is polynomially solvable for two groups. For two groups, the chapter proves that the minimum number of periods in which to complete all lectures without conflicts equals $\lceil T \rceil$, where $T$ is the optimal value of an $LP$-relaxation of the University timetabling problem. The $LP$-relaxation permits fractions of periods in feasible solutions, and thus it minimizes makespan for the preemptive open shop with multiprocessor operations. The proof also gives a polynomial-time algorithm for the University timetabling problem with two groups. The chapter then gives polynomial-time algorithms for makespan minimization in preemptive open shops with multiprocessor operations having any fixed, possibly higher than two, number of groups. The solutions can be rounded to provide approximate solutions guaranteeing fixed absolute errors for the University timetabling problem. The complexity of preemptive open shop scheduling with multiprocessor operations remains open for arbitrary number of groups.

Chapter 5 deals with concurrent open shop scheduling. Concurrent open shops permit processing more than one operation of the same job at a time—a key departure from traditional open shop that excludes such concurrent processing. Applications of concurrent open shop scheduling are broad and range from order scheduling, where different components of an order can be produced concurrently on dedicated machines to be delivered as a complete order, to airplane maintenance. Concurrent open shop scheduling naturally concentrates on total completion time- and due date-based objective functions. The chapter provides an in-depth review of the complexity of optimization of those objective functions that are notoriously difficult to optimize for the concurrent open shop. The chapter shows that permutation schedules include optimal schedules for regular objective functions and that the EDD (earliest due date) permutation minimizes maximum lateness. The permutation schedule has the same permutation of jobs on each machine. The chapter gives 2-approximation algorithm for total weighted completion time and shows that total completion time is $NP$-hard to approximate within factor $2 - \epsilon$ for any $\epsilon > 0$ assuming that the Unique Games Conjecture holds. It shows that both number of tardy jobs and total tardiness are $NP$-hard to approximate within factor $(1 - c) \ln m$ for any $c > 0$. Those results are based on recent inapproximability results for set cover and independent set problems in $r$-uniform hypergraphs. The chapter also reviews the existing literature on exact algorithms, and heuristics for concurrent open shop scheduling.

Chapter 6 considers open shop scheduling with simultaneity constraints. The constraints require that some operations be processed simultaneously at any time. We show that the constraints are capable of modeling edge coloring in arbitrary graphs, which in turn models many real-life problems. Motivated by applications in scheduling wireless networks, we consider the problem of covering the edges of a graph by a sequence of matchings subject to the constraint that each edge $e$ appears in at least a given fraction $r(e)$ of the matchings. It can be determined in polynomial time whether such a sequence of matchings exists or not; however, makespan minimization of the sequence is computationally intractable in general. We restrict our investigation to a special class of graphs, the so-called OLoP (*Overall Local Pooling*) graphs shown important in an online distributed wireless network scheduling setting. We also generalize the results to a larger class of graphs called GOLoP graphs. In particular, we show that deciding whether a given GOLoP graph has a matching sequence of length at most $k$ can be done in linear time. If the answer is affirmative, we show how to construct, in quadratic time, the matching sequence of length at most $k$. Finally, we prove that, for GOLoP graphs, the length of a shortest sequence does not exceed a constant times the least common denominator of the fractions $r(e)$. This leads to a pseudopolynomial-time algorithm for minimizing the length of the sequence. We show that the constant equals 1 for OLoP graphs and, following Seymour [12], conjecture that the constant is as small as 2 for general graphs. This conjecture holds for all graphs with at most 10 vertices.

Chapter 7 considers proportionate open shop scheduling. Two models of proportionate open shops have emerged in the literature: job-proportionate open shop, where all operations of each job have the same processing time; and machine-proportionate open shops, where all operations on each machine have the same processing time. The two models are equivalent for makespan minimization but not for other objective functions. Somewhat surprisingly, the makespan minimization remains NP-hard in the ordinary sense for three-machine job-proportionate open shops. A pseudopolynomial-time algorithm has been recently given for the three-machine case. The chapter reviews approximation algorithms for that case and presents $\frac{7}{6}$-algorithm. The chapter proposes a new approach to makespan minimization in machine-proportionate open shops with arbitrary number of machines $m$. The approach uses approximation algorithms for bin packing as subroutines to improve the solutions for the machine-proportionate open shops with $n < m$. The problem is solvable in polynomial time for $n \geq m$. The chapter considers total completion time minimization for machine-proportionate open shops. The complexity status of the problem has been a long-standing open question. The chapter shows that the problem is NP-hard for three machines and gives a polynomial-time algorithm for two machines. It remains, however, open whether the two-machine problem is polynomial with respect to a succinct input encoding. Finally, the chapter reviews sufficient conditions that make makespan minimization for ordered open shops, and open shop scheduling with maximal, dominated, and bottleneck machines polynomial.

Chapter 8 deals with multiprocessor open shop scheduling. Two models of open shops with multiprocessors have emerged in the literature: single-operation machine and multiple-operation machine models. The former replaces a single machine in each stage of open shop by parallel machines, typically identical, but limits processing of each operation to the machines from one stage. The latter permits processing each operation on a set of machines possibly from different stages. The research on those models has focused mainly on makespan minimization. For the preemptive case, the makespan minimization can be done by a two-phase approach. The approach first allocates operations to machines using McNaughton algorithm or linear programming depending on the model, thus creating an instance of preemptive open shop. The second phase then applies algorithms for makespan minimization for preemptive open shop to find the schedule for the instance. The chapter presents this approach for both models. The non-preemptive case is NP-hard in the strong sense. The chapter gives a 2-approximation algorithm that works for an arbitrary number of machines and reviews approximation schemes for a fixed number of stages. Finally, the chapter reviews the applications of multiprocessor open shop scheduling to health care and other areas.

Chapter 9 considers compact and cyclic compact open shop scheduling. The compact scheduling is motivated by timetabling, and the cyclic compact scheduling by scheduling in just-in-time systems. A compact schedule is a preemptive schedule that requires each job to be processed in a no-wait fashion (without idle time between its operations) and each machine to process operations in a no-idle fashion (without idle time between operations processed by the machine). The compact open shop schedules are traditionally recast as interval edge colorings of bipartite graphs $G$. Compact schedules do not always exist, which adds to the complexity of the makespan minimization problem. The chapter presents smallest instances for which compact schedules do not exist, and a number of tests to determine whether compact schedules exist or not depending on maximum degree $\Delta(G)$ of $G$. Such tests run in polynomial time for open shops with $\Delta(G) \leq 4$, but the decision problem becomes $NP$-complete in the strong sense for $\Delta(G) = 5$. The minimization of makespan over all compact schedules has polynomial-time solution for $\Delta(G) \leq 3$, and its complexity status remains open even for $(3, 4)$-biregular open shops, where every job has length equal to 3 and each machine has workload equal to 4. Deficiency of a schedule has been proposed as a measure of deviation of the schedule from compactness. However, the problem of minimizing deficiency is not only $NP$-hard in the strong sense but also the existing integer programming and constraint programming models can hardly solve problems of size $n + m = 10$ in reasonable time. The chapter presents polynomial-time algorithms for makespan minimization of cyclic compact schedules for open shops with $\Delta(G) \leq 4$, and sufficient conditions that make makespan minimization for $\Delta(G) \geq 5$ solvable in polynomial time.

Chapter 10 deals with no-wait scheduling of open shops. No-wait schedules model the lack of intermediate storage to store jobs between operations or lack of buffer for optical messages in optical networks. No-wait open shop scheduling does not rule out preemptions per se. The optimal preemptive and non-preemptive no-wait schedules may differ. We prove that preemptive no-wait scheduling to minimize makespan is $NP$-hard in the strong sense for two machines. We show polynomial-time tests to check whether or not there is a no-wait schedule with makespan $C_{\max} \leq 3$ or $C_{\max} \leq 4$. The existence of a polynomial-time test for $C_{\max} \leq 5$ is an open question even for 0–1 operations. We characterize instances with negative and affirmative answer to the last problem. The problem with optimal makespan $C_{\max} \leq 6$ is $NP$-hard in the strong sense even for open shops with all jobs of length 3 and each machine workload 6. This implies that no PTAS for no-wait makespan minimization exists unless $P = NP$. A PTAS, however, exists for two-machine no-wait makespan minimization. We show how to obtain no-wait schedules from cyclic compact schedules to further exploit the results of Chap. 9. We argue that the existing optimization and approximation algorithms have been developed under assumption that the operations are non-preemptive and, therefore, leave a room for research on such algorithms for preemptive case.

Chapter 11 reviews the applications of open shop scheduling. The preemptive open shop scheduling to minimize makespan finds surprisingly many prominent applications in telecommunications and information technology. Those include satellite-switched time-division multiple access (SS/TDMA) method used to allocate the communication bandwidth provided by a satellite link to carry traffic between earth stations; scheduling reconfigurable data centers; scheduling crossbar switches to guarantee 100% throughput for traffic with required traffic rates; multimessage unicasting and multicasting; file transfer scheduling; and scheduling and bandwidth allocation problem. The open shop scheduling also finds important applications in scheduling theory as a key component of the two-phase method for preemptive scheduling. The method uses algorithms, for instance, linear programming or maximum network flow, to determine an optimal instance (or instances) of the preemptive open shop in the first phase. The phase depends on a particular scheduling problem under consideration. The second phase uses preemptive open shop schedule with minimum makespan to find a feasible schedule for the preemptive scheduling problem.

Chapter 12 considers two-machine open shop scheduling with job-dependent time lags. The time lags model transportation delays, intermediate processes, or operations that are not constrained by resources but take time. The open shop scheduling with time lags can be viewed as a generalization of no-wait open shop scheduling. The chapter gives a new proof of $NP$-hardness of makespan minimization for two-machine open shop with time lags and unit-time operations. It presents an algorithm that runs in $O(n \log n)$ time for a special case where all time lags are distinct. The algorithm produces ideal schedules that minimize makespan and total completion time at the same time. The chapter then proves that the total completion time minimization is $NP$-hard in the strong sense for two-machine open shop with time lags and unit-time operations. This answers an open question from the literature. The chapter reviews special cases solvable in polynomial time and approximation algorithms for the makespan minimization problem with time lags, as well as inapproximability results for the problem with exact time lags. The chapter also discusses some open problems.

Special thanks go to my friends and colleagues, listed in random order, for their encouragement and support: Dominique de Werra (École Polytechnique Fédérale de Lausanne), Tamás Kis (Hungarian Academy of Sciences), Erwin Pesch (University of Siegen), Jan Węglarz and Jacek Błażewicz (Poznań University of Technology), Moshe Dror (University of Arizona), Krzysztof Giaro and Dariusz Dereniowski (Gdańsk University of Technology), Denis Trystram (Université Grenoble Alpes), Edward G. Coffman Jr. (Columbia University), and Michael Pinedo (New York University). I am particularly indebted to Camille C. Price, Series Editor, for her constant encouragement.

St. John's, NL, Canada Wieslaw Kubiak
August 2021

## References

1. M.M. Ahmadian, M. Khatami, A. Salehipour, T.C.E. Cheng, Four decades of research on the open-shop scheduling problem to minimize makespan. Eur. J. Oper. Res. (2021). https://doi.org/10.1016/j.ejor.2021.03.026
2. E. Anand, R. Panneerselvam, Literature review of open shop scheduling problems. Intell. Inform. Manag. **7**, 33–52 (2015)
3. J. Błażewicz, K. Ecker, E. Pesch, G. Schmidt, J. Węglarz, *Handbook on Scheduling: From Theory to Applications*. International Handbooks on Information Systems. (Springer, Berlin, 2007)
4. P. Brucker, *Scheduling Algorithms* (Springer, Berlin, 1995)

5. D. de Werra, Graph-theoretical models for preemptive scheduling, in *Advances in Project Scheduling*, ed. by R. Słowiński, J. Węglarz (Elsevier, Amsterdam, 1989), pp. 171–185
6. T. Gonzalez, Open shop scheduling, in *Handbook on Scheduling: Algorithms, Models, and Performance Analysis*, ed. by J.Y.-T. Leung (Chapman and Hall/CRC, 2004), pp. 6-1–6-14
7. T. Gonzalez, S. Sahni, Open shop scheduling to minimize finish time. J. ACM **23**, 665–679 (1976)
8. R.L. Graham, E.L. Lawler, J.K. Lenstra, A.H.G. Rinnooy Kan, Optimization and approximation in deterministic sequencing and scheduling: a survey. Ann. Discr. Math. **5**, 287–326 (1979)
9. M. Pinedo, L. Schrage, Stochastic shop scheduling: a survey, in *Deterministic and Stochastic Scheduling*, ed. by M.A.H. Dempster, J.K. Lenstra, A.H.G. Rinnooy Kan (Reidel, Dordrecht, 1982), pp. 181–196
10. M.L. Pinedo, *Scheduling: Theory, Algorithms, and Systems*, 5th edn. (Springer, Berlin, 2016)
11. C. Prins, Open shop scheduling, in *Production Scheduling*, chapter 10, (Wiley, London, 2010), pp. 271–300
12. P.D. Seymour, On multi-colourings of cubic graphs, and conjectures of Fulkerson and Tutte, in *Proc. London Math. Soc.(3)*, vol. 38 (CiteSeerX, 1979), pp. 423–460
13. V.S. Tanaev, Y.N. Sotskov, V.A. Strusevich, *Scheduling Theory: Multi-Stage Systems* (Kluwer Academic Publishers, Dordecht, 1994)
14. G.J. Woeginger, The open shop scheduling problem, in *35th Symposium on Theoretical Aspects of Computer Science (STACS 2018)*, ed. by R. Niedermeier, B. Vallée, Leibniz. International Proceedings in Informatics (Dagstuhl Publishing, 2018), pp. 4:1–4:12

# Contents

# Chapter 1
# Preliminaries

## 1.1 Open Shop

We begin by introducing notation, terminology, and definitions for traditional open shop scheduling. Those will be expanded in Chap. 2 to include open shops with additional resources, in Chap. 5 to include concurrent open shops, and in Chap. 6 to include open shops with simultaneity constraints. For the time being, we leave details to those chapters where the motivation for the extensions becomes more natural.

An open shop consists of set of machines (processors) $\mathcal{M} = \{M_1, \ldots, M_m\}$, and set of jobs $\mathcal{J} = \{J_1, \ldots, J_n\}$. Each job $J_i$ is made up of $m$ operations $O_{i,1}, \ldots, O_{i,m}$ where operation $O_{i,h}$ requires machine $M_h$ for its execution (processing). In a feasible schedule, the operations of a job can be executed in any order; however, executions of any two operations of the same job may not be done in parallel. Each machine can execute at most one job at a time. A job completes its execution as soon as all its operations have been executed. The completion time of job $J_i$ in a schedule $S$ is denoted by $C_i(S)$. To simplify notation, we omit the schedule from the notation whenever the schedule is clear from the context. An instance of an open shop is specified by an $n \times m$ non-negative real matrix

$$\mathbb{P} = \begin{bmatrix} p_{1,1} & \cdots & p_{1,m} \\ p_{2,1} & \cdots & p_{2,m} \\ \cdot & \cdots & \cdot \\ \cdot & p_{i,h} & \cdot \\ \cdot & \cdots & \cdot \\ p_{n,1} & \cdots & p_{n,m} \end{bmatrix}.$$

The convention assumed in this book is that the rows of $\mathbb{P}$ correspond to jobs, and the columns correspond to machines. The entry $p_{i,h}$ of $\mathbb{P}$ is processing time of

 
W. Kubiak, *A Book of Open Shop Scheduling*, International Series in Operations Research & Management Science 325,
https://doi.org/10.1007/978-3-030-91025-9_1

operation $O_{i,h}$. The sum of all entries in column $h$ equals workload of machine $M_h$, $h = 1, \ldots, m$,

$$L_h = \sum_{i=1}^{n} p_{i,h}. \tag{1.1}$$

The sum of all entries in row $i$ equals length of job $J_i$, $i = 1, \ldots, n$,

$$P_i = \sum_{h=1}^{m} p_{i,h}. \tag{1.2}$$

We define the maximum machine workload

$$L = \max_h \{L_h\}, \tag{1.3}$$

the longest job

$$P = \max_i \{P_i\}, \tag{1.4}$$

and the maximum degree

$$\Delta(\mathbb{P}) = \max\{L, P\}. \tag{1.5}$$

We shall also refer to $\Delta(\mathbb{P})$ as the degree of instance. We define the longest operation

$$p_{\max} = \max_{i,h}\{p_{i,h}\}. \tag{1.6}$$

For each instance $\mathbb{P}$, there is a *dual instance*, which is the transpose $\mathbb{P}^T$ of $\mathbb{P}$

$$\mathbb{P}^T = \begin{bmatrix} p_{1,1} & \cdots & p_{1,n} \\ p_{2,1} & \cdots & p_{2,n} \\ \cdot & \cdots & \cdot \\ \cdot & p_{h,i} & \cdot \\ \cdot & \cdots & \cdot \\ p_{m,1} & \cdots & p_{m,n} \end{bmatrix}.$$

Thus jobs of an instance become machines of its dual instance, and machines of the instance become jobs of its dual instance. The input size is $O(nm \log p_{max})$.

An instance may also include non-negative job release dates $r_1, \ldots, r_n$, job due dates $d_1, \ldots, d_n$, and job weights $w_1, \ldots, w_n$ for jobs $J_1, \ldots, J_n$, respectively. The open shop scheduling problem consists in finding an optimal feasible schedule, if any, for each instance of the problem. The optimality depends on objective function.

The objective functions used in this book include: makespan $C_{\max} = \max_i\{C_i\}$, maximum lateness $L_{\max} = \max_i\{C_i - d_i\}$, total completion time $\sum C_i$, total tardiness $\sum T_i = \sum \max\{0, C_i - d_i\}$, number of tardy jobs $\sum U_i$, where $U_i = 1$ if $C_i > d_i$, and $U_i = 0$ if $C_i \leq d_i$, or their weighted counterparts. A comprehensive list of objective functions used in scheduling as well as the complexity relationships between them can be found in the books on scheduling by Błażewicz et al. [8], Pinedo [25], and Brucker [10]. The well-known notation used for classification of scheduling problems introduced by Graham et al. [20] will be often used to describe open shop scheduling problems throughout the book. The notation has become a standard for scheduling problems and it is well-described in the books on scheduling (Błażewicz et al. [8], Pinedo [25], and Brucker [10]), and hence it will not be discussed in detail here.

The reader is referred to the books by Garey and Johnson [15], and Ausiello et al. [2] for the essential concepts and results on computational complexity and approximation algorithms and schemes used in this book.

Types of open shops discussed in this book include:

- open shop;
- open shop with additional resources;
- open shop with multiprocessor operations, where some operations require all machines in a set of machines for execution;
- open shop with simultaneity constraints;
- proportionate open shop, where either all operations of each job have the same processing time or all operations on each machine have the same processing time;
- multiprocessor open shop, where each machine is replaced by a set of identical parallel machines;
- compact open shop, where no waiting time between operations of jobs is allowed, and no idle time on machines is allowed;
- no-wait open shop, where no waiting time between operations of jobs is allowed;
- open shop with time lags, where a certain minimum time must elapse between operations of each job.

## 1.2 Open Shop Scheduling and Multigraph Edge Coloring

An early research on open shop scheduling originated from the basic class–teacher timetabling problem; see Gotlieb [19] and de Werra [12], where teachers correspond to machines and classes to jobs: teacher $M_h$ is required to teach class $J_i$ for $p_{i,h}$ periods of equal duration but not necessarily consecutive. Bondy and Murty [9] describe the problem as follows:

> ... in any one period, each teacher can teach at most one class, and each class can be taught by at most one teacher - this, at least, is our assumption Thus a teaching schedule for one period corresponds to a matching in the graph and, conversely, each matching corresponds to a possible assignments of teachers to classes for one period. Our problem, therefore, is

to partition the edges of $\mathcal{G}$ into as few matchings as possible or, equivalently, to properly colour the edges of $\mathcal{G}$ with as few colours as possible. Since $\mathcal{G}$ is bipartite, we know...

This description clearly defines the scheduling problem but also gives a method for its solution. The method is to recast open shop as a bipartite class–teacher multigraph and to search for a preemptive schedule where preemptions are permitted at integer points only. The solutions can then be viewed as collections of matchings in the bipartite multigraphs, each matching corresponding to a different color in edge coloring, which in turn corresponds to a different unit-time period in the schedule. Thus the focus becomes edge coloring of bipartite graphs. We now introduce a graph theoretic notation and terminology pertinent to this approach often used in this book.

All graphs considered in this book are finite. Let $G = (V, E)$ be a simple graph. For a vertex $v \in V$, let $N_G(v)$ denote the set of vertices in $G$ that are adjacent to $v$, i.e., the neighbors of $v$. $N_G(v)$ is called the *neighborhood of vertex* $v$. Whenever $G$ is clear from the context, we will drop the subscripts and write $N(v) = N_G(v)$. The *degree of a vertex* $v$, denoted by $\deg(v)$, is the number of neighbors of $v$, i.e., $\deg(v) = |N(v)|$. The set of edges incident to a vertex $v$ will be denoted by $\delta(v)$. $\Delta(G)$ denotes the *maximum degree* of $G$, i.e., $\Delta(G) = \max_{v\in V}\{\deg(v)\}$. For $X \subseteq V$, we denote by $G[X]$ the subgraph induced by $X$. We write $G - v$ for the subgraph obtained from $G$ by deleting a vertex $v$. Similarly, for $X \subseteq V$, we denote by $G - X$ the subgraph of $G$ obtained by deleting the set $X$, i.e., $G - X = G[V \setminus X]$. A *matching* in a graph $G = (V, E)$ is a set of pairwise non-adjacent edges. The set of all matchings in $G$ is denoted by $\mathcal{M}_G$, and the set of all matchings in $G$ containing an edge $e \in E(G)$ is denoted by $\mathcal{M}_G(e)$. For $u \in V(G)$, let $\mathcal{M}_G(u)$ denote the set of matchings that contain an edge incident with $u$, and for $v \in V(G)$, $v \neq u$, let $\mathcal{M}_G(u, v) = \mathcal{M}_G(u) \cap \mathcal{M}_G(v)$. If $M \in \mathcal{M}_G(u)$, we say that the matching $M$ *saturates* $u$.

A *multigraph* $H$ is a pair $(G, \mathrm{mp})$, where $G$ is a simple graph and $\mathrm{mp}\colon E(G) \to \mathbb{Z}_+$ is a function. The value of $\mathrm{mp}(e)$ is the *multiplicity of edge* $e$ in the multigraph $H$. We define $V(H) = V(G)$, $E(H) = E(G)$, and $\mathcal{M}_H = \mathcal{M}_G$. (Thus, to clarify, $\mathcal{M}_H$ does not contain two matchings that differ only in the choice of parallel edges.) For $v \in V(H)$, we define $\deg_H(v) = \sum_{e\in\delta(v)} \mathrm{mp}(e)$. Moreover, $\Delta(H) = \max_{v\in V(H)}\{\deg(v)\}$ denotes the maximum degree of $H$.

### 1.2.1 Job-Machine Bipartite Multigraph

Let $\mathbb{P}$ be an instance of open shop with all entries being non-negative integers. The rows correspond to jobs, and the columns correspond to machines. The bipartite multigraph $H = (G, \mathrm{mp})$ corresponding to $\mathbb{P}$ is defined as follows: $G = (\mathcal{J}, \mathcal{M}, E)$, where $E = \{(M_h, J_i) : M_h \in \mathcal{M}, J_i \in \mathcal{J}, \text{ and } p_{i,h} > 0\}$, with the edge $(M_h, J_i) \in E$ and multiplicity $\mathrm{mp}(M_h, J_i) = p_{i,h}$. To illustrate, consider an instance

$$\mathbb{P} = \begin{bmatrix} 1 & 1 \\ 3 & 1 \\ 1 & 4 \\ 1 & 1 \\ 1 & 2 \\ 2 & 1 \end{bmatrix},$$

with its corresponding bipartite multigraph shown in Fig. 1.1a. The two representations are equivalent for open shop scheduling problems with unit-time operations, denoted by $O|p_{ij} = 1 \ldots| \ldots$, or 0–1 operations, denoted by $O|p_{ij} \in \{0, 1\} \ldots | \ldots$, or the $O|\text{pmtn} \ldots | \ldots$. open shops with integral operation processing times and preemptions allowed at integral points only. For those problems, the two representations will be often used interchangeably following the literature convention.

### 1.2.2 Edge Coloring

An *edge coloring of a multigraph H* is a mapping $c\colon E(H) \to 2^{\mathbb{Z}_+}$ such that $|c(e)| = \text{mp}(e)$ for all $e \in E(H)$ and if $e_1, e_2 \in E(H)$ share a vertex, then $c(e_1) \cap c(e_2) = \emptyset$. Let $e_{max} = \max\{z|z \in c(e)\}$. If for an edge coloring $c$ we have $\max_{e \in E(H)}\{e_{max}\} \leq k$ for all $e \in E$, then we call $c$ a *k-edge coloring* of $H$. The smallest integer $k$ such that $H$ admits a $k$-edge coloring is called the *chromatic index* of $H$ and is denoted by $\chi'(H)$. For $k$-edge coloring $c$, define $E_i = \{e : i \in c(e)\}$, $i = 1, \ldots, k$. In edge coloring each $E_i$ is a matching. The collection $E_1, \ldots, E_k$ of matchings can be turned into a feasible schedule with $C_{\max} = k$. To see this, consider for instance a bipartite multigraph in Fig. 1.1a where a 10-edge coloring is also shown. Observe that the coloring is an interval edge coloring, i.e., all edges incident with any vertex are colored with colors making up an interval of integers. For instance, the five edges incident with a job-vertex $J_3$ are colored with colors in the interval [3, 7]. We will discuss interval edge coloring, or compact open shop, in more detail in Chap. 9. The edge coloring can be easily converted into a schedule by scheduling the jobs incident with the edges colored with color $i$ in the unit-time interval $[i - 1, i]$; see Fig. 1.1b. The makespan of the resulting schedule equals $C_{\max} = 10$, which is equal to the maximum vertex degree $\Delta = 10$ of the multigraph. The schedule is preemptive; however, a non-preemptive schedule can easily be obtained, and we leave it to the reader as an exercise. The following theorem is key to coloring edges of bipartite multigraphs.

**Theorem 1.1** *The edges of a bipartite multigraph G can be colored with*

$$\Delta(G) = \max\{L, P\} \tag{1.7}$$

*colors.*

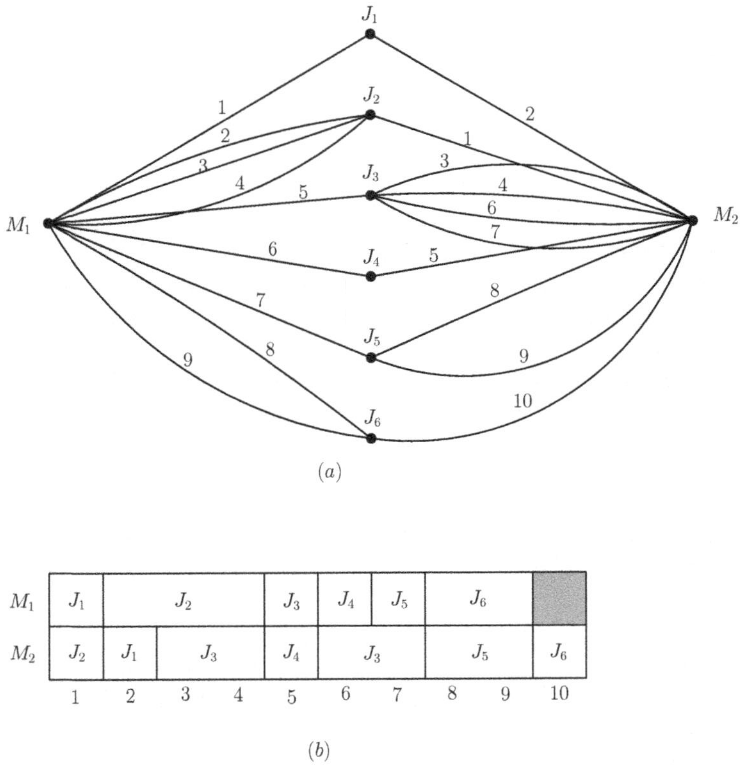


**Fig. 1.1** (**a**) bipartite multigraph corresponding to the open shop instance $\mathbb{P}$ and its 10-edge coloring; (**b**) schedule corresponding to the edge coloring

This is the famous König's edge coloring theorem (or König's line coloring theorem), König [23]. Remarkably, the coloring using $\Delta(G)$ colors can be found in polynomial time; see Cole et al. [11], Gabow and Kariv [14], de Werra [12], and Gonzalez and Sahni [18].

### 1.2.3 General Multigraph

It is well-known that $\chi'(H)$ can be written as the optimal value of the following integer linear program:

$$\chi'(H) = \chi'(G, \mathrm{mp}) = \min_{\mathbf{w} \in \mathbb{Z}_+^{\mathcal{M}_G}} \quad \sum_{M \in \mathcal{M}_H} w(M) \tag{1.8}$$

$$\text{subject to} \quad \sum_{M \in \mathcal{M}_H(e)} w(M) = \text{mp}(e) \text{ for all } e \in E(H).$$

The variable $w(M)$ from a solution vector $\mathbf{w}$ is the number of matchings in $\mathcal{M}_H$ used to construct a solution.

A *fractional edge coloring* of a graph $G$ is a mapping $f : \mathcal{M}_G \to \mathbb{R}_+$ such that $\sum_{M \in \mathcal{M}_G(e)} f(M) = 1$ for every edge $e$. If for a fractional edge coloring $f$ we have $\sum_{M \in \mathcal{M}_G} f(M) \leq k$, we call $f$ a *fractional k-edge coloring*. The smallest integer $k$ such that $G$ admits a fractional $k$-edge coloring is called the *fractional chromatic index* of $G$ and is denoted by $\chi'_f(G)$; see Schrijver [26].

A *rate function* (for a graph $G$) is a function $r : E(G) \to \mathbb{Q} \cap (0, 1]$. Similar to the degree of a vertex, for any $v \in V(G)$, we will write $r(v) = \sum_{e \in \delta(v)} r(e)$. We will think of rate functions as a continuous versions of multiplicity functions mp. With this in mind, we introduce the following continuous version of the fractional chromatic index. For a graph $G$ and a weight function $\gamma : E(G) \to \mathbb{R}_+$ (we will usually take $\gamma$ to be either a multiplicity function mp or a rate function $r$), define

$$\chi'_f(G, \gamma) = \min_{\mathbf{w} \in \mathbb{R}_+^{\mathcal{M}_G}} \sum_{M \in \mathcal{M}_G} w(M) \tag{1.9}$$

$$\text{subject to} \quad \sum_{M \in \mathcal{M}_G(e)} w(M) = \gamma(e) \text{ for all } e \in E(G).$$

Given a multigraph $H = (G, \text{mp})$, define

$$t(H) = \max \left\{ \frac{2|E(H')|}{|V(H')| - 1} : H' \text{ is an induced subgraph of } H, |V(H')| \text{ is odd}, |V(H')| \geq 3 \right\}.$$

We have the following fundamental result for the fractional chromatic index of any multigraph.

**Theorem 1.2 (Edmonds [13])** $\chi'_f(H) = \max\{\Delta(H), t(H)\}$, *for every multigraph* $H$.

### *1.2.4 Open Shop Scheduling and Doubly Stochastic Matrices*

A square matrix is a doubly stochastic matrix if it is a non-negative matrix where each row sums up to 1 and each column sums up to 1; see Berge [7], Asratian et al. [1], Horn and Johnson [22], Bapat and Raghavan [4], and Marshall and Olkin [24] for introduction to doubly stochastic matrices.

A square matrix $P$ is a permutation matrix if exactly one entry in each row and each column equals 1 and all other entries are 0; see Horn and Johnson [22]. The following theorem known as Birkhoff–von Neumann theorem (see Horn and Johnson [22]) holds.

**Theorem 1.3** *An $n \times n$ matrix $Q$ is doubly stochastic if and only if it is a convex combination of permutation matrices, i.e., there exist $q$ permutation matrices $P_1, \ldots, P_q$ and positive real numbers $\lambda_1, \ldots, \lambda_q$ such that*

$$Q = \lambda_1 P_1 + \ldots + \lambda_q P_q \tag{1.10}$$

*and*

$$\lambda_1 + \ldots + \lambda_q = 1. \tag{1.11}$$

*Moreover, $q \leq n^2 - n + 1$.*

The theorem offers an immediate solution to makespan minimization for preemptive open shops where

(1) the number of machines $m$ equals the number of jobs $n$,
(2) all machine workloads are equal and equal to all job lengths.

Processing times of operations can be real numbers, and preemptions are allowed at any points. As an example of application of the theorem, consider the following open shop:

$$\mathbb{P} = \begin{bmatrix} 1 & 2 & 3 \\ 2 & 3 & 1 \\ 3 & 1 & 2 \end{bmatrix}.$$

We have $L_1 = L_2 = L_3 = P_1 = P_2 = P_3 = 6$, and $n = m = 3$. We get the following doubly stochastic matrix for the instance $\mathbb{P}$

$$Q = \frac{1}{6}\mathbb{P} = \begin{bmatrix} \frac{1}{6} & \frac{1}{3} & \frac{1}{2} \\ \frac{1}{3} & \frac{1}{2} & \frac{1}{6} \\ \frac{1}{2} & \frac{1}{6} & \frac{1}{3} \end{bmatrix}.$$

By the Birkhoff–von Neumann theorem, $Q$ is a convex combination of permutation matrices

$$Q = \frac{1}{6}\begin{bmatrix} 1\,0\,0 \\ 0\,0\,1 \\ 0\,1\,0 \end{bmatrix} + \frac{1}{3}\begin{bmatrix} 0\,1\,0 \\ 1\,0\,0 \\ 0\,0\,1 \end{bmatrix} + \frac{1}{2}\begin{bmatrix} 0\,0\,1 \\ 0\,1\,0 \\ 1\,0\,0 \end{bmatrix},$$

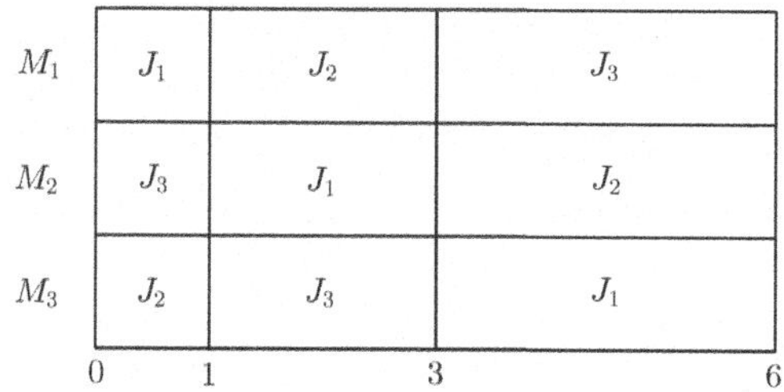


**Fig. 1.2** Schedule corresponding to the convex combination of permutation matrices for a doubly stochastic $Q$

where $\lambda_1 = \frac{1}{6}$, $\lambda_2 = \frac{1}{3}$, and $\lambda_3 = \frac{1}{2}$. Each permutation matrix $P_i$ in the convex combination represents a perfect matching of $n$ machines with $n$ jobs, or an assignment of jobs to machines. The multiplier $6 \times \lambda_i$ equals the duration of that assignment in an open shop schedule corresponding to the convex combination. The schedule has makespan $6(\lambda_1 + \lambda_2 + \lambda_3) = C_{\max} = 6$. The schedule is shown in Fig. 1.2. Observe that only the switch from one permutation matrix to the other can possibly generate preemption in the schedule obtained for the convex combination. Thus the number of distinct points in time when preemptions occur is bounded by the number of permutation matrices $q$ in the convex combination. The combination is not unique. We return to the number of permutation matrices $q$ required by the convex combination in Chap. 3 where we also show that the makespan minimization for any, not just directly meeting the conditions (1) and (2), preemptive open shop can be reduced to the application of Birkhoff–von Neumann theorem.

Gonzalez and Shani [18] consider a job–machine bipartite graph $(\mathcal{J}, \mathcal{M}, E)$ with weights $p_{i,h} > 0$ for edge $(J_i, M_h) \in E$ rather than the multiplicity that by definition needs to be integer. Their algorithm, though designed to minimize makespan of preemptive schedules, can be viewed as an algorithm for finding a convex combination in the Birkhoff–von Neumann theorem. The algorithm runs in $O(|E|(\min\{|E|, m^2\} + m \log n))$ time, which is strongly polynomial. This may not be the case for some polynomial-time algorithms for edge coloring in multigraphs. Gonzalez [16] later proposed another algorithm for the problem that runs in $O(|E| + \min\{m^4, n^4, |E|^2\})$ time.

Gonzalez and Shani [18] were the first who, to the best of the author's knowledge, used the term open shop problem. Gonzalez [17] later provided the following explanation:

> Gonzalez and Shani [18] introduced the open shop problem back in 1974 to model several real-world applications that did not quite fit under the flow shop model.

### *1.2.5 Vector Rearrangement Theorem*

The following theorem, also known as Compact Vector Summation Theorem, is used in the development of polynomial-time algorithms for the makespan minimization of special cases of the non-preemptive open shop problem, $O||C_{\max}$ in Chap. 3. The theorem is discussed in-depth in Steinitz [28], Bárány [5] and [6], Grinberg and Sevast'janov [21], and Woeginger [29].

**Theorem 1.4** *Let* $\mathbf{x}_1, \ldots, \mathbf{x}_n$ *be vectors in* $\mathbb{R}^d$ *such that*

$$\sum_{i=1}^{n} \mathbf{x}_i = 0 \tag{1.12}$$

*and*

$$\|\mathbf{x_i}\| \leq 1 \textit{ for } i = 1, \ldots, n, \tag{1.13}$$

*where* $\|\cdot\|$ *is a vector norm. There exists a permutation* $\pi$ *of the vectors such that*

$$\|\mathbf{x}_{\pi(1)} + \cdots + \mathbf{x}_{\pi(i)}\| \leq d \textit{ for } i = 1, \ldots, n. \tag{1.14}$$

*The permutation can be calculated in* $O(n^2 d^2)$ *steps.*

For a symmetric norm, the $d$ can be replaced by $d - 1 + \frac{1}{d}$ in (1.14); see Banaszczyk [3] and Sevast'janov and Banaszczyk [27].

## References


1. A.S. Asratian, T.M.J. Denley, R. Häggkvist, *Bipartite Graphs and Their Applications*. Cambridge Tracts in Mathematics (131) (Cambridge University Press, Cambridge, 1998)
2. G. Ausiello, P. Crescenzi, G. Gambosi, V. Kann, A. Marchetti-Spaccamela, M. Protasi, *Complexity and Approximation: Combinatorial Optimization Problems and Their Approximability Properties* (Springer, Berlin, 1999)
3. W. Banaszczyk, The Steinitz constant of the plane. J. Reine Angew. Math. **373**, 218–220 (1987)
4. R.B. Bapat, T.E.S. Raghavan, *Nonnegative Matrices and Applications* (Cambridge University Press, Cambridge, 1997)
5. I. Bárány, A vector-sum theorem and its application to improving flow shop guarantees. Math. Oper. Res. **6**, 445–452 (1981)
6. I. Bárány, On the power of linera dependencies, in *Building Bridges Between Math and Computer Science*, ed. by M. Grötschel, G.O.H. Katona (Springer, 2008), pp. 31–45
7. C. Berge, *The Theory of Graphs* (Dover Publications, New York, 2001)
8. J. Błażewicz, K. Ecker, E. Pesch, G. Schmidt, J. Węglarz, *Handbook on Scheduling: From Theory to Applications*. International Handbooks on Information Systems (Springer, Berlin, 2007)
9. J.A. Bondy, U.S.R. Murty, *Graph Theory with Applications* (North-Holland Publishing, Amsterdam, 1976)
10. P. Brucker, *Scheduling Algorithms* (Springer, Berlin, 1995)
11. E. Cole, K. Ost, S. Schirra, Edge-coloring bipartite multigraphs in $O(E \log D)$. Combinatorica **21**, 5–12 (2001)
12. D. de Werra, On some combinatorial problems arising in scheduling. INFOR **8**, 165–175 (1970)

13. J. Edmonds, Maximum matching and a polyhedron with 0, 1-vertices. J. Res. Nat. Bur. Stand. **69**(1–2), 125–130 (1965)
14. H.N. Gabow, O. Kariv, Algorithms for edge coloring bipartite graphs and multigraphs. SIAM J. Comput. **11**, 117–129 (1982)
15. M.R. Garey, D.S. Johnson, *Computers and Intractability: A Guide to the Theory of NP-Completeness* (W. H. Freeman, San Francisco, 1979)
16. T. Gonzalez, A note on open shop preemptive schedules. IEEE Trans. Comput. **C-28**, 782–786 (1979)
17. T. Gonzalez, Open shop scheduling, in *Handbook on Scheduling: Algorithms, Models, and Performance Analysis*, ed. by J.Y.-T. Leung (Chapman and Hall/CRC, 2004), pp. 6-1–6-14
18. T. Gonzalez, S. Sahni, Open shop scheduling to minimize finish time. J. ACM **23**, 665–679 (1976)
19. C. Gotlieb, The construction of class-teacher time-tables, in *Proc. IFIP Congress 62, Munich* (North-Holland, Amsterdam, 1963)
20. R.L. Graham, E.L. Lawler, J.K. Lenstra, A.H.G. Rinnooy Kan, Optimization and approximation in deterministic sequencing and scheduling: a survey. Ann. Discr. Math. **5**, 287–326 (1979)
21. V.S. Grinberg, S.V. Sevast'janow, Value of the Steinitz constant (in Russian). Funktsional. Anal. i Prilozhen. (Functional Anal. Appl.) **14**, 125–126 (1980)
22. R.A. Horn, C.R. Johnson, *Matrix Analysis*, 2nd edn. (Cambridge University Press, Cambridge, 2013)
23. D. König, Über graphen und ihre anwendung auf determinantentheorie und mengenlehre. Math. Ann. **77**, 453–465 (1916)
24. A.W. Marshall, I. Olkin, *Inequalities: Theory of Majorization and Its Applications* (Academic Press, New York, 1979)
25. M.L. Pinedo, *Scheduling: Theory, Algorithms, and Systems*, 5th edn. (Springer, Berlin, 2016)
26. A. Schrijver, *Combinatorial Optimization. Polyhedra and Efficiency*, vol. A (Springer, Berlin, 2003)
27. S.V. Sevast'janow, W. Banaszczyk, To the Steinitz lemma in coordinate form. Discrete Math. **169**, 145–152 (1997)
28. E. Steinitz, Bedingt konvergente reihen und convexe systeme. J. Reine Angew. Math. **143**, 128–175 (1913)
29. G.J. Woeginger, The open shop scheduling problem, in *35th Symposium on Theoretical Aspects of Computer Science (STACS 2018)*, ed. by R. Niedermeier, B. Vallée. Leibniz International Proceedings in Informatics (Dagstuhl Publishing, 2018), pp. 4:1–4:12

# Chapter 2
# Makespan Minimization for Two-Machine Open Shops

## 2.1 Introduction

The two-machine open shop makespan minimization problem, $O2||C_{\max}$, is one of the most gracious scheduling problems. It has been studied in the seminal papers on open shop scheduling by Gonzalez and Sahni [11] and Pinedo and Schrage [16]. Detailed presentations of linear-time algorithms for the problem given in these two papers can be found in books by Pinedo [17], Błażewicz et al. [2], and Brucker [4], and in a book chapter by Gonzalez [10].

The two main observations that follow from those algorithms are that the problems $O2||C_{\max}$ and $O2|\text{pmtn}|C_{\max}$ have the same value of minimum makespan for any instance and that the value equals

$$C_{\max} = \max\{L, P\}. \tag{2.1}$$

That is, the minimum makespan equals either maximum machine workload or the length of the longest job, whichever is greater. This holds regardless of preemptions being allowed or not. de Werra [6] further observes that the general $n$-job two-machine open shop problem reduces to just 3-job problem in linear time. This leads to another linear-time algorithm for the problem. The algorithm will be presented in Sect. 2.2. Those features make the two-machine open shop makespan minimization problem quite unique in scheduling theory.

The importance of the problem has recently been further underlined by the operation of non-adjacent vertex cloning by which some real-life networks may evolve, in particular non-adjacent vertex cloning in wireless networks; see Chap. 6. Through such cloning a pre-existing vertex of degree 2 is cloned, and the clone becomes adjacent to both neighbors of the pre-existing vertex. The non-adjacent vertex cloning is shown in Fig. 2.1 where vertex $v$ in Fig. 2.1a is cloned in vertices $v_1$, $v_2$, $v_3$, $v_4$, and $v_5$ in Fig. 2.1b.



W. Kubiak, *A Book of Open Shop Scheduling*, International Series in Operations Research & Management Science 325,
https://doi.org/10.1007/978-3-030-91025-9_2

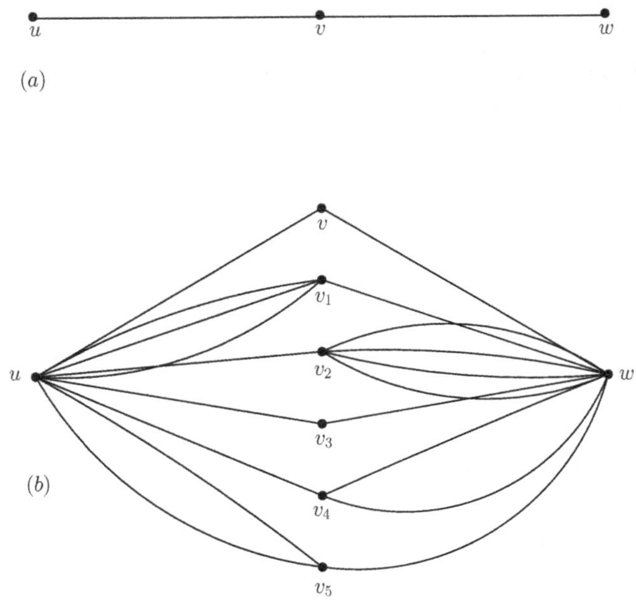


**Fig. 2.1** (**a**) vertex $v$ with $\deg(v) = 2$; (**b**) clones $v_1$, $v_2$, $v_3$, $v_4$, and $v_5$

The clones in Fig. 2.1b make up the following two-machine open shop with $n = 6$ jobs

$$\mathbb{P} = \begin{bmatrix} 1 & 1 \\ 3 & 1 \\ 1 & 4 \\ 1 & 1 \\ 1 & 2 \\ 2 & 1 \end{bmatrix}.$$

It is to be expected that some of those desirable features of $O2||C_{\max}$ problem disappear when we extend it to include additional characteristics of operations. The main challenge of the extension is to add the characteristics so that they stay relevant enough for the open shop scheduling yet they preserve some key features of $O2||C_{\max}$ at the same time. Section 2.3 presents an extension of the problem $O2||C_{\max}$. The extension requires operations to be allocated a certain number of units of an additional renewable resource, besides a machine, in order to be processed. All the units are released once the operation is completed in order to be used by other operations. This generalized problem can be solved in $O(n^3)$ time, which is shown in Sect. 2.5. The main observation is that the generalization has the same value of minimum makespan regardless of whether preemptions are being allowed or not. This preserves a key feature of the optimal solutions to $O2||C_{\max}$.

## 2.2 A Linear-Time Algorithm for Two-Machine Open Shop

In this section we show that the problem with $n$ jobs reduces to the problem with three jobs; de Werra [6]. The three jobs represent a partition of the set $\mathcal{J}$ of $n$ jobs into three disjoint subsets. Thus once the optimal schedule for the three jobs has been found, the partition can be used to find the optimal schedule for the original $n$-job instance. We now present the details of this algorithm. Take any order of jobs $1, \ldots, n$, let

$$\alpha = \max\{L_1 = \sum_{j=1}^{n} p_{j,1}, L_2 = \sum_{j=1}^{n} p_{j,2}, \max_j\{P_j = p_{j,1} + p_{j,2}\}\}. \tag{2.2}$$

Determine the smallest $i$ such that

$$\sum_{j=1}^{i}(p_{j,1} + p_{j,2}) \geq \alpha. \tag{2.3}$$

Since $\alpha \leq \sum_{j=1}^{n}(p_{j,1}+p_{j,2})$, such an $i$ exists. Denote it by $i^*$. Let $A = \{1, \ldots, i^*-1\}$, $B = \{i^* + 1, \ldots, n\}$, and $C = \{i^*\}$. Define three jobs: $A$ with processing time $p_{A,1} = \sum_{j=1}^{i^*-1} p_{j,1}$ and $p_{A,2} = \sum_{j=1}^{i^*-1} p_{j,2}$ on $M_1$ and $M_2$, respectively, $B$ with processing time $p_{B,1} = \sum_{j=i^*+1}^{n} p_{j,1}$ and $p_{A,2} = \sum_{j=i^*+1}^{n} p_{j,2}$ on $M_1$ and $M_2$, respectively, and $C$ with processing times $p_{C,1} = p_{i^*,1}$ and $p_{C,2} = p_{i^*,2}$ on $M_1$ and $M_2$, respectively. We use the same notation for the sets as for their corresponding jobs, but this should not cause any confusion. We show that a schedule that minimizes makespan for the three jobs can be easily converted into a schedule that minimizes makespan for the original $n$-job instance. We first show that

$$\alpha = \max\{p_{A,1} + p_{B,1} + p_{C,1}, p_{A,2} + p_{B,2} + p_{C,2}, p_{C,1} + p_{C,2}\}. \tag{2.4}$$

We observe that $L_1 = p_{A,1}+p_{B,1}+p_{C,1}$, and $L_2 = p_{A,2}+p_{B,2}+p_{C,2}$. Thus (2.4) holds, for $\max\{L_1, L_2\} \geq \max_j\{P_j\}$. Suppose that $\max\{L_1, L_2\} < P_{j^*} = \alpha$ for some $j^*$. If $j^* = i^*$, then (2.4) holds. Otherwise, $j^* > i^*$, which leads to contradiction because by (2.3) it implies

$$2\alpha \leq \sum_{j=1}^{i^*}(p_{j,1} + p_{j,2}) + (p_{j^*1} + p_{j^*2}) \leq L_1 + L_2 < 2\alpha.$$

Thus $i^* = j^*$ and (2.4) holds. Let us now find an optimal schedule for the jobs $A$, $B$, and $C$. For $\max\{L_1, L_2\} \leq p_{C,1} + p_{C,2} = \alpha$, the schedule in Fig. 2.2 is optimal.

Now assume that $\alpha = \max\{L_1, L_2\} > p_{C,1} + p_{C,2}$. For $L_1 \geq L_2$, one of the schedules in Figs. 2.3, 2.4, and 2.5 is optimal if the following condition is met:

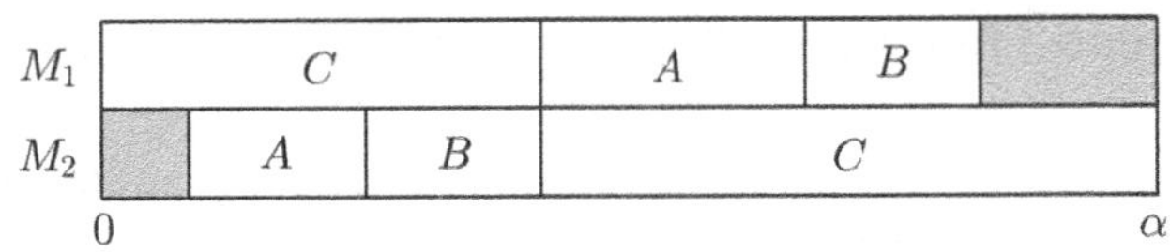


**Fig. 2.2** An optimal schedule for $\max\{L_1, L_2\} \le p_{C,1} + p_{C,2} = \alpha$

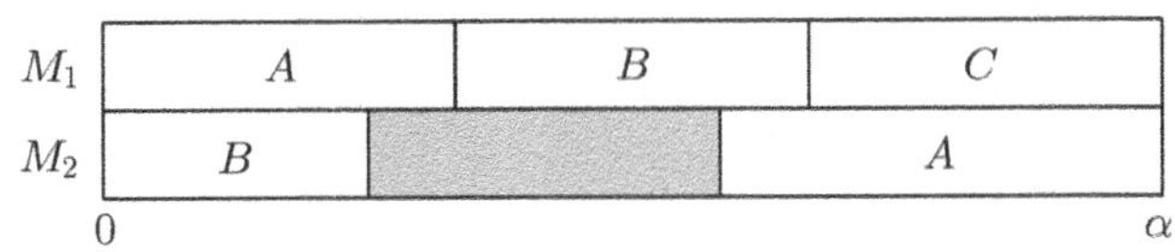


**Fig. 2.3** An optimal schedule for $p_{A,1} \ge p_{B,2}$ and $p_{C,1} \le p_{A,2}$. The job $C$ can be scheduled anywhere in the shaded interval on $M_2$

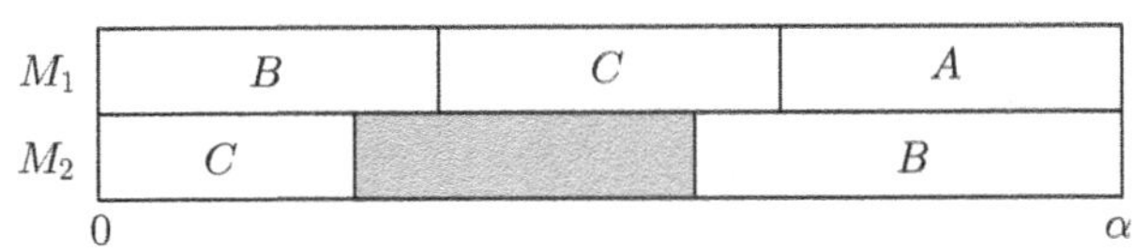


**Fig. 2.4** An optimal schedule for $p_{B,1} \ge p_{C,2}$ and $p_{A,1} \le p_{B,2}$. The job $A$ can be scheduled anywhere in the shaded interval on $M_2$

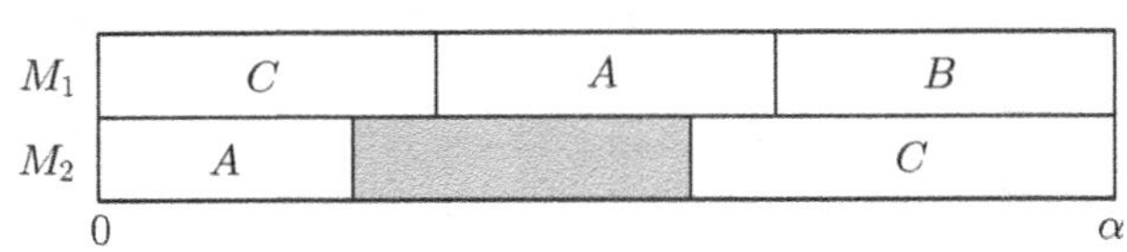


**Fig. 2.5** An optimal schedule for $p_{C,1} \ge p_{A,2}$ and $p_{B,1} \le p_{C,2}$. The job $B$ can be scheduled anywhere in the shaded interval on $M_2$

($p_{A,1} \ge p_{B,2}$ and $p_{C,1} \le p_{A,2}$) or ($p_{B,1} \ge p_{C,2}$ and $p_{A,1} \le p_{B,2}$) or ($p_{C,1} \ge p_{A,2}$ and $p_{B,1} \le p_{C,2}$). If this condition is not met, i.e., the following condition is met: ($p_{A,1} < p_{B,2}$ or $p_{C,1} > p_{A,2}$) and ($p_{B,1} < p_{C,2}$ or $p_{A,1} > p_{B,2}$) and ($p_{C,1} < p_{A,2}$ or $p_{B,1} > p_{C,2}$), then ($p_{A,1} < p_{B,2}$ and $p_{C,1} < p_{A,2}$ and $p_{B,1} < p_{C,2}$), which contradicts the assumption $L_1 \ge L_2$, or ($p_{C,1} > p_{A,2}$ and $p_{B,1} > p_{C,2}$ and $p_{A,1} > p_{B,2}$) in which case the optimal schedule is shown in Fig. 2.6. For $L_2 \ge L_1$, optimal schedules are obtained by the symmetry between $M_1$ and $M_2$.

To illustrate the algorithm run, let us consider the instance $\mathbb{Q}$ below:

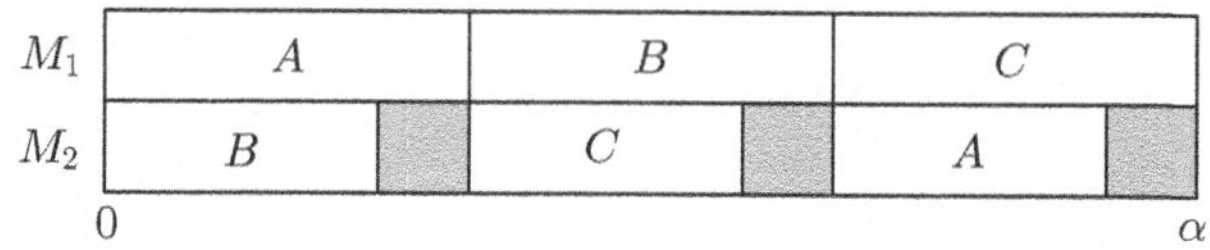


**Fig. 2.6** An optimal schedule for $p_{C1} > p_{A2}$ and $p_{B1} > p_{C2}$ and $p_{A1} > p_{B2}$

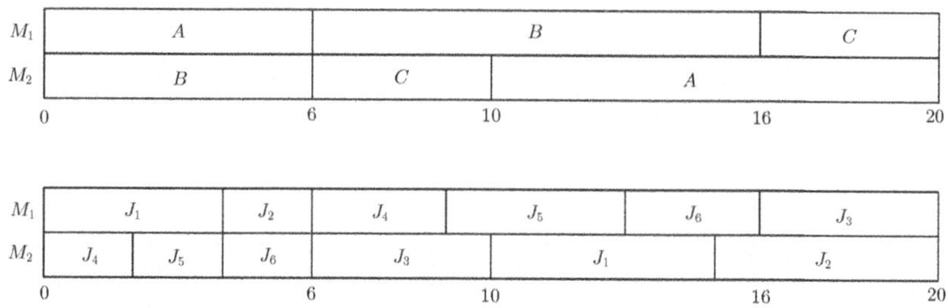


**Fig. 2.7** An optimal schedule for the instance $\mathbb{Q}$. The schedule is obtained from the partition into jobs $A$, $B$, and $C$

$$\mathbb{Q} = \begin{bmatrix} 4 & 5 \\ 2 & 5 \\ 4 & 4 \\ 3 & 2 \\ 4 & 2 \\ 3 & 2 \end{bmatrix}.$$

We have $\alpha = 20$ for $\mathbb{Q}$. For the order $J_1, \ldots, J_6$ of jobs, we have $A = \{J_1, J_2\}$, $B = \{J_4, J_5, J_6\}$, and $C = \{J_3\}$. The processing times of jobs $A$, $B$, and $C$ are equal: $p_{A,1} = 6$, $p_{A,2} = 10$, $p_{B,1} = 10$, $p_{B,2} = 6$, and $p_{C,1} = p_{C,2} = 4$. Thus the schedule in Fig. 2.3 is optimal for the three jobs $A$, $B$, and $C$; see Fig. 2.7. Observe that a reduction to two jobs is also possible for this instance. Take $D = \{J_1, J_5, J_6\}$ and $E = \{J_2, J_3, J_4\}$ for which the processing times are equal: $p_{D,1} = 11$, $p_{D,2} = 9$, $p_{E,1} = 9$, and $p_{E,2} = 11$. The optimal schedule is shown in Fig. 2.8.

A natural question arises whether the number of jobs in the reduction can always be reduced to two. The answer unfortunately is negative since there may be instances of $n$ jobs such that $\sum_{i=1}^{n}(p_{i1} + p_{i2}) = 2\alpha$ and such that for any subset $A$ of the jobs either $\sum_{i\in A}(p_{i,1}+p_{i,2}) < \alpha$ or $\sum_{i\in A}(p_{i,1}+p_{i,2}) > \alpha$. Thus either $\sum_{i\in \mathcal{J}\setminus A}(p_{i,1}+p_{i,2}) > \alpha$ or $\sum_{i\in A}(p_{i,1} + p_{i,2}) > \alpha$ for any subset $A$ of $\mathcal{J}$. Therefore either the job $\mathcal{J} \setminus A$ or job $A$ would be too long to guarantee makespan $\alpha$ for the two-job instance. However $\alpha$ is an optimal makespan for $\mathcal{J}$. This proves that the reduction to a two-job instance is not always possible. The problem to decide whether there is a reduction to a two-job instance or not is shown NP-complete by Gribkovskaia et al. [13]. Soper [20] uses similar ideas in yet another linear-time algorithm for the $O2||C_{\max}$.

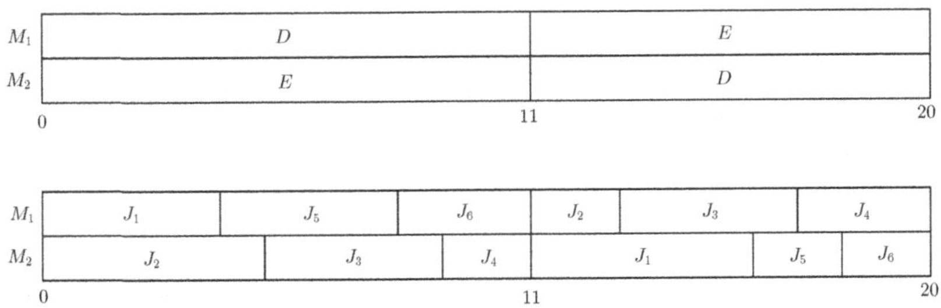


**Fig. 2.8** An optimal schedule for the instance $\mathbb{Q}$. The schedule is obtained from the partition into jobs $D$ and $E$

Van den Akker et al. [24], see also Shakhlevich and Strusevich [19], consider the two-machine open shop problem with bounds $C_1$ and $C_2$ on the completion time of machine $M_1$ and $M_2$, respectively. They show necessary and sufficient conditions for a feasible schedule that meets the bounds to exist. The conditions can be checked in linear time, and a feasible schedule, if any, can be computed in linear time.

## 2.3 Open Shop with Additional Renewable Resources

We now define the two-machine open shop scheduling problems with additional renewable resources we alluded to in the introduction. There are two machines $M_1$ and $M_2$ and $l$ renewable resource types $R_1, \ldots, R_l$; $0 \leq s_\ell$ units of resource type $R_\ell$ are available at any time. The number $s_\ell$ is called resource capacity of resource type $R_\ell$. Operation $O_{i,h}$ needs machine $M_h$ and $r_\ell(O_{i,h}) \leq s_\ell$ units of resource type $R_\ell$ at any time during its execution. The number $r_\ell(O_{i,h})$ is called resource requirement of operation $O_{i,h}$. In a feasible schedule each job is processed by at most one machine at a time, each machine processes at most one operation at a time, and the total number of resource requirements of operations processed simultaneously does not exceed resource capacity for any resource type at any time. Błażewicz et al. [3] propose a three-dot notation, res$\ldots$, to extend the notation of Graham et al. [12] to include scheduling problems with additional renewable resources. The first dot represents an arbitrary number of resource types, the second dot arbitrary resource capacities, and the third dot arbitrary resource requirements. The arbitrary here means that those numbers are part of the problem input. By replacing any of the dots by a positive integer, we make the corresponding part of the input constant for all instances. For example, the notation res1.. denotes all instances with a single additional resource type, and the other two dots mean that the capacity of that additional resource type is a part of the input and so are the resource requirements of operations. In Sect. 2.4 we consider the two-machine problem $O2|\text{res}\ldots,\text{pmtn}|C_{\max}$ with preemptions where the number of additional resources, their capacities, and operation's resource requirements are all part of the input. Section 2.5 considers the problem $O2|\text{res}1..|C_{\max}$ with a single resource type.

The resource requirements of each operation $O$ are represented by a vector $[r_1(O), \ldots, r_l(O)]$, where $0 \leq r_\ell(O) \leq s_\ell$ for $\ell = 1, \ldots, l$, and the resource capacities of resource types $R_1, \ldots, R_l$ by a vector $[s_1, \ldots, s_l]$. Though we will use the vector representation of resource requirements and capacities to express resource constraints in the following sections, it is worth mentioning that other equivalent representations, e.g., conflict graphs or agreement graphs, are possible for two-machine open shops in particular to express those constraints. Namely, the resource requirement vectors define an operation conflict graph $G = (V, E)$ for two-machine open shops with additional resources where $V$ is the set of all operations, and $(O_{i,h}, O_{j,k}) \in E$ if and only if $i = j$ or there is $\ell$ such that $r_\ell(O_{i,h}) + r_\ell(O_{j,k}) > s_\ell$. On the other hand, for an operation conflict graph $G = (V, E)$ for operations in $V$ there are $|V|$ resource types each with capacity 1. The resource type $R_v$ corresponds to the vertex $v$ of $G$. The vertex $v$ has a 0–1 vector of resource requirements where the resource requirements are set to 1 for the resource types corresponding to the vertices in $N(v) \cup \{v\}$, where $N(v)$ is the neighborhood of $v$, and they are set to 0 for the resources corresponding to the vertices in $V \setminus (N(v) \cup \{v\})$. Thus operations $u$ and $v$ can be processed simultaneously if and only if they are not neighbors (they are not adjacent) in $G$. If they were, $u$ would request one unit of resource $R_u$ and one unit of resource $R_v$, and $v$ would request one unit of resource $R_u$ and one unit of resource $R_v$. Thus the total request for both $R_u$ and $R_v$ would equal 2, which exceeds resource capacity 1. The two representations of resource constraints are equivalent. Tellache and Boudhar [22] consider a similar model of constraints. They study a job conflict graph $G = (V, E)$ (rather than the operation conflict graph), where $V$ is the set of jobs and $(J_i, J_j) \in E$ if and only if jobs $J_i$ and $J_i$ cannot be processed simultaneously. For a job conflict graph $G = (V, E)$, Tellache et al. [23] define a job agreement graph that is simply $\bar{G} = (V, \bar{E})$. Therefore clearly the translations between different representations of the resource constraints can easily be done in polynomial time, which allows for translations between the complexity results as well.

## 2.4 A Network Flow Algorithm for $O2|\text{res} \ldots, \text{pmtn}|C_{\max}$

Following Jurisch and Kubiak [14], we show a polynomial-time algorithm for the $O2|\text{res} \ldots, \text{pmtn}|C_{\max}$ problem in this section. For a given instance of the problem, we define a network $G = (V, A)$ as follows:

- The set of vertices, $V$, consists of a source $s$, vertices representing the operations $O_{1,2}, \ldots, O_{n,2}$ on machine $M_2$, vertices representing the operations $O_{1,1}, \ldots, O_{n,1}$ on machine $M_1$, and sink $t$.
- The set of directed arcs with capacities, $A$, consists of arcs $(s, O_{i2})$ with capacities $p_{i,2}$, $i = 1, \ldots, n$, arcs $(O_{i,2}, O_{j,1})$, $i, j = 1, \ldots, n$, with unlimited

capacity $\infty$ if $O_{i,2}$ and $O_{i,1}$ can be processed in parallel, i.e., if $i \neq j$ and $r_\ell(O_{i,2}) + r_\ell(O_{j,1} \leq s_\ell$ for all $\ell = 1, \ldots, l$, and arcs $(O_{j,1}, t)$ with capacities $p_{j,1}$, $j = 1, \ldots, n$.

Let $f(i, j)$ be the flow through $(i, j) \in A$ in a feasible solution to the max-flow problem defined by $G$. We define a feasible preemptive schedule to the corresponding open shop problem as follows:

- Schedule the operations $O_{1,1}, \ldots, O_{n,1}$ on machine $M_1$ in any order.
- Schedule $f(O_{i,2}, O_{j,1})$ time units of operation $O_{i2}$ in parallel with $O_{j,1}$, $i, j = 1, \ldots, n$.
- If there are time units of an operation $O_{i,2}$ left, i.e., if $\sum_{j=1}^{n} f(O_{i,2}, O_{j1}) < p_{i2}$, then schedule $p_{i,2} - \sum_{j=1}^{n} f(O_{i,2}, O_{j,1})$ time units of $O_{i,2}$ on $M_2$ after the completion of the last $M_1$ operation.

This procedure converts any feasible solution to the network max-flow problem $G$ into a feasible solution to $O2|\text{res} \ldots, \text{pmtn}|C_{\max}$ with the makespan

$$C_{\max} = \sum_{i=1}^{n} p_{i,1} + \sum_{i=1}^{n} [p_{i,2} - \sum_{j=1}^{n} f(O_{i,2}, O_{j,1})]. \tag{2.5}$$

Thus when $\sum_{i=1}^{n} \sum_{j=1}^{n} f(O_{i,2}, O_{j,1})$ is maximized in $G$, $C_{\max}$ of the open shop schedule is minimized.

**Lemma 2.1** *An optimal schedule for $O2|res \ldots, pmtn|C_{\max}$ can be found in $O(n^3)$ time. The number of preemptions in the schedule does not exceed $n^2$.*

## 2.5 An Algorithm for $O2|\text{res } 1\,..|C_{\max}$

We now focus on the non-preemptive case of the problem with a single resource, $O2|\text{res } 1\,..|C_{\max}$. We show that any preemptive schedule obtained in Sect. 2.4 can be turned into a non-preemptive schedule with the same makespan, and the conversion can be done in polynomial time. To streamline the presentation of the conversion and the algorithm for the non-preemptive case, let us simplify the notation introduced in Sect. 2.3 and make simple observations about the preemptive schedules first. For simplicity, we denote the resource by $R$ and assume that $s$ units of $R$ are available at any time. Operation $O_{i,h}$, $i = 1, \ldots, n$; $h = 1, 2$, requires $r(O_{i,h})$ units of $R$ all the time during it execution. Without loss of generality, we let $r(O_{1,1}) \geq r(O_{2,1}) \geq \cdots \geq r(O_{n,1})$. We shall solve this two-machine open shop problem with a single resource, i.e., $O2|\text{res } 1\,..|C_{\max}$, in two steps:

*Step 1:* Solve $O2|\text{res } 1\,.., \text{pmtn}|C_{\max}$ with a max-flow algorithm (Lemma 2.1) to obtain schedule $S$.

*Step 2:* Convert the resulting schedule $S$ into a schedule of $O2|\text{res } 1\,..|C_{\max}$ with the same makespan.

To simplify the presentation of Step 2, we make two assumptions about schedule $S$. First, without loss of generality, we may assume that neither machine is idle in $S$. We can easily meet this condition by adding dummy operations on either machine, if necessary. Second, again without loss of generality, we may assume that $S$ meets the following conditions:

- No operation on $M_1$ is preempted.

(2.6)

- The operations on $M_1$ are scheduled in descending order of their resource requirements.

(2.7)

Thus, schedule $S$ determines time slots and their orders. The time slots correspond to time interval occupied by exactly one operation on $M_1$. By the $i$th time slot, or simply slot $I_i$ of $S$, we mean the time interval $[\sum_{j=1}^{i-1} p_{j,1}, \sum_{j=1}^{i} p_{j,1}]$. The following theorem gives the details of Step 2.

**Theorem 2.1** *Any feasible schedule for $O2|\text{res } 1\,..,pmtn|C_{\max}$ can be converted into a feasible schedule for $O2|\text{res } 1\,..|C_{\max}$ with the same makespan.*

***Proof*** The proof consists of Lemmas 2.2 and 2.3 and a recursive procedure Process-Time-Slot. Lemma 2.2 shows how to reduce the number of preempted $M_2$-operations scheduled both in slot $I_i$ and in some slots $I_j$, $j > i$ to two. It also gives a good characterization of the case with exactly two such operations in $I_i$. The lemma is as follows. □

**Lemma 2.2** *Let $S$ be a schedule for which both (2.6) and (2.7) hold. The $S$ can be converted into a schedule $S'$ for which both (2.6) and (2.7) still hold; moreover,*

- $C_{\max}(S') = C_{\max}(S)$.

- *For any $i$, there are at most two $M_2$-operations scheduled in slot $I_i$ and in some slots $I_j$ with $j > i$.*

(2.8)

- *If there are exactly two operations $O_{a,2}$, $O_{b,2}$, $a < b$, scheduled both in slot $I_i$ and in some slots $I_j$ with $j > i$, then $b > i + 1$ and $O_{b,2}$ is continued in slots $I_{i+1}, \ldots, I_{b-1}$ only, and $O_{b,2}$ is the only operation scheduled on $M_2$ in these slots.*

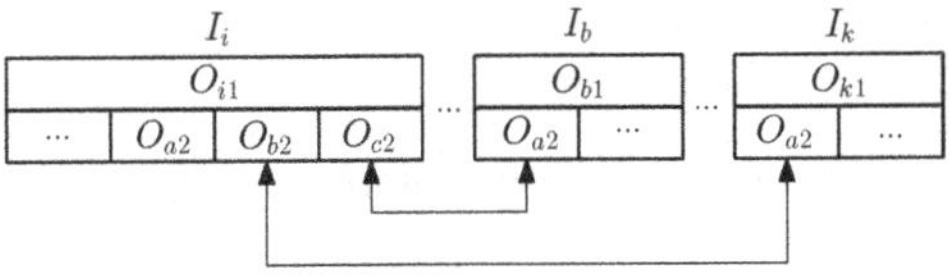


**Fig. 2.9** Proof of Lemma 2.2: reduction of the number of preempted $M_2$-operations in $I_i$

(2.9)

***Proof*** The proof is by induction. Let $S$ be a schedule that meets conditions (2.6) and (2.7), and let $i$ be the smallest index such that (2.8) or (2.9) is not met for slot $I_i$. We construct a schedule $S'$ that meets (2.6)–(2.9) for $I_i$ without making any change to $I_j$ with $j < i$. First, assume that there are at least three $M_2$-operations $O_{a,2}$, $O_{b,2}$, and $O_{c,2}$ scheduled both in slot $I_i$ and in some slots $I_j$ with $j > i$. We show that it is then possible to exchange parts of $O_{b,2}$ and $O_{c,2}$ from $I_i$ with parts of $O_{a,2}$ from slots $I_j$, $j > i$ until either:

(1) $O_{a,2}$ is no longer in slots $I_j$ with $j > i$; or
(2) $O_{b,2}$ or $O_{c,2}$ is no longer in $I_i$.

In either case, the number of $M_2$-operations that occur both in $I_i$ and in $I_j$ with $j > i$ decreases by at least 1. Assume that a part of $O_{a,2}$ is scheduled in slot $I_k$ with $k > i$. Denote by:

- $q_b$ the length of the piece of $O_{b,2}$ in $I_i$,
- $q_c$ the length of the piece of $O_{c,2}$ in $I_i$, and
- $q_a$ the length of the piece of $O_{a,2}$ in $I_k$.

We proceed as follows:

- If $k = b$, then exchange $\min\{q_c, q_a\}$ time units of $O_{c,2}$ from $I_i$ with the same number of time units of $O_{a,2}$ in $I_k$. If $q_a \geq q_c$, then (2) is met.
- If $k \neq b$, then exchange $\min\{q_b, q_a\}$ time units of $O_{b,2}$ from $I_i$ with the same number of time units of $O_{a,2}$ in $I_k$. If $q_a \geq q_b$, then (2) is met.

This procedure is shown in Fig. 2.9. Note that the exchanges are feasible since we have $r(O_{1,1}) \geq \ldots \geq r(O_{n,1})$. We apply this procedure to any part of $O_{a,2}$ in slots $I_j$ with $j > i$ until eventually either (1) or (2) is met. Then, the number of $M_2$-operations scheduled in both $I_i$ and $I_j$ with $j > i$ decreases by at least 1. The procedure is repeated for $I_i$ until the number of such $M_2$-operations is reduced to at most 2.

Now, assume that there are exactly two $M_2$-operations $O_{a,2}$, $O_{b,2}$, $a < b$, scheduled both in slot $I_i$ and in slots $I_j$ with $j > i$. We show that it is possible to exchange parts of $M_2$-operations between slots $I_i, \ldots, I_n$ until one of the following holds:

(3) Either $O_{a,2}$ is not processed in any slot $I_j$ with $j > i$ or $O_{b,2}$ is no longer processed in $I_i$.

(4) Either $O_{b,2}$ is not processed in any slot $I_j$ with $j > i$ or $O_{a,2}$ is no longer processed in $I_i$.
(5) Each of the two operations $O_{a,2}$ and $O_{b,2}$ is processed in $I_i$ as well as in $I_j$ with $j > i$, but in $I_{i+1}, \ldots, I_{b-1}$ only $O_{b,2}$ is processed on $M_2$, and $O_{b,2}$ is not processed in any slot $I_j$ with $j \geq b$.

In cases (3) and (4), the number of operations processed in $I_i$ and in $I_j$ with $j > i$ is reduced to 1 or 0. In all three cases (2.8) and (2.9) hold for slot $I_i$. In case (5), (2.8) and (2.9) also hold for $I_{i+1}, \ldots, I_{b-1}$. The exchange works as follows.

If $b < i$ (or $a < i$), then we can exchange parts of $O_{b,2}$ (or $O_{a,2}$) in $I_i$ with parts of $O_{a,2}$ (or $O_{b,2}$) in slots $I_j$ with $j > i$ until (3) (or (4)) holds.

Now, assume that $i < a < b$. We show that then we can modify slots $I_i, \ldots, I_n$ step by step until either (3) or (4) or (5) is met. First we attempt to extend $O_{a,2}$ in $I_i$ in such a way that (3) is met. This is done as follows:

First, exchange parts of $O_{b,2}$ scheduled in $I_i$ with parts of $O_{a,2}$ scheduled in slots $I_j$ with $j > i$, $j \neq b$ (see Fig. 2.10a) until either:

- (3) is met and consequently (2.8) and (2.9) hold for $I_i$ or
- no more parts of $O_{a,2}$ are scheduled in slots $I_j$ with $j > i$, $j \neq b$. Thus, a part of $O_{a,2}$ is scheduled in $I_b$.

Again, these exchanges are feasible since $r(O_{1,1}) \geq \ldots \geq r(O_{n,1})$. Now, assume that an operation $O_{x,2}$ with $x \neq b$ is processed in a slot $I_j$ with $i < j < b$. In this case we shift parts of:

- $O_{b,2}$ from $I_i$ to $I_j$,
- $O_{x,2}$ from $I_j$ to $I_b$,
- $O_{a,2}$ from $I_b$ to $I_i$,

as shown in Fig. 2.10b until either:

- (3) is met and consequently (2.8) and (2.9) hold for $I_i$ or
- there are no operations $O_{x,2}$ with $x \neq b$ scheduled in $I_j$ with $i < j < b$.

Again, it is easy to see that these shifts are feasible. Now, assume that parts of $O_{b,2}$ are processed in slots $I_j$ with $j > b$ (if they were not, then (5) would be met). In this case, we attempt to extend $O_{b,2}$ in $I_i$ until (4) is met. This is done as follows:

We exchange parts of $O_{a,2}$ scheduled in $I_i$ with parts of $O_{b,2}$ scheduled in slots $I_j$ with $j > b$ (see Fig. 2.10c) until either:

- (4) is met and consequently (2.8) and (2.9) hold for $I_i$ or
- no more parts of $O_{b,2}$ are scheduled in slots $I_j$ with $j > b$. In this case, (5) is met (see Fig. 2.10d). Thus, (2.8) and (2.9) hold for slots $I_i, \ldots, I_{b-1}$. □

Lemma 2.3 shows when and how a slot and a feasible partial schedule without preemptions and idle time can be combined into another feasible schedule without preemptions and idle time. The lemma is as follows.

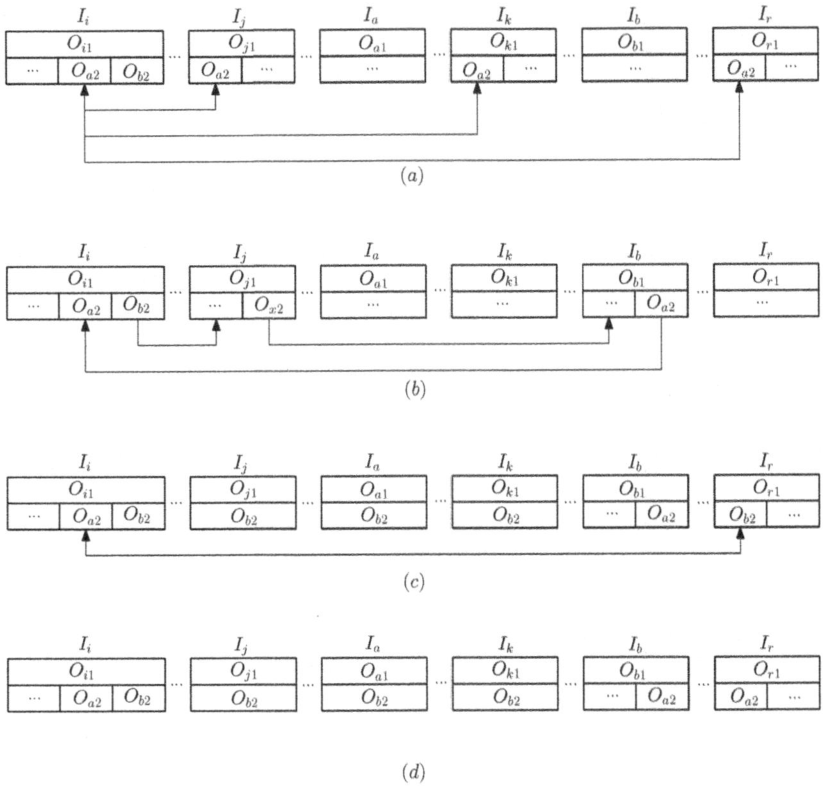


**Fig. 2.10** Proof of Lemma 2.2: the case of exactly two preempted $M_2$-operations in $I_i$

**Lemma 2.3** *Let $S$ be a feasible partial schedule without preemptions and idle time. Let $I_i$ be a slot not in $S$ that meets the following conditions:*

- *For each operation $O_{k,2}$ scheduled in $I_i$ and each operation $O_{l1}$ scheduled in $S$, we have $r(O_{k,2}) + r(O_{l,1}) \leq s$.*
- *There is at most one operation $O_{a,2}$ processed both in $I_i$ and in $S$.*

*Then $I_i$ and $S$ can be merged into a feasible schedule $S'$ without preemptions and idle time.* □

***Proof*** First, assume that, in $S$, the operation $O_{a,2}$ starts or finishes together with an operation on $M_2$. Then, we can easily modify $S$ in such a way that $O_{a,2}$ starts the whole schedule $S$. By scheduling $O_{a,2}$ at the end of $I_i$, we obtain a feasible schedule $S'$ without preemptions and idle time as shown in Fig. 2.11a.

Now, we assume that, in $S$, operation $O_{a,2}$ is scheduled in parallel with only one operation $O_{x,1}$ on $M_1$. If $O_{a,1}$ is scheduled after $O_{a,2}$ in $S$, then we can move $O_{a,2}$ to the left by successively exchanging it with its preceding operations until it either

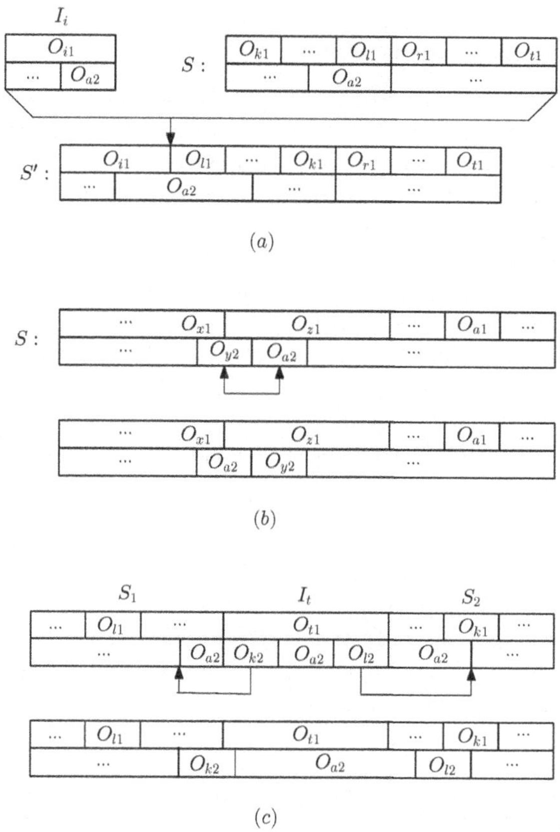


**Fig. 2.11** Proof of Lemma 2.3, insertion of slot $I_i$ in schedule $S$

starts or finishes together with an operation on $M_1$, or it is scheduled in parallel with at least two operations on $M_1$. Observer that $O_{a,2}$ can potentially be scheduled in parallel with all $M_1$-operations in $S$ except $O_{a,1}$ (see Fig. 2.11b). If $O_{a1}$ is not scheduled after $O_{a,2}$ in $S$, then move $O_{a,2}$ to the right. Thus, we may assume that $O_{a,2}$ is scheduled in parallel with at least two operations $O_{x,1}$, $O_{z,1}$ in $S$ and that it neither starts nor finishes together with an operation on $M_1$. We obtain a feasible schedule by merging $S$ and $I_i$ as follows. Let $S_1$ be the part of $S$ that finishes with $O_{x,1}$ on $M_1$, and let $S_2$ be the part of $S$ that starts with $O_{z,1}$ on $M_1$ (observe that $O_{a,2}$ finishes $S_1$ and starts $S_2$ on $M_2$). We insert $I_i$ between $S_1$ and $S_2$ and apply the following procedure to all operations $O_{y,2} \neq O_{a,2}$ scheduled in $I_i$ (Fig. 2.11c):

- If $O_{y,1}$ is scheduled in $S_1$, then we insert $O_{y,2}$ into $S_2$ immediately after $O_{a,2}$ and shift the intermediate operations on $M_2$ to the left.
- Otherwise, we insert $O_{y,2}$ into $S_1$ immediately before $O_{a,2}$ and shift the intermediate operations on $M_2$ to the right.

Note that these shifts are feasible due to the assumption that $r(O_{k,2}) + r(O_{l,1}) \leq s$ for all $O_{k,2}$ scheduled in $I_i$ and all $O_{l,1}$ scheduled in $S$. After moving all operations

$O_{y,2} \neq O_{a,2}$ either to the left or to the right, we obtain a feasible schedule $S'$ without preemptions and idle time as required. □

We use Lemmas 2.2 and 2.3 in the following recursive procedure for converting a preemptive schedule into a non-preemptive schedule with the same makespan. It returns a feasible non-preemptive schedule made of the time slots $I_j, \ldots, I_n$ if called with $i = j$. Observe that we call the procedure recursively with index $i + 1$ if we reduce the number of $M_2$-operations that are processed both in slot $I_i$ and in some slots $I_j$ with $j > i$ to 1 or 0. Otherwise, due to Lemma 2.3 the time slots $I_{j+1}, \ldots, I_{b-1}$ have a special structure: only operation $O_{b,2}$ is processed on $M_2$, and moreover $O_{b,2}$ is not processed in any slot $I_j$ with $j \geq b$. In this case, we first insert $I_i$ into the non-preemptive schedule made of the slots $I_b, \ldots, I_n$, and then we insert the slots $I_{i+1}, \ldots, I_{b-1}$ one after another. The insertion can be done easily because the properties given in Lemma 2.3 are met. □

The following theorem is a direct consequence of Lemma 2.1 and Theorem 2.1.

**Theorem 2.2** *The problem $O2|res\ 1\ldots|C_{\max}$ can be solved in $O(n^3)$ time.*

***Proof*** The max-flow algorithm computes an optimal preemptive schedule in $O(n^3)$ time. The number of operations processed both in slot $I_i$ and in slots $I_j$ with $j > i$ can be reduced to at most 2 in time $O(n^2)$. It takes $O(n)$ time to change the schedule so that it meets (2.8) and (2.9) for a given slot $I_i$. Finally, the insertion of one time slot into a feasible partial schedule takes $O(n)$ time. Since the number of slots is $n$, an overall complexity is $O(n^3)$. □

**Procedure** Process-time-slot($i$)

**begin**
- Exchange operations on $M_2$ between time slots $I_i, I_{i+1}, \cdots, I_n$ such that the number $P$ of $M_2$-operations processed both in $I_i$ and in some slots $I_j$ with $j > i$ is minimal (Proof of Lemma 2.2);
- **if** $P \leq 1$ **then**
  - S:=Process-Time-Slot ($i + 1$);
  - S:= Insert $I_i$ into $S$ (Proof of Lemma 2.3);
- **else**
  - ($P = 2$; let $O_{b,2}$ with $b > i$ be the operation scheduled in $I_i, I_{i+1}, \ldots, I_{b-1}$(Proof of Lemma 2.2))
  - S:=Process-Time-Slot($b$);
  - FOR $j := i$ TO $b - 1$ DO
  - S:= Insert $I_j$ into $S$ (Proof of Lemma 2.3);
- **end**
- return($S$);

**end**

To illustrate the working of the procedure Process-Time-Slot, let us consider an instance shown in Table 2.1 with $s = 10$. Figure 2.12a shows a preemptive schedule $S$ for the instance that meets conditions (2.8) and (2.9). The time slots of $S$ are shown

**Table 2.1** An instance of $O2|\text{res}1..|C_{max}$

| | Processing times | | Resource requirements | |
|---|---|---|---|---|
| Job | $M_1$ | $M_2$ | $M_1$ | $M_2$ |
| 1 | 4 | 5 | 10 | 3 |
| 2 | 2 | 5 | 8 | 0 |
| 3 | 4 | 4 | 7 | 0 |
| 4 | 3 | 2 | 5 | 2 |
| 5 | 4 | 2 | 2 | 9 |
| 6 | 3 | 2 | 0 | 5 |

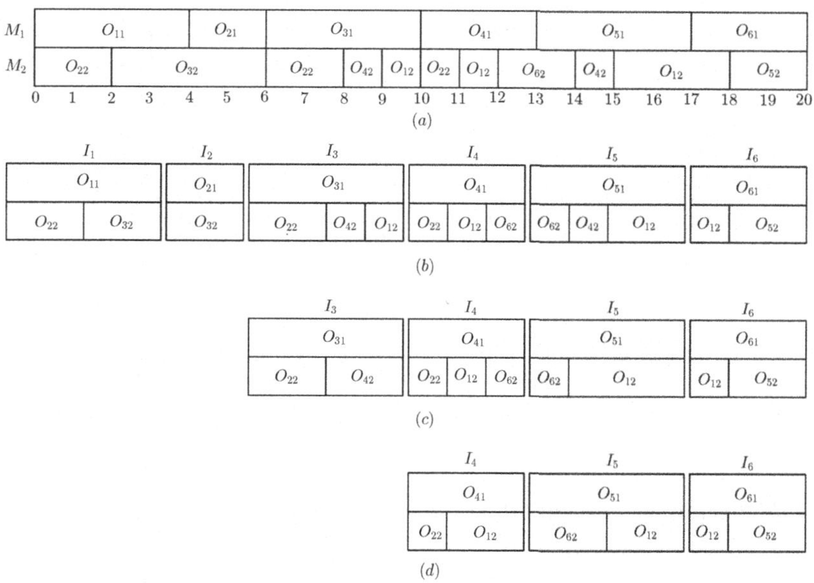

**Fig. 2.12** Reduction of the number of preempted $M_2$-operations in time slots in the example

in Fig. 2.12b. We cannot reduce the number of preempted $M_2$-operations processed in slot $I_1$ to 1, but the condition (2.9) is satisfied for $I_1$. Thus, we skip the slot $I_2$ and continue with $I_3$. We extend $O_{4,2}$ by moving one unit of $O_{1,2}$ to slot $I_5$; see Fig. 2.12c. We continue with slot $I_4$ and extend $O_{1,2}$ there; see Fig. 2.12d. After this step, all slots meet the conditions in Lemma 2.2 and we can start to build up the non-preemptive schedule recursively. Figure 2.13a shows the schedule obtained by combining the slots $I_5$ and $I_6$. Next, slots $I_4$ and $I_3$ are inserted; see Fig. 2.13b and c. Now we have to skip $I_2$ and insert $I_1$; see Fig. 2.13d. Finally, we obtain a feasible non-preemptive schedule by inserting slot $I_2$; see Fig. 2.13e. Figure 2.14 shows how the demand for the resource changes over time in the schedule from Fig. 2.13e.

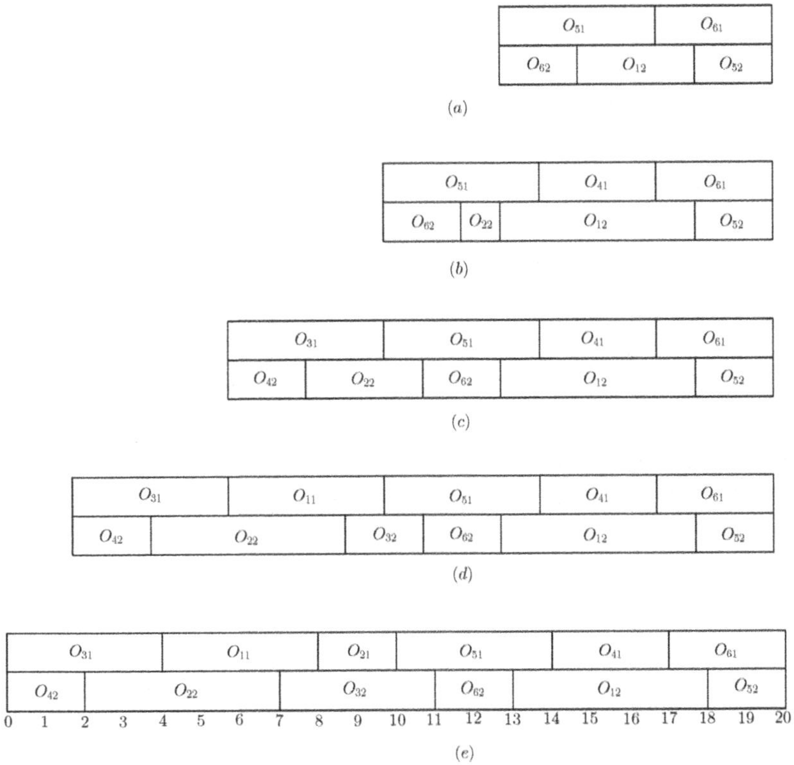


**Fig. 2.13** Recursive buildup of the schedule in the example

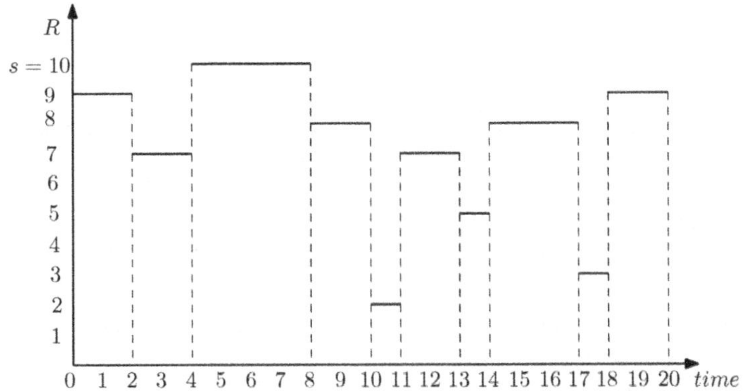


**Fig. 2.14** The number of units of the additional resource required by the optimal schedule in Fig. 2.13e

## 2.6 Open Problems

Let $\mathcal{R}_1$ be the class of operation conflict graphs $G = (V, E)$ corresponding to the instances of $O2|\text{res}1..|C_{\max}$. We show in Sect. 2.5 that for any $G \in \mathcal{R}_1$ optimal preemptive and non-preemptive solutions have the same makespan. The vertex set $V$ of $G \in \mathcal{R}_1$ can be partitioned into two disjoint sets $V_1 = \{O : R(O) \leq \frac{s}{2}\}$ and $V_2 = \{O : R(O) > \frac{s}{2}\}$. The set $V_2$ is a clique in $G$, and the set $V_1$, however, is *not* an independent set since it may include both operations of the same job, which creates a conflict (an edge in $G$). In addition, edges may exist between the vertices in $V_1$ and $V_2$. The following question remains open.

**Problem 2.1** Is there a class $\mathcal{R}$, $\mathcal{R}_1 \subset \mathcal{R}$, of operation conflict graphs $G = (V, E)$ for which optimal preemptive and non-preemptive solutions to the two-machine open shop have the same makespan?

**Problem 2.2** Characterize the class of all operation conflict graphs $G = (V, E)$ for which optimal preemptive and non-preemptive solutions to the two-machine open shop have the same makespan.

The two-machine open shop problem with *two* or more resource types benefits from preemptions. The optimal schedules with preemptions can be *shorter* than those without preemptions. Jurisch and Kubiak [14] show that the problem $O2|\text{res}211|C_{\max}$ with *two* resource types of capacity 1 each is NP-hard, and the problem $O2|\text{res}.11|C_{\max}$ with *arbitrary* number of resource types of capacity 1 each is NP-hard in the strong sense. The open shop problem with additional resources has been studied by Błażewicz et al. [1], Cochand et al. [5], de Werra et al. [9], and de Werra and Błażewicz [7, 8]. A review of some of those results can be found in Kubiak et al. [15]. Recent complexity results for two-machine open shop with additional resources can be found in Shabtay and Kaspi [18], Tellache and Boudhar [22], and Tellache et al. [23].

## Problems

**2.1** The network flow algorithm for multiple resources works in $O(n^3)$ time. Can the time be reduced to $n^2$ for the network for a single resource type?

**2.2** Prove that the problem to decide whether there is a reduction to two jobs in the algorithm in Sect. 2.5 is $NP$-complete.

**2.3** Define conflict and agreement graphs for the instance in Table 2.1.

**2.4** Prove that the problem $O2|\text{res}211|C_{\max}$ is NP-hard.

**2.5** Prove that the problem $O2|\text{res}.11|C_{\max}$ is NP-hard in the strong sense.

## References


1. J. Błażewicz, W. Cellary, R. Słowiński, J. Węglarz, *Scheduling Under Resource Constraints—Deterministic Models* (J. C. Baltzer AG, Basel, Switzerland, 1986)
2. J. Błażewicz, K. Ecker, E. Pesch, G. Schmidt, J. Węglarz, *Handbook on Scheduling: From Theory to Applications*. International Handbooks on Information Systems. (Springer, Berlin, 2007)
3. J. Blazewicz, J.K. Lenstra, A.H.G. Rinnooy Kan, Scheduling subject to resource constraints: classification and complexity. Discrete Appl. Math. **5**, 11–24 (1983)
4. P. Brucker, *Scheduling Algorithms* (Springer, Berlin, 1995)
5. D. Cochand, D. de Werra, R. Słowiński, Preemptive scheduling with staircase and piecewise linear resource availability. Z. Opns. Res. **33**, 297–313 (1989)
6. D. de Werra, Graph-theoretical models for preemptive scheduling, in *Advances in Project Scheduling*, ed. by R. Słowiński, J. Węglarz (Elsevier, Amsterdam, 1989), pp. 171–185
7. D. de Werra, J. Błażewicz, Some preemptive open shop problems with a renewable or a nonrenewable resources. Discrete Appl. Math. **35**, 205–219 (1992)
8. D. de Werra, J. Błażewicz, Addendum: some preemptive open shop scheduling problems with a renewable or a nonrenewable resources. Discrete Appl. Math. **35**, 103–104 (1993)
9. D. de Werra, J. Błażewicz, W. Kubiak, A preemptive open shop scheduling problem with one resource. Oper. Res. Letts. **10**, 9–15 (1991)
10. T. Gonzalez, Open shop scheduling, in *Handbook on Scheduling: Algorithms, Models, and Performance Analysis*, ed. by J.Y.-T. Leung (Chapman and Hall/CRC, 2004), pp. 6-1–6-14
11. T. Gonzalez, S. Sahni, Open shop scheduling to minimize finish time. J. ACM **23**, 665–679 (1976)
12. R.L. Graham, E.L. Lawler, J.K. Lenstra, A.H.G. Rinnooy Kan, Optimization and approximation in deterministic sequencing and scheduling: a survey. Ann. Discr. Math. **5**, 287–326 (1979)
13. I.V. Gribkovskaia, C.-Y. Lee, V.A. Strusevich, D. de Werra, Three is easy, two is hard: open shop sum-bath scheduling problem refined. Oper. Res. Lett. **34**, 456–464 (2006)
14. B. Jurisch, W. Kubiak, Two-machine open shops with renewable resources. Opns. Res. **45**, 544–552 (1997)
15. W. Kubiak, C. Sriskandarajah, K. Zaras, A note on the complexity of openshop scheduling problems. INFOR **29** (1991)
16. M. Pinedo, L. Schrage, Stochastic shop scheduling: a survey, in *Deterministic and Stochastic Scheduling*, ed. by M.A.H. Dempster, J.K. Lenstra, A.H.G. Rinnooy Kan (Reidel: Dordrecht, 1982), pp. 181–196
17. M.L. Pinedo, *Scheduling: Theory, Algorithms, and Systems*, 5th edn. (Springer, Berlin, 2016)
18. D. Shabtay, M. Kaspi, Minimizing the makespan in open-shop scheduling problems with a convex resource consumption function. Naval Res. Logist. **53**, 204–216 (2006)
19. N.V. Shakhlevich, V.A. Strusevich, Two machine open shop scheduling problem to minimize an arbitrary regular penalty function. Eur. J. Oper. Res. **70**, 391–404 (1993)
20. A.J. Soper, A cyclical search for the two machine flow shop and open shop to minimize finish time. J. Sched. **18**, 311–314 (2015)
21. V.S. Tanaev, Y.N. Sotskov, V.A. Strusevich, *Scheduling Theory: Multi-Stage Systems* (Kluwer Academic Publishers, Dordrecht, 1994)
22. N.E.H. Tellache, M. Boudhar, Open shop scheduling problems with conflict graphs. Discrete Appl. Math. **227**, 103–120 (2017)
23. N.E.H. Tellache, M. Boudhar, F. Yalaoui, Two-machine open shop problem with agreement graph. Theor. Comput. Sci. **796**, 154 –169 (2019)
24. M. van den Akker, H. Hoogeveen, G.J. Woeginger, The two-machine open shop problem: to fit or not to fit, that is the question. Oper. Res. Lett. **31**, 219–224 (2003)
25. G.J. Woeginger, The open shop scheduling problem, in *35th Symposium on Theoretical Aspects of Computer Science (STACS 2018)*, ed. by R. Niedermeier, B. Vallée Leibniz. International Proceedings in Informatics (Dagstuhl Publishing, 2018), pp. 4:1–4:12

# Chapter 3
# General Open Shop Scheduling

## 3.1 Complexity of Makespan Minimization

This section considers the makespan minimization for non-preemptive open shops scheduling problem, $O||C_{\max}$. It shows that the problem is NP-hard in the strong sense even when limited to the instances with the maximum degree of 4, i.e., the instances with maximum machine workload not exceeding 4, and the longest job being not longer than 4. The decision problem to determine whether there is a schedule with makespan not exceeding 4 or not for such instances is NP-complete in the strong sense. We assume integer processing times in this section unless explicitly stated otherwise. The instances with maximum degree of 3 are easier to solve; the decision problem to determine whether there is a schedule with makespan not exceeding 3 or not can be solved in $O(n^3)$ time.

Williamson et al. [81] prove the following complexity result for general open shop.

**Theorem 3.1** *The problem $O||C_{\max} \leq 4$ is NP-complete in the strong sense even for operation processing time values limited to* 0, 1, *or* 2.

***Proof*** The reduction is from a satisfiability problem. Let set $V = \{x_1, \ldots, x_v\}$ of variables and set $C = \{c_1, \ldots, c_u\}$ of clauses make up an instance of MONOTONE-NOT-ALL-EQUAL-3SAT problem; Garey and Johnson [33]. Let variable $x_i$ occur $n_i \geq 1$ times in the clauses of $C$. We have $n_i$ *assignment* jobs $J_{i,1}, \ldots, J_{i,n_i}$, and $2n_i$ *assignment* machines $M_A(i, 1), \ldots, M_A(i, n_i)$ and $M_B(i, 1), \ldots, M_B(i, n_i)$ in the open shop instance for variable $x_i$. The assignment job $J_{i,j}$ has two operations of length 2 each, one processed on $M_A(i, j)$ and the other on $M_B(i, j)$. Hence any feasible schedule with makespan 4 permits only two possible schedules for job $J_{i,j}$ shown in Fig. 3.1. In order to ensure that the schedules for all assignment jobs $J_{i,1}, \ldots, J_{i,n_i}$ are consistent in any feasible schedule with makespan 4, the

 
W. Kubiak, *A Book of Open Shop Scheduling*, International Series in Operations Research & Management Science 325,
https://doi.org/10.1007/978-3-030-91025-9_3

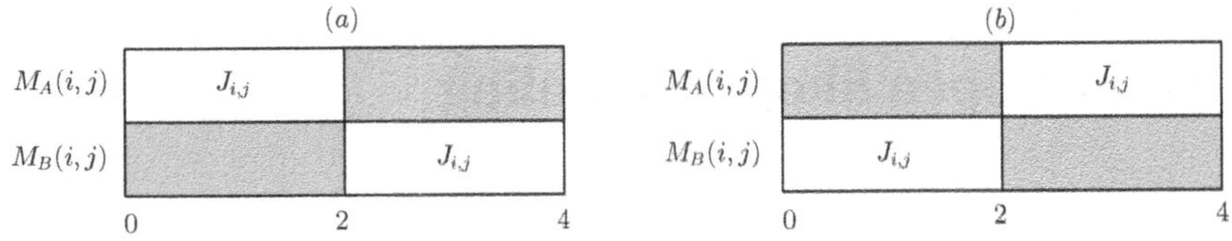


**Fig. 3.1** Two possible schedules for assignment job $J_{ij}$

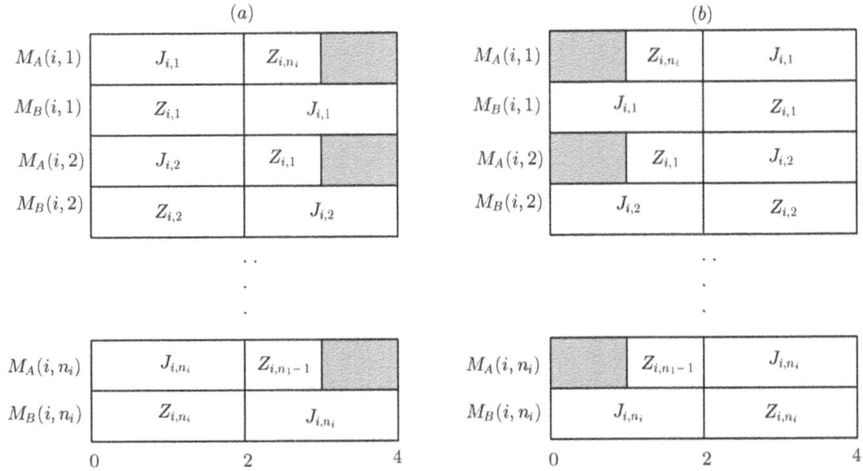

**Fig. 3.2** (**a**) $T$-configuration of $x_i$ and (**b**) $F$-configuration of $x_i$. Observe that the consistency jobs can be processed either in the interval [2, 3] or in the interval [3, 4] in $T$-configuration, and either in the interval [0, 1] or in the interval [1, 2] in $F$-configuration

open shop instance also includes $n_i$ *consistency* jobs $Z_{i,1}, \ldots, Z_{i,n_i}$ for variable $x_i$. The consistency job $Z_{i,j}$ has two operations: one of length 2 processed on assignment machine $M_B(i, j)$ and the other of length 1 processed on assignment machine $M_A(i, j \mod n_i + 1)$. Thus any feasible schedule with makespan 4 permits only two possible schedules of jobs $J_{i,1}, \ldots, J_{i,n_i}$ and $Z_{i,1}, \ldots, Z_{i,n_i}$ on machines $M_A(i, 1), \ldots, M_A(i, n_i)$ and $M_B(i, 1), \ldots, M_B(i, n_i)$ shown in Fig. 3.2. The schedule in Fig. 3.2a will be called $T$-configuration of $x_i$ and the schedule in Fig. 3.2b will be called $F$-configuration of $x_i$. Therefore, any schedule of makespan 4 selects either $T$-configuration or $F$-configuration for $x_i$. Thus we can decide the truth assignment for $x_i$ based on the configuration for $x_i$ selected by the schedule. Moreover, either configuration for $x_i$ leaves $n_i$ unit-time intervals, one on each machine $M_A(i, 1), \ldots, M_A(i, n_i)$, for jobs other than the assignments and consistency jobs. The $T$-configurations leave them in the interval [2, 4] and the $F$-configurations leave them in the interval [0, 2]. Those unit-time intervals are left for the *clause* jobs $C_1, \ldots, C_u$ that correspond to the clauses $c_1, \ldots, c_u$, respectively. The clause job $C_k$ corresponds to the clause $c_k = (x_i \vee x_j \vee x_l)$. Now, let $i'$, $j'$, and $l'$ be the numbers of times variables $x_i, x_j, x_l$, respectively, occur in the clauses

$c_1, \ldots, c_k$. The job $C_k$ consists of three unit-time operations to be processed on machines $M_A(i, i')$, $M_A(j, j')$, and $M_A(l, l')$. Thus, in any feasible schedule with makespan 4 a clause job $C_k$ has at least one operation scheduled in the interval [0, 2] and at least one in the interval [2, 4]. Therefore, $c_k$ has at least one true variable and at least one false variable in the truth assignment selected by the schedule. This proves that if the schedule with makespan 4 exists for the open shop instance, then there exists a truth assignment for $V$ such that each clause in $C$ has at least one true variable and at least one false variable.

Now, for a truth assignment for $V$ such that each clause in $C$ has at least one true variable and at least one false variable, we pick the $T$-configuration for true $x_i$, and $F$-configuration for false $x_i$. The clause $c_k = (x_i \vee x_j \vee x_l)$ has either two true variables and one false or one true variable and two false variables in the truth assignment, thus the job $C_k$ can either process two of its operations in the interval [2, 4] and one in the interval [0, 2] or process one of its operations in the interval [2, 4] and two in the interval [0, 2] on machines $M_A(i, i')$, $M_A(j, j')$, and $M_A(l, l')$. Therefore, each clause job completes by 4 in the schedule and the whole schedule has makespan 4. This completes the proof. □

Observe that the proof is done for the instances with at most *three* operations per job, and at most *three* operations per machine. Kononov et al. [48] have further strengthened this result by showing that the problem remains NP-complete even if limited to the instances with at most *two* operations per job and at most *three* operations per machine. To close the gap, we now give a polynomial-time algorithm to test whether for an instance with at most *two* operations per job and at most *two* operations per machine there is a schedule with makespan $C_{\max} \leq 4$.

**Theorem 3.2** *The problem $O||C_{\max} \leq 4$ with at most two operations per job and at most two operations per machine is polynomial.*

***Proof*** Consider the bipartite multigraph $G = (\mathcal{J}, \mathcal{M}, E)$ for the instances with operation processing time values limited to 0, 1, or 2, and with at most *two* operations per job and at most *two* operations per machine. Since $C_{\max} \leq 4$, we have $\Delta(G) \leq 4$. Replace each multiedge with multiplicity 2 in $G$ by a single edge. The resulting bipartite graph $H$ has $\Delta(H) \leq 2$, and hence $H$ is a collection of even cycles and paths that can be colored with two colors 1 and 2. Delete the edges colored with colors 1 or 2 in $H$ from $G$. The resulting graph $G'$ has $\Delta(G') \leq 2$ since each multiedge with multiplicity 2 in $G$ becomes a single edge in $G'$. Hence $G'$ is a collection of even cycles and paths that can be colored with two colors 3 and 4. Therefore we can obtain a schedule with $C_{\max} \leq 4$. Now suppose that operations with processing times 3 and 4 are also permitted. Each operation with processing time 4 does not share its machine with any other operation, and it is the only operation of some job if $C_{\max} \leq 4$. Hence those operations are scheduled on their own machines and can be removed from an instance without loss of generality. Each operation with processing time 3 belongs to a job with its other operation, if any, having processing time 1, and can only share its machine with an operation, if any, having processing time 1. Thus the multiedge with multiplicity 3 in $G$ can be replaced with a single edge as it has been done for each edge with multiplicity 2 in $G$. The algorithm will then proceed as for the instances with processing times 0, 1, or 2 and color one of the multiple edges with multiplicity 3 with either 1 or 2, and the remaining two with 3 and 4. □

Consider now the problem $O||C_{\max}$ with operation processing time values limited to 0, 1, or 2, and $C_{\max} \leq 3$. Williamson et al. [81] showed that the problem can be solved in polynomial time. We give a different proof of that fact here.

**Theorem 3.3** *The problem $O||C_{\max} \leq 3$ with operation processing time values limited to* 0, 1, 2, *or* 3 *is polynomial.*

***Proof*** The bipartite multigraph $G = (\mathcal{J}, \mathcal{M}, E)$ for the instances with $C_{\max} \leq 3$ has maximum degree $\Delta(G) \leq 3$. We first observe that a schedule with $C_{\max} \leq 3$ exists if and only if there is 3-edge coloring of $G$ with colors 1, 2, or 3 where each multiedge with multiplicity 2 has one of its edges colored with color 2. We leave the proof of the observation as an exercise; see Problem 3.5. Without loss of generality, we may assume that each multiedge in $G$ has multiplicity 1 or 2. Observe that the multiedges, if any, with multiplicity 3 are colored with colors 1, 2, and 3 in any 3-edge coloring of $G$. Also, any vertex incident with such a multiedge is not incident with any other multiedge in $G$.

Thus it remains to test whether 3-edge coloring of $G$ with colors 1, 2, or 3 where each multiedge with multiplicity 2 has one of its edges colored with color 2 exists or not. We now show such a test that runs in $O(n^3)$ time. Let $X$ be the set of all vertices of degree 3 incident with three edges having multiplicity 1 each in $G$. Let $Y$ be the set of all vertices incident with multiedges having multiplicity 2 in $G$. The sets $X$ and $Y$ are disjoint, and they partition the set of all vertices of degree 3 in $G$. Let $M_0$ be the set of all edges, if any, incident with a vertex in $X$ and a vertex in $Y$ from $G$. We now turn the bipartite multigraph $G$ into a simple bipartite graph $H$ with weighted edges as follows. Each multiedge with multiplicity 2 in $G$ is turned into an edge with weight 2 in $H$. Let $F$ be the set of all these edges in $H$. Since $\Delta(G) \leq 3$, no two different edges in $F$ share a vertex in $H$. Thus $F$ is a matching in $H$, and $|Y| = 2|F|$. Each edge with multiplicity 1 in $G$ that is not in $M_0$ is turned into an edge in $H$ with weight equal to the number of vertices in $X$ incident with the edge. Any edge in $M_0$ is turned into an edge in $H$ with weight equal to 0.

We now show that a maximum weighted matching in $H$ has weight at least $|X| + |Y|$ if and only if there is a 3-edge coloring of $G$ with colors 1, 2, or 3 where each multiedge with multiplicity 2 has one of its edges colored with color 2.

Suppose that the maximum weighted matching $M$ in $H$ has weight at least $|X| + |Y|$. The set of vertices incident with the edges in $M$ includes $X \cup Y$ since each vertex in the set $X \cup Y$ adds at most 1 to the weight and any vertex outside of the set adds 0. The set $X \cup Y$ is precisely the set of all vertices of degree 3 in $G$. Furthermore, $F \subseteq M$. Color the edges in $M$ with color 2 in $G$ and remove them from $G$. The resulting multigraph has $\Delta \leq 2$, and thus by König's edge-coloring theorem, its edges can be colored with two colors 1 and 3. This gives the required 3-edge coloring of $G$ since each multiedge of multiplicity 2 in $G$ has one of its edges colored with 2. Now suppose that there is 3-edge coloring of $G$ with colors 1, 2, or 3 where each multiedge with multiplicity 2 has one of its edges colored with color 2. Let $M$ be the set of edges colored with color 2 in $G$. Thus $M$ is a matching in $H$ with $F \subseteq M$ by the property of the coloring. Moreover, for each vertex in $X$ there is an edge in $M$ incident with that vertex but not with a vertex in $Y$. Thus $M$ a matching with total weight $|X| + |Y|$ in $H$ as required.

Therefore the test consists in checking if a maximum weighted matching in a bipartite $H$ has weight at least $|X| + |Y|$ or not. This can be done in polynomial time by the Hungarian method for instance. □

The test given in the proof is not a polynomial-time algorithm for the problem $O||C_{\max}$ on the instances of degree 3. The design of such an algorithm is left as Problem 3.6.

## 3.2 Approximate Solutions

This section considers approximation algorithms for makespan minimization for non-preemptive open shops with more than two machines. A greedy algorithm produces dense schedules that may be longer than the optimal schedules but no more than twice longer; see Sect. 3.2.1. A conjecture proved for up to $m = 8$ machines claims that those dense schedules actually guarantee the factor $2 - \frac{1}{m}$ rather than 2. On the other hand, Theorem 3.1 implies that no polynomial-time algorithm can reduce the factor below $\frac{5}{4}$; see Sect. 3.2.2. No polynomial-time algorithm that guarantees approximation between $\frac{5}{4}$ inclusive and 2 exclusive (or $2 - \frac{1}{m}$ if the conjecture holds) has been found so far. For the problem with fixed number of machines, a PTAS exists; see Sect. 3.2.3. This implies that for any fixed $m$ and $\epsilon$ a polynomial-time $(1 + \epsilon)$-approximation algorithm exists. However, the algorithm can still be impractical since the polynomial that bounds running time of the PTAS is *not* a polynomial in either $m$ or $\frac{1}{\epsilon}$. The question whether a pseudopolynomial-time algorithm for the problem with fixed $m$ exists or not is open as is the existence of a Fully Polynomial-Time Approximation Scheme (FPTAS) for the problem.

### *3.2.1 A Greedy 2-Approximation Algorithm*

We have the following theorem due to Bárány and Fiala [9].

**Theorem 3.4** *A greedy algorithm for $O||C_{\max}$ produces schedules that are shorter than twice optimal makespan.*

***Proof*** Start with an empty schedule that has all machines available from $t = 0$ on. For each operation $O_{i,h}$ not yet scheduled, determine the earliest $t$ such that

1. no job is processed on $M_h$ at $t$, and
2. job $J_i$ is not processed on any machine in $\mathcal{M} \setminus \{M_h\}$ at $t$.

The greedy algorithm proposed in Bárány and Fiala [9] works as follows. Select a not yet scheduled operation $O_{i,h}$ with the smallest $t$, break ties arbitrarily, and schedule it to start at $t$ on $M_h$. Continue until all operations are scheduled. The schedule $S$ thus obtained is feasible since by (1) no operation can start on machine $M_h$, $h = 1, \ldots, m$, at $t$ when the machine is occupied by processing another operation at $t$, and by (2) no operation of job $J_i$ can start at $t$ if some other operation of that job is being processed at $t$. Let $O_{i,h}$ be the last operation scheduled on machine $M_h, h = 1, \ldots, m$, in $S$. We have $C^h_{\max} \leq L_h + P_i - p_{i,h}$, $p_{i,h} > 0$. To prove this, we observe that $O_{ih}$ cannot start earlier in $S$ since either (1) or (2) does not hold for any $t$ earlier than the start time $s$ of $O_{i,h}$ is $S$, which means that for any $t < s$ either a job is processed on $M_h$ or job $J_i$ is processed on some machine in $\mathcal{M} \setminus \{M_h\}$. Thus $s \leq L_h - p_{i,h} + P_i - p_{i,h}$, which proves $C^h_{\max} \leq L_h + P_i - p_{i,h}$. Clearly we have the following lower bounds on the optimal makespan $C^*_{\max} \geq L_h$ and $C^*_{\max} \geq P_i$. Hence $C^h_{\max}/C^*_{\max} \leq 2 - p_{i,h}/C^*_{\max} < 2$ for each $h = 1, \ldots, m$. Thus the greedy algorithm produces schedule $S$ with makespan $C_{\max}$ such that $C_{\max}/C^*_{\max} < 2$ for each instance of $O||C_{\max}$. □

The schedules obtained by the greedy algorithm of Bárány and Fiala [9] are referred to as *dense schedules*. Chen and Strusevich [20] conjecture that those schedules are not longer than $(2 - \frac{1}{m})C^*_{\max}$ for any instance of $O||C_{\max}$, where $C^*_{\max}$ is optimal makespan for the instance. They prove the conjecture for $m = 3$ and show instances for which dense schedules are not shorter than $(2 - \frac{1}{m})C^*_{\max}$. Chen and Yu [21] prove the conjecture for $m = 4$. Chen and Yu [24] and [25] and Chen [22] prove it for $m = 5, 6$. Recently Chen et al. [23] prove it for $m = 7, 8$.

### 3.2.2 No $(\frac{5}{4} - \epsilon)$-Approximation Algorithm Exists Unless P=NP

Williamson et al. [81] prove the following inapproximability result. The result is a direct consequence of Theorem 3.1.

**Theorem 3.5** *If $P \neq NP$, then no polynomial-time algorithm for $O||C_{\max}$ exists with the worst case ratio less than $\frac{5}{4}$.*

***Proof*** Consider the set $\mathcal{I}$ of open shop instances defined in the proof of Theorem 3.1. The problem $\Pi$ defined by $\mathcal{I}$ and the question whether $I \in \mathcal{I}$ has a schedule with makespan not exceeding 4 or not is $NP$-complete, which follows immediately from the proof of Theorem 3.1. Suppose for contradiction that there is a polynomial-time algorithm $A$ such that $C^A_{\max}/C^*_{\max} < 5/4$ for any instance of $O||C_{\max}$. Thus, in particular, $C^A_{\max}/C^*_{\max} < 5/4$ for any instance of $\Pi$. The algorithm $A$ can be used to solve $\Pi$ as follows. If $C^A_{\max} \leq 4$ for instance $I$, then the answer for $I$ is affirmative.

Otherwise, if $C^A_{\max} > 4$ for $I$, then, since all processing times in $I$ are integer, we have $C^A_{\max} \geq 5$ and integer. Thus, since $C^*_{\max} > 4C^A_{\max}/5$, we get $C^*_{\max} > 4$ and the answer for $I$ is negative. Since $C^A_{\max}$ can be computed in polynomial time for each $I \in \mathcal{I}$, we have $\Pi$ in $P$. This implies $P = NP$ since $\Pi$ is $NP$-complete and gives contradiction. □

### 3.2.3 Approximation for Fixed Number of Machines $Om||C_{\max}$ and $m = 3, 4, \ldots$

Sevastianov and Woeginger [71] (see also Woeginger [82]) give a PTAS for $Om||C_{\max}$ where the number of machines is not a part of the input data. Theorem 3.1 implies that there is no PTAS for the problem with $m$ being a part of the input data if $P \neq NP$, i.e., for the problem $O||C_{\max}$.

## 3.3 A Scheme for Polynomially Solvable Cases

We begin by informally describing the main idea behind one of the most frequently studied polynomial cases first. Suppose that all machines have the same workload $L_{\max}$. We may conjecture that an instance with sufficiently large ratio $\frac{L_{\max}}{p_{\max}}$ (i.e., sufficiently many jobs) has a schedule with optimal makespan $C_{\max} = L_{\max}$ and that the schedule can be found in polynomial time. We may further conjecture that the same holds for some lower bound $\frac{L_{\max}}{p_{\max}} \geq b(m)$ common for all instances with the same number of machines $m$. Next we could claim that if the conjecture holds, then the problem $O||C_{\max}$ limited to the instances that satisfy the condition $\frac{L_{\max}}{p_{\max}} \geq b(m)$ is polynomial. This idea was proposed by Bárány and Fiala [9] and further developed in Fiala [32] and Sevast'janov [69, 72]. The conjecture may hold for different bounds $b(m)$, and different polynomial-time algorithms for optimal schedules. Thus the idea actually proposes a scheme to obtain cases of the problem $O||C_{\max}$ that are polynomial.

We now give the details of an instance of this scheme proposed by Sevast'janov [72]. We assume $L_{\max} \geq mp_{\max}$ and $m \geq 3$ (recall that the problem $O2||C_{\max}$ is polynomial for $m = 2$). If the former condition does not hold, then there is no polynomial-time algorithm that minimizes makespan over all instances that satisfy $L_{\max} < mp_{\max}$ unless $P = NP$. This follows from the proof of Theorem 3.1, where $L_{\max} = 4$, $p_{\max} = 2$, and $m = 3u$, $u \geq 1$. Since $L_{\max} \leq np_{\max}$, we have $n \geq m$.

Thus our search for polynomial-time algorithms needs to be limited to instances with $L_{\max} \geq mp_{\max}$. Without loss of generality, we assume that

$$L_h = L \text{ for } h = 1, \ldots, m \tag{3.1}$$

and

$$p_{i,h} \leq 1 \text{ for } i = 1, \ldots, n \text{ and } h = 1, \ldots, m, \tag{3.2}$$

for each instance with $L_{\max} \geq mp_{max}$. If (3.1) is not met by machine $M_h$, i.e., $L_h < L_{\max}$, then we can increase processing times in column $h$ of processing time matrix $\mathbb{P}$ by splitting $L_{\max} - L_h$ into $n$ numbers $\epsilon_{i,h}$ such that $p_{i,h} + \epsilon_{i,h} \leq p_{\max}$. This can easily be done since $np_{\max} \geq L_{\max}$. The increase in operation processing times may increase job's processing time to at most $mp_{\max}$, which however does not exceed $L_{\max}$ by assumption. Moreover, it can be done in $O(nm)$ time. Finally, we can take the processing time $\frac{1}{p_{\max}}\mathbb{P}$, where $\mathbb{P}$ meets (3.1) to ensure that (3.2) holds. Thus $L = \frac{L_{\max}}{p_{\max}}$. Now, for each job $J_i$, $i = 1, \ldots, n$, define a vector in $\mathbb{R}^{m-1}$

$$\mathbf{x}_i = (p_{i,1} - p_{i,m}, \ldots, p_{i,h} - p_{i,m}, \ldots, p_{i,m-1} - p_{i,m}). \tag{3.3}$$

By (3.1) and (3.2), we have

$$\mathbf{x}_1 + \ldots + \mathbf{x}_n = 0$$

and

$$\|\mathbf{x}_i\| \leq 1 \text{ for } i = 1, \ldots, n,$$

for the vectors $\mathbf{x}_1, \ldots, \mathbf{x}_n$, where the symmetric norm $\| \cdot \|$ is defined as follows:

$$\|\mathbf{x} = (x_1, \ldots, x_n)\| = \max_{\ell,j}\{|x_\ell|, |x_\ell - x_j|\}. \tag{3.4}$$

Thus the vectors in (3.3) satisfy the conditions of the Vector Rearrangement Theorem 1.4. By this theorem, there is a permutation $\mathbf{x}_{\pi(1)}, \ldots, \mathbf{x}_{\pi(n)}$ of those vectors such that

$$\|\mathbf{x}_{\pi(1)} + \ldots + \mathbf{x}_{\pi(k)}\| \leq (m-1) - 1 + \frac{1}{m-1} \text{ for } k = 1, \ldots, n. \tag{3.5}$$

The bound on the right-hand side is given in Banaszczyk [6]. The permutation $\pi$ for that bound can be calculated in $O(n^2m^2)$ time; see Banaszczyk and Sevast'janow [73]. We now consider schedule $S$ that schedules all jobs without idle time, and in the same order $J_{\pi(1)}, \ldots, J_{\pi(n)}$ on each machine; see Fig. 3.3 (all figures in this proof are for $m = 4$, which suffices to illustrate the main ideas). The schedule $S$ is likely infeasible since it permits jobs to be processed simultaneously on more than one machine at a time, and thus it needs to be modified in order to remove this infeasibility. The modification, however, needs to be done without increasing the schedule makespan beyond $L$. We now describe such a modification and give a sufficient condition that makes the resulting schedule feasible. Both the modification

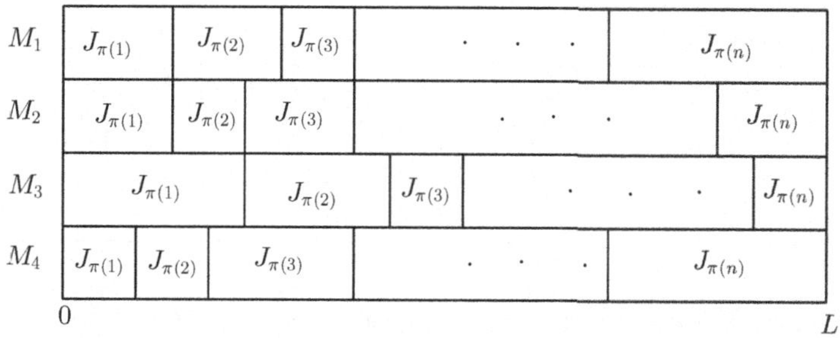


**Fig. 3.3** Schedule $S$ for permutation $\pi$ obtained by the Vector Rearrangement Theorem

and the condition rely on (3.5). Since

$$\mathbf{x}_1 + \ldots + \mathbf{x}_k = \Big( \sum_{j=1}^{k} (p_{\pi(j),1} - p_{\pi(j),m}), \ldots, \sum_{j=1}^{k} (p_{\pi(j),m-1} - p_{\pi(j),m}) \Big), \tag{3.6}$$

the condition (3.5) ensures

$$\Big| \sum_{j=1}^{k} (p_{\pi(j),h} - p_{\pi(j),h+1}) \Big| \le m - 2 + \frac{1}{m-1} \text{ for } k = 1, \ldots, n \text{ and } h = 1, \ldots, m-1, \tag{3.7}$$

and

$$\Big| \sum_{j=1}^{k} (p_{\pi(j),m} - p_{\pi(j),1}) \Big| \le m - 2 + \frac{1}{m-1} \text{ for } k = 1, \ldots, n. \tag{3.8}$$

Thus we get the following upper bound on the difference between the completion of $J_{\pi(k)}$ on $M_h$ and its start on $M_{h+1}$ in $S$, which is equal to the completion time of job $J_{\pi(k-1)}$ on $M_{h+1}$ in $S$

$$C_{h,\pi(k)} - C_{h+1,\pi(k-1)} = \sum_{j=1}^{k-1} (p_{\pi(j),h} - p_{\pi(j),h+1}) + p_{\pi(k),h} \le m - 1 + \frac{1}{m-1}, \tag{3.9}$$

for $h = 1, \ldots, m-1$ and $k = 1, \ldots, n$, and

$$C_{m,\pi(k)} - C_{1,\pi(k-1)} = \sum_{j=1}^{k-1} (p_{\pi(j),m} - p_{\pi(j),1}) + p_{\pi(k),m} \le m - 1 + \frac{1}{m-1}. \tag{3.10}$$

We take $C_{h,\pi(0)} = 0$, $h = 1, \ldots, m$, in those formulas. Now define

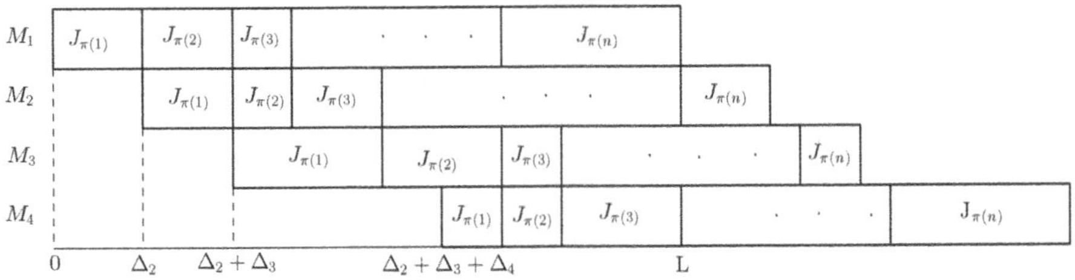


**Fig. 3.4** Schedule $S'$ obtained by delaying $S$ on machines $M_2, \ldots, M_m$

$$\Delta_h = \max_k \{C_{h-1,\pi(k)} - C_{h,\pi(k-1)}, 0\} \tag{3.11}$$

for $h = 2, \ldots, m$. For positive $\Delta_h$, $h = 2, \ldots, m$, by delaying the start of the schedule $S$ by $\Delta_h$ on $M_h$ with respect to the start of the schedule $S$ on $M_{h-1}$ we make each job $J_i$ on $M_h$ to start *not earlier* than the completion of $J_i$ on $M_{h-1}$. Observe that any further such delay preserves this condition so we could use a longer delay, if needed, to obtain a feasible schedule. For $\Delta_h = 0$, each job $J_i$ starts *not earlier* on $M_h$ than it completes on $M_{h-1}$ in $S$. By these observations, we get that starting $S$ at

$$\Delta_2 + \ldots + \Delta_h \tag{3.12}$$

instead of 0 on $M_h$, $h = 2, \ldots, m$ produces a feasible schedule $S'$; see Fig. 3.4.

The schedule $S'$, however, has makespan $L + \Delta_2 + \ldots + \Delta_m$. In order to reduce the makespan we observe that the delays created idle interval $[0, \Delta_2 + \ldots + \Delta_h]$ on $M_h$ at the cost of extending the schedule to the interval $[L, L + \Delta_2 + \ldots + \Delta_h]$ on $M_h$, $h = 2, \ldots, m$; see Fig. 3.4. The idea now is to move the operations from $[L, L+\Delta_2+\ldots+\Delta_h]$ back to $[0, \Delta_2+\ldots+\Delta_h]$ on machine $M_h$, $h = 2, \ldots, m$, and thus to reduce the makespan back to $L$. There are two obstacles to implementing this idea. First, there may be a *straddling operation* that starts at $L-\delta_h$, $0 < \delta_h < 1$, and completes after $L$ on $M_h$ in $S'$; such a straddling operation should not be preempted in the process and thus we need to start the schedule on $M_h$ even later in order to align the start time of this straddling operation with $L$. That would require a further delay by $\delta_h$ on each $M_h, \ldots, M_m$ to avoid preemptions. Thus, starting $S'$

$$\Delta_2 + \ldots + \Delta_h + \delta_2 + \ldots + \delta_h \tag{3.13}$$

on $M_h$, $h = 2, \ldots, m$, results in a feasible schedule that is aligned at $L$ on each machine. Second, and more difficult to overcome, obstacle is to ensure that the schedule obtained after moving the operations from $[L, L + \Delta_2 + \ldots + \Delta_h + \delta_2 + \cdots + \delta_h]$ back to $[0, \Delta_2+\ldots+\Delta_h+\delta_2+\cdots+\delta_h]$ on machine $M_h$, $h = 2, \ldots, m$ (see Fig. 3.5) is feasible. Though this obstacle can possibly be overcome in a number of ways, we show the one used in Sevast'janov [72], which is based on the following condition sufficient to ensure feasibility of the resulting schedule:

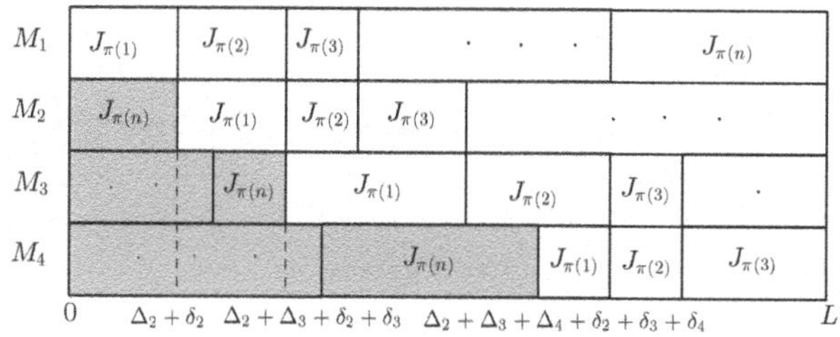


**Fig. 3.5** If $L - \Delta_1 \geq \Delta_2 + \ldots + \Delta_m + \delta_2 + \ldots + \delta_m$, the schedule is feasible

$$L - \Delta_1 \geq \Delta_2 + \ldots + \Delta_m + \delta_2 + \ldots + \delta_m, \tag{3.14}$$

where

$$\Delta_1 = \max_k \{C_{m,\pi(k)} - C_{1,\pi(k-1)}, 0\}. \tag{3.15}$$

The key is to observe that delaying the start of schedule $S$ on $M_1$ with respect to the start of schedule $S$ by $\Delta_1$ on $M_m$ makes each job $J_i$ on $M_1$ to start *not earlier* than the completion of $J_i$ on $M_m$. Thus moving the jobs from the interval $[L, L + \Delta_2 + \ldots + \Delta_m + \delta_2 + \ldots + \delta_m]$ to the interval $[0, \Delta_2 + \ldots + \Delta_m + \delta_2 + \ldots + \delta_m]$ on $M_m$ and keeping their order as in $S$ results in a feasible schedule, provided that (3.14) is met. Now consider machine $M_{m-1}$. Move the jobs from the interval $[L, L + \Delta_2 + \ldots + \Delta_{m-1} + \delta_2 + \ldots + \delta_{m-1}]$ to the interval $[0, \Delta_2 + \ldots + \Delta_{m-1} + \delta_2 + \ldots + \delta_{m-1}]$ on $M_{m-1}$, and keep their order as in $S$. By the construction of $S'$, each job $J_i$ starts on $M_m$ not earlier than it completes on $M_{m-1}$ in $S'$, and it does so after the move. Moreover any job $J_i$ in $[0, \Delta_2 + \ldots + \Delta_m + \delta_2 + \ldots + \delta_m]$ completes on $M_m$ by its start on $M_1$. Thus by induction, we extend the feasible schedule to include machine $M_{m-1}$, and the remaining machines. At the end, we obtain a feasible schedule with makespan $L$, provided that (3.14) holds, which by (3.9), (3.10), and (3.11) gives

$$L \geq m^2 + \frac{1}{m-1}. \tag{3.16}$$

Therefore we obtained an optimal schedule with makespan $C_{\max} = L_{\max}$ for any instance of the original open shop that meets the following condition:

$$\frac{L_{\max}}{p_{\max}} \geq m^2 + \frac{1}{m-1}. \tag{3.17}$$

It takes $O(n^2m^2)$ steps to calculate the permutation $\pi$ that ensures (3.9), (3.10), and (3.11), and all remaining steps required to turn $S$ into a feasible schedule with makespan $L$ require $O(nm)$ steps. The condition (3.17) can be easily verified in $O(nm)$ steps. Therefore the overall complexity is $O(n^2m^2)$. We summarize these results in the following theorem.

**Table 3.1** Lower bounds $b(m)$ and the computational complexity of the scheme, $m' = 2^{\lceil \log_2 m \rceil}$ in the table

| Reference | Lower bound $b(m)$ on the ratio $\frac{L_{\max}}{p_{\max}}$ | Complexity |
|---|---|---|
| Bárány and Fiala [9] | $m^2 + 2m - 1$ | $O(n^2m^3)$ |
| Bárány and Fiala [9] | $8m' \log_2 m' + 5m'$ | $O(nm^3)$ |
| Fiala [32] | $16m' \log_2 m' + 5m'$ | $O(n^2m^3)$ |
| Sevast'janov [69, 72] | $\frac{16}{3}m \log_2 m' + \frac{61}{9}m - 7.4$ | $O(nm^2 \log_2 m)$ |
| Sevast'janov [72] | $m^2 - 1 + \frac{1}{m-1}$ | $O(n^2m^2)$ |

**Theorem 3.6** *The problem $O||C_{\max}$ limited to instances that satisfy the condition $\frac{L_{\max}}{p_{\max}} \geq (m^2 + \frac{1}{m-1})$ can be solved to optimality in $O(n^2m^2)$ time. The optimal solution has makespan $C_{\max} = L_{\max}$.*

Other than the $m^2 + \frac{1}{m-1}$, lower bounds $b(m)$ on the ratio $\frac{L_{\max}}{p_{\max}}$ also guarantee polynomial-time algorithms for $O||C_{\max}$ if limited to the instances with $\frac{L_{\max}}{p_{\max}} \geq b(m)$. For all those instances, the optimal makespan equals $L_{\max}$. The summary of those bounds and the complexities of the corresponding polynomial-time algorithms are given in Table 3.1. The reader is referred to Sevastianov [70] for a survey of application of the Vector Rearrangement Theorem to scheduling.

## 3.4 Other Objective Functions

Achugbue and Chin [1] show that the minimization of total completion time for two-machine open shop, $O2||\sum C_i$, is NP-hard in the strong sense. They also show that the shortest processing time (SPT) schedules guarantee $m$-approximation of optimal schedules. Hoogeveen et al. [45] show that the problem $O||\sum C_i$ does not have a PTAS unless $P = NP$. Lenstra et al. [55] and [56] show that the maximum lateness minimization for two-machine open shops, $O2||L_{max}$, is NP-hard in the strong sense. Graham et al. [39] show that the makespan minimization for the problem with release dates, $O2|r_i|C_{\max}$, is NP-hard in the strong sense, and it remains NP-hard even for two distinct release dates $O2|r_i \in \{0, r\}|C_{\max}$. Lu and Posner [61] show that the average-case complexity for the latter problem is polynomial when operation processing times are random variables with distributions coming from certain classes of probability distributions. Józefowska et al. [47] show that the minimization of number of tardy jobs is NP-hard in the ordinary sense for a common due date, $O2|d_i = d|\sum U_i$. They also give a dynamic programming algorithm running in time $O(nd^2)$ for the problem. Koulamas and Kyparisis [49] develop polynomial-time algorithms for some special cases of $O2|d_i = d|\sum U_i$.

## 3.5 Open Shop Scheduling with All Unit-Time Operations

In this section we consider open shop scheduling problem with all operations of a job being unit-time. This implies that each job has exactly $m$ unit-time operations, and no operations are missing.

### *3.5.1 Open Shop Scheduling and Scheduling on Identical Parallel Processors*

The idea of using scheduling of jobs on parallel identical processors to scheduling open shops was introduced by Kubiak et al. [53, p. 288], and later applied to various open shop scheduling problems by Brucker et al. [17] and Timkovsky [79]. We now describe the idea in detail. Consider a problem $\mathcal{P}$ of scheduling of $n$ jobs on $m$ parallel identical processors; see Błażewicz et al. [10] for terminology and notation for scheduling on parallel identical processors. We limit the problem to instances with integer values of the numerical data such as release dates, processing times, due dates, deadlines but not necessarily the weights in the objective function. We also permit preemptions of jobs. Let $\mathcal{P}_m$ be the problem $\mathcal{P}$ limited to the instances with all processing times equal to $m$, and with $m$ processors. Let $\mathcal{S}_I$ be the set of all feasible schedules with preemptions at integer points for an instance $I$ of $\mathcal{P}_m$. A schedule $S \in \mathcal{S}_I$ is called *open shop*-like schedule if each job is scheduled for exactly one unit of time on each of the $m$ processor. We have the following theorem.

**Theorem 3.7** *For each schedule $S \in \mathcal{S}_I$, there is an open-shop like schedule $S^O \in \mathcal{S}_I$ with the same value of objective function as $S$. The schedule $S^O$ can be found in $O((n + (C_{\max} - n)m)m \log m)$ steps, where $C_{\max}$ is the makespan of $S$.*

***Proof*** Consider schedule $S \in \mathcal{S}_I$ with makespan $C_{\max}$. To streamline discussion, assume $C_{\max} = n$ for the time being. We return to the case $C_{\max} > n$ later. Consider a simple bipartite graph $G = (\mathcal{J}, Y, E)$ for $S$, where $\mathcal{J} = \{J_1, \dots, J_n\}$ is the set of $n$ jobs, $Y = \{1, \dots, n\}$ is the set on $n$ unit-time slots $\{[0, 1], \dots, [n-1, n]\}$, and $(J_i, j) \in E$ if and only if $J_i$ is scheduled in $[j-1, j]$ in $S$. The graph $G$ is $m$-regular since by definition of $I$ there are exactly $m$ different unit-time slots where $J_i$ is scheduled in $S$, thus $\deg(J_i) = m$ for each $J_i$, and since there are exactly $m$ different jobs scheduled in each unit-time slot $[j-1, j]$, thus $\deg(j) = m$. By König's edge coloring theorem, there are $m$ disjoint matchings $M_1, \dots, M_m$ that cover all edges in $E$, i.e., $E = M_1 \cup \dots \cup M_m$. Moreover, since $G$ is $m$-regular, each matching $M_i$ is perfect, i.e., covers all jobs in $\mathcal{J}$ and all unit-time slots in $Y$. Therefore, we obtain a schedule $S^O$ where the schedule on processor $i$ is the matching $M_i$, $i = 1, \dots, m$. The schedule $S^O$ is open-shop like since $M_i$ is a perfect matching of jobs in $\mathcal{J}$ and unit-time slots in $Y$. Besides the schedule $S^O$ schedules each job in the same unit-time slots as in $S$. Thus each job is scheduled on at most one processor at a time. Therefore, $S^O$ preserves the feasibility, and the value of objective function of

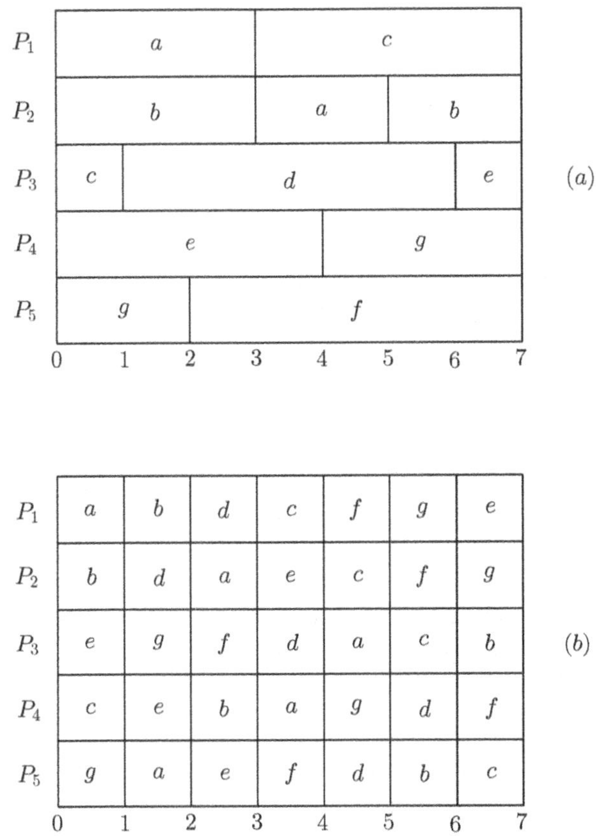


**Fig. 3.6** (**a**) Schedule $S$ for parallel processors and (**b**) open-shop like schedule corresponding to $S$

$S$. The $S^O$ can be obtained by the edge-coloring algorithm of Cole et al. [28] in $O(|E| \log m) = O(nm \log m)$ steps. Let us turn now to the case $C_{\max} > n$. Let $c$ be the total number of unit-time slots idle in the interval $[0, C_{\max}]$ on all $m$ processors in $S$. Since $c + nm = mC_{\max}$, $c$ is a multiple of $m$. Add $c$ dummy jobs to $\mathcal{J}$, and extend $Y = \{1, \ldots, C_{\max}\}$ by the set $\{C_{\max} + 1, \ldots, C_{\max} + \frac{(m-1)c}{m}\}$. Add $c$ edges connecting the dummy jobs with the unit-time slots in $Y$ to increase the degree of each vertex in $Y$ to $m$, and finally add $(m-1)c$ edges between the dummy jobs and the unit-time slots $\{C_{\max} + 1, \ldots, C_{\max} + \frac{(m-1)c}{m}\}$ to make the resulting bipartite graph $m$-regular. We can now repeat the argument used for $C_{\max} = n$ earlier in the proof. Since $|E| = m(n+c)$, the $S^O$ can be obtained by the edge-coloring algorithm in $O((n + (C_{\max} - n)m)m \log m)$ steps. □

The approach described in the proof of Theorem 3.7 is illustrated in Fig. 3.6, where Fig. 3.6a shows a schedule $S$ with $C_{\max} = 7$ of $n = 7$ jobs $a, b, c, d, e, f, g$ each with processing time equal to 5 on $m = 5$ identical parallel processors $P_1, \ldots, P_5$. Figure 3.6b shows an open-shop like schedule $S^O$ for $S$. The start and completion times are the same in both $S$ and $S^O$.

The open-shop like schedules $S^O$ obtained from schedules $S$ for the problem $\mathcal{P}_m$ solve the corresponding open shop problem $O_m$ on $m$ machines. The corresponding open shop problem has the same number of jobs $n$, each job with exactly $m$ unit-time operations, and the objective function as $\mathcal{P}_m$. Moreover by the proof of Theorem 3.7, those open-shop like schedules preserve the completion times of jobs, and thus they preserve the value of any objective function of job completion times. On the other hand, any feasible schedule for the open shop problem $O_m$ is also feasible for $\mathcal{P}_m$. Therefore, we can use the existing polynomial-time algorithms for non-preemptive or preemptive scheduling on identical parallel processors, if any, to solve the corresponding open shop scheduling problems. We show this approach for a number of open shop scheduling problems in the following sections. Observe that the $NP$-hardness of $\mathcal{P}$ does not imply the $NP$-hardness of $\mathcal{P}_m$. Thus the $NP$-hardness of $\mathcal{P}$ does not imply the $NP$-hardness of $O_m$.

### 3.5.2 Horn's Algorithm for $P|pmtn, p_i = m, r_i|L_{\max}$ Turned into an Algorithm for $O|p_{ij} = 1, r_i|L_{max}$

Horn [46] gives a polynomial-time algorithm for the minimization of maximum lateness on identical parallel machines, $P|\text{pmtn}, r_i|L_{\max}$. Each job $J_i$ has release date $r_i$, due date $d_i$, and processing time $p_i$. We assume that all data are integer in this section. The algorithm tests for a given integer $L$ whether there is a feasible preemptive schedule that schedules each job $J_i$ in the interval $[r_i, d_i + L]$. We assume non-negative $L$ for the time being, and we later show that the assumption is made without loss of generality. Also without loss of generality, we may assume that $r_i < d_i + L$ for each job $J_i$ since otherwise the test is obviously negative. The test first orders the $2n$ ends: $r_1, \dots, r_n, d_1 + L, \dots, d_n + L$ of those intervals in ascending order to determine all distinct ends and their order

$$e_1 < \cdots < e_\ell, \tag{3.18}$$

where $\ell > 1$. For each interval $[e_k, e_{k+1}]$, $k = 1, \dots, \ell - 1$, let $E_k$ be the set of jobs $J_i$ such that $r_i \leq e_k$ and $e_{k+1} \leq d_i + L$, i.e., $[e_k, e_{k+1}] \subseteq [r_i, d_i + L]$. Thus any part $x_{i,k} \geq 0$ of job $J_i \in E_k$ can be scheduled in $[e_k, e_{k+1}]$, provided that it is not longer than $e_{k+1} - e_k$. However the total of all parts $x_{i,k}$ of jobs $J_i \in E_k$ scheduled in $[e_k, e_{k+1}]$ may not exceed $m(e_{k+1} - e_k)$, which is the processing capacity the $m$ processors provide in the interval $[e_k, e_{k+1}]$. Finally, all parts of job $J_i$ need to sum up to the job's processing time $p_i$. Observe that by the construction, for each job $J_i$ there are $a(i) \leq b(i)$ such that $J_i \in E_k$ for each $k = a(i), \dots, b(i)$ and $J_i \notin E_k$ for each $k \notin [a(i), b(i)]$. We summarize those constraints in the following system $F$ of linear inequalities:

$$0 \leq x_{i,k} \leq e_{k+1} - e_k \quad J_i \in E_k, \; k = 1, \dots, \ell, \; i = 1, \dots, n;$$

$$0 = x_{i,k} \qquad J_i \notin E_k,\ k = 1, \ldots, \ell,\ i = 1, \ldots, n;$$

$$\sum_{J_i \in E_k} x_{i,k} \leq m(e_{k+1} - e_k) \qquad k = 1, \ldots, \ell - 1;$$

$$\sum_{k=a(i)}^{b(i)} x_{i,k} = p_i \qquad i = 1, \ldots, n;$$

The feasibility of $F$ for a given $L$ can be tested in $O(n^3)$ steps by the max-flow algorithm given in Malhotra et al. [64]. The test, if positive, gives integer values of $x_{i,k}$, $i = 1, \ldots, n$, $k = 1, \ldots, \ell$ since all lower and upper bounds on the flows in $F$ are integer. A binary search in the interval $[0, \sum_i p_i]$ suffices to find the $L_{\max}$. That is easy to see for the instances with $\max_i\{r_i + p_i - d_i\} \geq 0$, which result in non-negative $L_{\max}$. For an instance $I$ with $\max_i\{r_i + p_i - d_i\} < 0$, consider an instance $I_r$ with $r_i' = r_i + r$, where $r = -\max_i\{r_i + p_i - d_i\}$, and otherwise the same as $I$. Thus we have $\max_i\{r_i' + p_i - d_i\} \geq 0$, and consequently non-negative $L_{\max}'$ for $I_r$. Moreover, $L_{\max}' = r + L_{\max}$. Therefore a binary search for $L_{\max}'$ in the interval $[0, \sum_i p_i]$ finds $L_{\max}$. Thus regardless of the type of instance for $P|\text{pmtn}, r_i|L_{\max}$, we need a binary search in $[0, \sum_i p_i]$ for $L_{\max}$ and hence $O(\log \sum_i p_i)$ tests. For $p_i = m$, $i = 1, \ldots, n$, the number of tests for the open shop is $O(\log nm)$, and the algorithm requires $O(n^3 \log nm)$ steps to solve $P|\text{pmtn}, p_i = m, r_i|L_{\max}$ with $m$ processors. Table 3.2 gives an instance of $P|\text{pmtn}, p_i = m, r_i|L_{\max}$ with $n = 7$ jobs and $m = 4$ machines. Figure 3.7 gives a solution to $F$ for $L = 3$, and a schedule for the solution, where

$$e_1 = 0 < e_2 = 1 < e_3 = 2 < e_4 = 3 < e_5 = 6 < e_6 = 7 < e_7 = 8 < e_8 = 9,$$

and

$$x_{1,4} = x_{2,4} = x_{4,4} = x_{5,4} = x_{7,4} = 2 \text{ and } x_{3,4} = x_{6,4} = 1$$

are parts of jobs $J_1, J_2, J_4, J_5, J_7, J_3$, and $J_6$, respectively, in the interval $[e_4, e_5] = [3, 6]$. The parts become processing times in an instance of the problem $P|\text{pmtn}|C_{\max}$, which can be solved to optimality by McNaughton algorithm [66]. The constraints in the system $F$ guarantee that the optimal makespan equals $e_5 - e_4 = 3$. The schedule in Fig. 3.7 is *not* an open-shop like, but it can be turned into one by Theorem 3.7. Figure 3.8 gives such an open-shop like schedule solving the corresponding instance of $O|p_{ij} = 1, r_i|L_{max}$. The schedule in Fig. 3.8 has exactly the same start and completion times for each job as the schedule in Fig. 3.7. This also follows immediately from the proof of Theorem 3.7. Therefore we obtained a polynomial-time algorithm for $O|p_{ij} = 1, r_i|L_{max}$ that runs in $O(n^3 \log nm)$ time.

**Table 3.2** An instance of $P|\text{pmtn}, r_i|L_{\max}$ with $m = 4$ machines

| Job ($i$) | Release dates ($r_i$) | Due dates ($d_i$) | Processing time ($p_i$) |
|---|---|---|---|
| 1 | 0 | 3 | 4 |
| 2 | 2 | 5 | 4 |
| 3 | 1 | 4 | 4 |
| 4 | 3 | 6 | 4 |
| 5 | 1 | 3 | 4 |
| 6 | 1 | 4 | 4 |
| 7 | 2 | 5 | 4 |

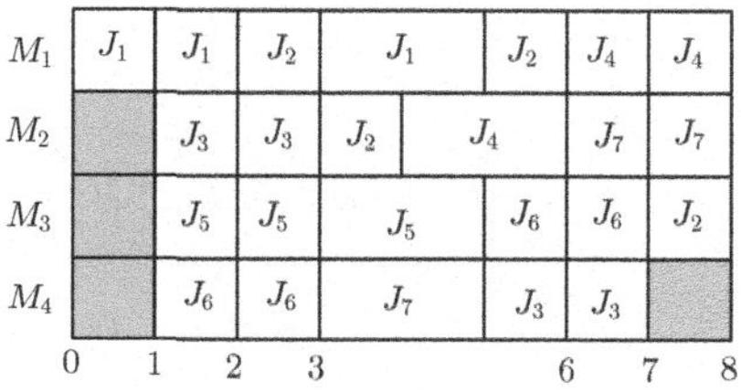


**Fig. 3.7** A schedule for the instance of $P|\text{pmtn}, r_i|L_{\max}$ in Table 3.2

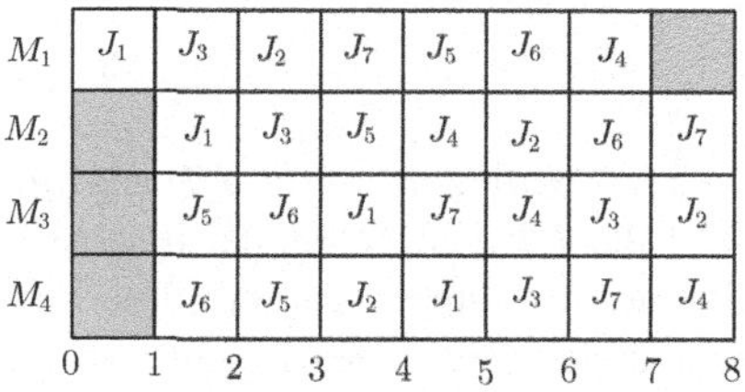


**Fig. 3.8** A schedule for the corresponding instance of $O|p_{ij} = 1, r_i|L_{max}$

### 3.5.3 *Algorithm for $P|pmtn, p_i = m, r_i|\sum C_i$ Turned into an Algorithm for $O|p_{ij} = 1, r_i|\sum C_i$*

The problem $P|\text{pmtn}, p_i = p, r_i|\sum C_i$ was studied in Baptiste et al. [8]. They prove that schedules $S$ that complete each job on machine $M_1$, that is, the schedules with $C_i = C_{i,1}, i = 1, \ldots, n$, where $C_i$ is the completion time of job $J_i$ in $S$ and $C_{i,1}$ the completion time of job $J_i$ on machine $M_1$ in $S$, include optimal schedules. This feature helps to conveniently express the objective function $\sum C_i$ as $\sum C_{i,1}$. Since all machines are identical, any order $M_1, \ldots, M_m$ of them can be selected and fixed by the algorithm at its start. The jobs are not preempted on a single machine though they can be preempted between the machines; in other words, there is a single interval, perhaps *empty*, for each job on each machine. The disjoint intervals on each machine $h$ follow the same order: the interval $[S_{1,h}, C_{1,h}]$ for $J_1$ first, followed by the interval $[S_{2,h}, C_{2,h}]$ for $J_2$ second, ..., and followed by the interval $[S_{n,h}, C_{n,h}]$ for $J_n$ last. We assume the same order of jobs determined by the order $r_1 \leq \cdots \leq r_n$ on each machine. For a job $J_i$ to observe its release date $r_i$, none of the job's $m$ intervals may start prior to $r_i$. To meet this constraint, the order of machines comes handy. Baptiste et al. [8] show that for the optimality it suffices to consider each job's $J_i$ disjoint intervals $[S_{i,m}, C_{i,m}], \ldots, [S_{i,1}, C_{i,1}]$ on machines $M_m, \ldots, M_1$, respectively, in ascending order $S_{i,m} \leq \cdots \leq S_{i,1}$ of their start times on machines $M_m, \ldots, M_1$, respectively. Observe that $S_{i,h} = S_{i,h-1}$ is possible for some machine $M_h$, which means that the job $J_i$ occupies an empty interval on $M_h$. Therefore job $J_i$ respects its release date $r_i$ as long as $r_i \leq S_{i,m}$. These features guarantee that the following linear program $LP$ with all non-negative variables finds optimal intervals for each job on each machine:

$$\min \sum_i C_{i,1}$$

$$\text{s. t} \qquad r_i \leq S_{i,m} \quad i = 1, \ldots, n;$$

$$\sum_h (C_{i,h} - S_{i,h}) = p \quad i = 1, \ldots, n;$$

$$S_{i,h} \leq C_{i,h} \quad i = 1, \ldots, n, \ h = 1, \ldots, m;$$

$$C_{i,h} \leq S_{i,h-1} \quad i = 1, \ldots, n, \ h = 2, \ldots, m;$$

$$C_{i,h} \leq S_{i+1,h} \quad i = 1, \ldots, n-1, \ h = 1, \ldots, m.$$

The coefficients of the $LP$ are limited to the values $-1$, 0, or 1; thus, by Tardos [78] the $LP$ can be solved in $O(n^5 m^5)$ steps. The solution, if it includes values that are not integer, can be turned into an integer solution with the same value of total completion time by the network-flow approach given in Baptiste et al. [8]. It is worth observing that though the $LP$ solves the problem $P|\text{pmtn}, p_i = p, r_i|\sum C_i$ with equal processing times, it does *not* solve the problem $P|\text{pmtn}, r_i|\sum C_i$ with arbitrary processing times, which is shown to be NP-hard in the strong sense in Baptiste et al. [8].

Table 3.3 gives an instance of $P|\text{pmtn}, p_i = m, r_i|\sum C_i$ with $n = 10$ jobs and $m = 5$ machines. Figure 3.9 gives a solution to the $LP$, and a schedule for the solution. The jobs follow order $J_1, \ldots, J_{10}$ on each machine, though some jobs, like $J_6$ and $J_7$, are scheduled in empty interval (12, 12) on machine $M_1$ for the instance. The job $J_5$ is scheduled in non-empty intervals ($S_{5,3} = 7, C_{5,3} = 8$) on machine $M_3$, ($S_{5,2} = 8, C_{5,2} = 11$) on machine $M_2$, and ($S_{5,1} = 11, C_{5,1} = 12$) on machine $M_1$. The job $J_7$ is scheduled in non-empty intervals ($S_{7,5} = 7, C_{7,5} = 8$) on machine $M_5$, ($S_{7,4} = 8, C_{7,4} = 11$) on machine $M_4$, and ($S_{7,3} = 11, C_{7,3} = 12$) on machine $M_3$. The job $J_5$ completes at $C_5 = C_{5,1} = 12$, and the job $J_7$ completes at $C_7 = C_{7,3} = C_{7,2} = C_{7,1} = 12$. All jobs, except for job $J_9$, complete at $r_i + m$, and $J_9$ completes at $r_9 + m + 1$; thus, the schedule in Fig. 3.9 minimizes the total completion time. We leave the proof of that as an exercise; see Problem 3.2. The schedule in Fig. 3.9 is *not* an open-shop like, but it can be turned into one by Theorem 3.7. Figure 3.10 shows such an open-shop like schedule solving the

**Table 3.3** An instance of $P|\text{pmtn}, p_i = p, r_i|\sum C_i$ with $m = 5$ machines

| Job ($i$) | Release dates ($r_i$) | Processing time ($p_i = p$) |
|---|---|---|
| 1 | 0 | 5 |
| 2 | 2 | 5 |
| 3 | 3 | 5 |
| 4 | 6 | 5 |
| 5 | 7 | 5 |
| 6 | 7 | 5 |
| 7 | 7 | 5 |
| 8 | 10 | 5 |
| 9 | 10 | 5 |
| 10 | 13 | 5 |

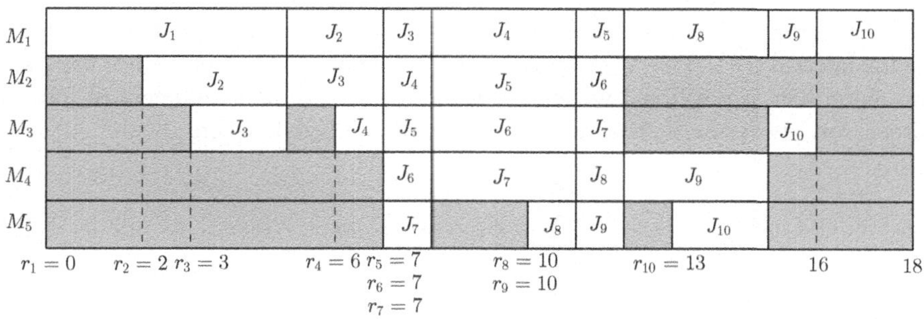


**Fig. 3.9** A schedule for the instance of $P|\text{pmtn}, p_i = p, r_i|\sum C_i$ in Table 3.3

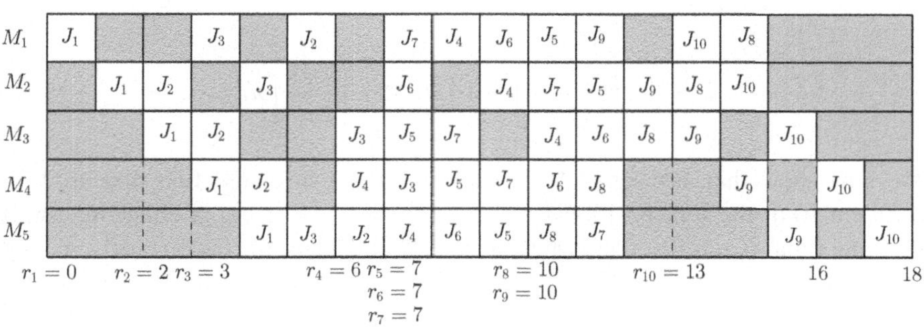


**Fig. 3.10** A schedule for the corresponding instance of $O|p_{ij} = 1, r_i|\sum C_i$

corresponding instance of $O|p_{ij} = 1, r_i|\sum C_i$. The schedule in Fig. 3.9 has exactly the same start and completion times for each job as the schedule in Fig. 3.10, which also follows immediately from the proof of Theorem 3.7. Therefore we obtained a polynomial-time algorithm for $O|p_{ij} = 1, r_i|\sum_i C_i$ that runs in $O(n^5m^5)$ time.

### 3.5.4 Algorithm for $P|pmtn, p_i = m|\sum C_i$ Turned into an Algorithm $O|p_{ij} = 1|\sum w_i C_i$

We begin by considering the problem $P|\text{pmtn}, p_i = m|\sum C_i$ with $m$ parallel identical processors. By McNaughton [66], preemptions of jobs do not reduce total completion time and thus it suffices to solve $P|p_j = m|\sum C_i$ to solve $P|\text{pmtn}, p_i = m|\sum C_i$. To find an optimal solution to $P|p_j = m|\sum C_i$, we match the following $|\mathcal{J}|$ positional weights

$$\underbrace{m,\ldots,m}_{m-positions},\ldots,\underbrace{km,\ldots,km}_{m-positions},\ldots,\underbrace{\left\lfloor\frac{|\mathcal{J}|}{m}\right\rfloor m,\ldots,\left\lfloor\frac{|\mathcal{J}|}{m}\right\rfloor m}_{m-positions},\underbrace{\left\lceil\frac{|\mathcal{J}|}{m}\right\rceil m,\ldots,\left\lceil\frac{|\mathcal{J}|}{m}\right\rceil m}_{r-positions}, \text{ where}$$

$|\mathcal{J}| = \lfloor\frac{|\mathcal{J}|}{m}\rfloor + r$ with the $|\mathcal{J}|$ jobs. Any matching will do since all jobs have the same processing time. The job $J_i$ matched with positional weight $\ell m$ is then scheduled in the interval $[(\ell-1)m, \ell m]$ on any of the $m$ identical processors. Since there are no more than $m$ positional weights of value $\ell m$, a feasible assignment of jobs to processors always exist. Thus we obtain a feasible schedule where

$$\sum_{i=km+1}^{(k+1)m} C_i = (k+1)m \tag{3.19}$$

where $C_{mk+1}, \ldots, C_{(m+1)k}$ are completion times of jobs matched with $mk+1$-st, $\ldots$, $(k+1)m$-st positional weights, respectively, $k = 0, \ldots, \left\lfloor\frac{|\mathcal{J}|}{m}\right\rfloor$. Thus the total completion time equals

$$\left\lfloor\frac{|\mathcal{J}|}{m}\right\rfloor\left(\left\lfloor\frac{|\mathcal{J}|}{m}\right\rfloor + 1\right)m^2/2 + \left\lceil\frac{|\mathcal{J}|}{m}\right\rceil mr. \tag{3.20}$$

We now show that any schedule $S'$ such that the $\ell$-th completion time in $S'$ is different from the $\ell$-th positional weight, $C_\ell \neq \lfloor\frac{\ell}{m}\rfloor m + m$, is either infeasible or suboptimal. Consider the earliest such $\ell$ in $S'$. Thus we have exactly $m$ jobs in each interval $[0, m], \ldots, [(\lfloor\frac{\ell}{m}\rfloor - 1)m, \lfloor\frac{\ell}{m}\rfloor m]$ each of length $m$. Therefore the job that completes at $C_\ell$ cannot start before $\lfloor\frac{\ell}{m}\rfloor m$ and thus it cannot finish before $\lfloor\frac{\ell}{m}\rfloor m + m$. Therefore $C_\ell > \lfloor\frac{\ell}{m}\rfloor m + m$. This schedule $S'$, however, cannot be optimal. Thus we proved not only that the algorithm provides an optimal schedule but also that each optimal schedule is a matching of the jobs with the positional weights. The positional weights are also completion times of jobs. By Theorem 3.7, we obtain an optimal solution to $O|p_{ij} = 1|\sum C_i$; see also Adiri and Amit [2]. Consider now $O|p_{ij} = 1|\sum w_i C_i$. To get an optimal schedule for this weighted completion time problem, order the jobs in non-increasing order of their weights $w_1 \geq w_2 \geq \cdots \geq w_{|\mathcal{J}|}$ and match them with positional weights so that the position weight in position $\ell$ is matched with the job in position $\ell$. The resulting solution minimizes the weighted total completion time by the rearrangement inequality of Hardy, Littlewood, and Polya [44].

Let us return to the problem $P|\text{pmtn}, p_i = m|\sum C_i$. The solution presented earlier minimizes the total completion time but not the makespan, which equals $\left\lceil \frac{|\mathcal{J}|}{m} \right\rceil m$ while the minimum makespan equals $|\mathcal{J}|$. It was shown by Adiri and Amit [2] that an ideal feasible schedule (see Coffman et al. [27]), which minimizes total completion time and makespan at the same time, exists for each instance of $P|\text{pmtn}, p_i = m|\sum C_i$ and thus for each instance of $O|p_{ij} = 1|\sum C_i$. Such ideal schedule can be obtained by completing each of the last $r$ jobs by $m - r$ time units earlier, and by delaying $m-r$ jobs by $r$ units of time each. Thus all jobs complete by $n$, and the total completion time decreases by total $r(m-r)$ and increases by $(m-r)r$ with respect to the original schedule, which still minimizes total completion time. Thus the resulting schedule, if feasible, is ideal. We leave the proof of feasibility as an exercise; see Problem 3.3. We illustrate this approach to obtaining feasible ideal schedules by an example with $n = 7$ and $m = 4$. Figure 3.11a gives an optimal solution to $P|p_i = m|\sum C_i$ with total completion time equal to 40 and $C_{\max} = 8$, and Fig. 3.11b gives an optimal solution to $P|\text{pmtn}, p_i = m|\sum C_i$ with the same value of total completion time but minimum makespan 7. The key to the approach

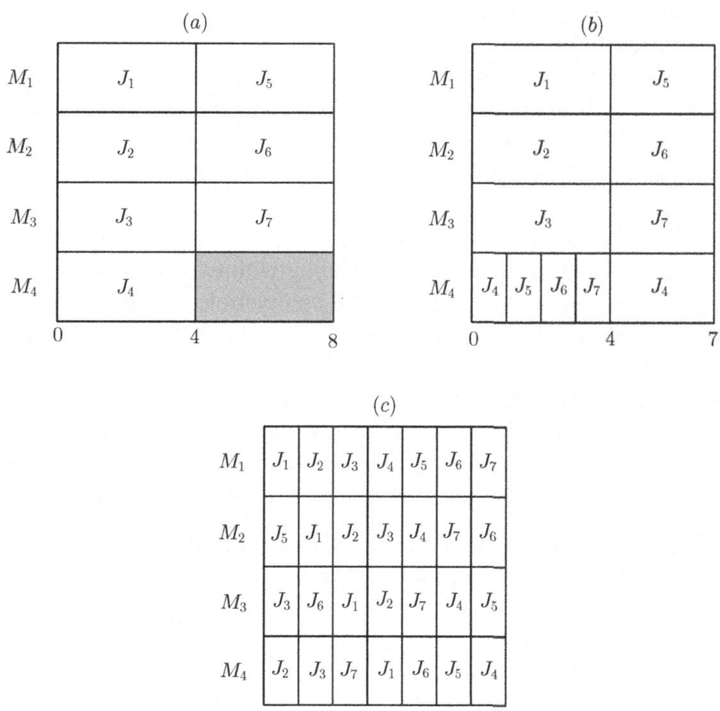


**Fig. 3.11** An example of ideal schedule for $O|p_{ij} = 1|\sum C_i$

is that the schedule in Fig. 3.11b is obtained from this in Fig. 3.11a by delaying the job $J_4$ by $r = 3$ units of time on $M_4$, which creates a gap of length 3 on $M_4$ $(m - r = 1)$, and speeding up the completion of jobs $J_5$, $J_6$, and $J_7$ $(r = 3)$ by one unit of time each. The gap on machine $M_4$ is then used to schedule one time unit of each job $J_4$, $J_5$, $J_6$, and $J_7$ to make the resulting ideal schedule feasible. Generally, the McNaughton [66] wrap-around rule ensures the feasibility of the changes. Figure 3.11c gives a resulting ideal schedule for $O|p_{ij} = 1|\sum C_i$.

Observe that only two numbers: $n$ and $m$ specify an instance of the problem $O|p_{ij} = 1|\sum C_i$. Thus the input of the problem is of size $O(\log n + \log m)$. Therefore an algorithm running in time $O(n)$, for instance, is no longer polynomial but rather pseudopolynomial for the succinct input encoding; see Kubiak [52] and Grigoriev [40] for other scheduling problems with succinct input encoding. However, the starting and completion times of each operation in a schedule can be computed in polynomial time even for the succinct encoding of the input data. We leave the proof of this as an exercise; see Problem 3.3.

The problem $O|p_{ij} = 1|\sum w_i C_i$ is not ideal. To see this, take an instance with $n = 5$ jobs where four jobs have weights 5 each, and one job has weight equal to 1, $m = 4$. The optimal solution has total weighted completion time equal to $4 \times 4 \times 5 + 8 \times 1 = 88$, while any schedule with $C_{\max} = 5$ has total weighted completion time at least $4 \times 5 + 3 \times 5 \times 5 + 5 = 100$.

Atay et al. [5] consider the problem $O|p_{ij} = 1|\sum C_i$ as a point of departure for their unit open shops scheduling game. The game is a cooperative game subject to an initial schedule. A coalition, i.e., a subset of jobs, can possibly improve its total completion time in comparison to the initial schedule by means of admissible rearrangements. Then the improvement results in a positive maximal cost saving if the rearrangement is done optimally. The authors prove that the core of the game is not empty.

In this context of bi-criteria open shop scheduling, we point out that Kyparisis and Koulamas [54] study open shop scheduling with two and three machines and arbitrary processing times to minimize total completion time subject to minimum makespan. They identify some polynomial cases for that NP-hard in the strong sense problem. Masuda and Ishii [65] consider both makespan and maximum lateness in a bi-criteria preemptive open shop scheduling for two machines. The maximum lateness can be minimized in polynomial time (see Sect. 3.7.2) and the makespan in linear time. The authors characterize non-dominated solutions whenever both criteria cannot be optimized at the same time. We have the following problem.

**Problem 3.1** Characterize non-dominated solutions for the Masuda and Ishii [65] bi-criteria open shop scheduling problem with $m > 2$ machines.

### 3.5.5 *Other Open Shop Scheduling Problems with All Unit-Time Operations*

Kravchenko [50] proves that the problem $O|p_{ij} = 1, r_i|\sum U_i$ is NP-hard in the strong sense. Baptiste [7] presents a dynamic program that runs in polynomial time to solve the problem $Om|p_{ij} = 1, r_i|\sum w_i U_i$ with fixed number of machines. Liu and Bulfin [60] show that the problems $O|p_{ij} = 1|\sum U_i$ and $O|p_{ij} = 1|\sum T_i$ are polynomial. Brucker et al. [18] (see also Brucker [14]) prove that the problem $Om|p_{ij} = 1|\sum w_i U_i$ is polynomial; however, the complexity status of the problem $O|p_{ij} = 1|\sum w_i U_i$ remains open.

## 3.6 Open Shops with 0–1 Operations

The previous section showed that open shop scheduling with unit-time operations is polynomial for a broad range of objective functions. This is due to a large degree to the fact that each job not only has the same number of unit-time operations as any other job but, more importantly, this number equals the number of machines $m$, i.e., no operation is missing. The introduction of missing operations with processing times 0 (we refer to such open shops as open shops with 0–1 operations) changes the open shop scheduling problem's complexity significantly. The problem becomes NP-hard for most objective functions though the makespan minimization remains polynomial. The open shop scheduling problem with missing operations often remains NP-hard even for all jobs having the same number of unit-time operation, which is, however, less than $m$. We discuss the complexity in the following subsections.

### 3.6.1 *Makespan Minimization: Problem $O|p_{ij} = 0, 1|C_{\max}$*

Each instance of the problem $O|p_{ij} = 0, 1|C_{\max}$ can be represented by a simple bipartite graph $G = (\mathcal{J}, \mathcal{M}, E)$, where $\mathcal{J}$ is the set of jobs, $\mathcal{M}$ is the set of machines, and $(J_j, M_h) \in E$ if and only if job $J_j$ has a unit-time operation on machine $M_h$. Thus by König's edge-coloring theorem (see Theorem 1.1), we have $C_{\max} = \{\max_j\{P_j\}, \max_h\{L_h\}\} = \Delta(G)$, and the makespan can be computed in $O(nm)$ time. The $\Delta(G)$-edge coloring can then be computed in time $O(|E| \log \Delta(G))$ by edge-coloring algorithm of Cole et al. [28]. Since $|E| \leq nm$ and $\Delta(G) \leq \max\{n, m\}$, we get an algorithm that runs in time $O(nm \log \max\{n, m\})$ for $O|p_{ij} = 0, 1|C_{\max}$. Thus we obtain the following theorem.

**Theorem 3.8** $O|p_{ij} = 0, 1|C_{\max}$ *can be solved in time* $O(nm \log \max\{n, m\})$.

A polynomial-time algorithm for $O|p_{ij} = 0, 1|C_{\max}$ can be also derived from algorithms for the problem $O|pmtn|C_{\max}$; see Gonzalez and Sahni [37] or Gonzalez [35].

### 3.6.2 Total Completion Time: Problem $O|p_{ij} = 0, 1|\sum C_i$

Gonzalez [36] shows that allowing missing operations renders the total completion time minimization NP-hard in the strong sense. He proves the following theorem.

**Theorem 3.9** $O|p_{ij} = 0, 1|\sum C_i$ *is NP-hard in the strong sense.*

***Proof*** The transformation is from 3-COLORING OF 4-REGULAR GRAPHS: Given a 4-regular graph $G = (N, E)$. Can the vertices of $G$ be colored using three different colors so that any two vertices connected by an edge in $E$ are colored with different colors? The question can be re-phrased as follows: Can $N$ be partitioned into disjoint subsets $N_1$, $N_2$, and $N_3$ so that $|e \cap N_\ell| \leq 1$ for $\ell = 1, 2, 3$, $e \in E$? The problem is NP-complete in the strong sense; see Gonzalez [36].

We now describe an instance of $O|p_{ij} = 0, 1|\sum C_i$ corresponding to $G$. We start with a general overview of the instance. Each job in the instance will have exactly five unit-time operations on some machines—on which will depend mainly on $G$—and it will be missing on all other machines. The number of jobs is three times the number of machines $m$, and each machine's workload will be exactly fifteen in the instance. The focus will be on such schedules for the instance that execute each job *entirely* either in the interval $[0, 5]$ or $[5, 10]$ or $[10, 15]$. We call those schedules *5-10-15* schedules. It is easy to verify that any *5-10-15* schedule for the instance results in mean flow time $F = 10$. Furthermore, any schedule with $F = 10$ for the instance is a *5-10-15* schedule. The proof of the latter claim is not trivial and quite technical; in order to streamline the NP-hardness proof, we leave its details for later.

Here are the details of the instance. The instance has an *interval selector* for each vertex $v_i \in N$. The interval selector for $v_i$ consists of five machines $s_1^i, s_2^i, s_3^i, s_4^i$, and $s_5^i$ and fourteen jobs $j_1^i, \ldots, j_{14}^i$. Each job of the selector has a unit-time operation on each of the five machines and it is missing on any other machine. The following is the *key* purpose of the selector in a *5-10-15* schedule: In a *5-10-15* schedule $S$, for each interval selector there are exactly five disjoint unit-time slots—one on each of the five machines of the selector—in the interval $[0, 15]$ where jobs that do *not* belong to the selector are executed. Moreover all five unit-time slots fall *entirely* in either the interval $[0, 5]$ or $[5, 10]$ or $[10, 15]$ in $S$. Accordingly, we say that the interval selector *selects* interval $[0, 5]$ or $[5, 10]$ or $[10, 15]$ for $v_i$ in $S$.

Furthermore, the instance has a *main* component that consists of $|E|$ machines $e_1, \ldots, e_{|E|}$, one machine for each edge of $E$. For each vertex $v_i \in N$, there are five *vertex*-jobs $a_i, b_i, c_i, d_i$, and $f_i$ each having unit-time operations on four machines $e_{i_1}, e_{i_2}, e_{i_3}$, and $e_{i_4}$ corresponding to the edges in $E$ incident with $v_i$. Moreover each job $a_i, b_i, c_i, d_i$, and $f_i$ has one unit-time operation on machines $s_1^i, s_2^i, s_3^i, s_4^i$, and $s_5^i$ of the interval selector for $v_i$. Those operations along with the feature of the interval selector for $v_i$ mentioned earlier ensure that all five jobs $a_i, b_i, c_i, d_i$, and $f_i$ will be executed *entirely* in either $[0, 5]$ or $[5, 10]$ or $[10, 15]$, and the selection of the interval is consistent with the selection made by the interval selector for $v_i$ in a *5-10-15* schedule $S$.

Finally, there are five copies of the main component and five copies of each interval selector in the instance. To avoid excessive notation, we drop the notation required to distinguish the copies. The five copies of the main component are linked together by the *edge*-jobs $g_j, h_j, x_j, y_j$, and $z_j$ for each edge $e_j$. Each of these jobs has a unit-time operation on processor $e_j$ in each copy. To summarize, there are $|E| + 5n$ processors, and $|E| + 5n + 14n$ jobs in each of the five copies. Thus $m = 5|E| + 25n$ processors and $5|E| + 95n$ jobs altogether in the instance. Since we have $|E| = 2n$ for 4-regular graphs, the instance has $m = 35n$ processors and $105n$ jobs. Recall that each job has exactly five unit-time operations.

Suppose that $N$ can be partitioned into disjoint subsets $N_1$, $N_2$, and $N_3$ so that $|e \cap N_\ell| \leq 1$ for $\ell = 1, 2, 3$ and $e \in E$. Let $E_\ell$ be the set of edges in $E$ that are not incident to the vertices in $N_\ell$, i.e., $E_\ell = \{e \in E : N_\ell \cap e = \emptyset\}$. The sets $E_1$, $E_2$, and $E_3$ are pairwise disjoint, and $E_1 \cup E_2 \cup E_3 = E$. Keeping this in mind, a *5-10-15* schedule can be obtained as follows: the vertex-jobs corresponding to the vertices in $N_\ell$ are assigned to machines in $E \setminus E_\ell$ in $[5(\ell - 1), 5\ell]$ for $\ell = 1, 2, 3$ (these machines correspond to the edges in $E$ that are incident with some node in $N_\ell$) in each of the five copies. The edge-jobs corresponding to edges in $E_\ell$ are assigned to machines in $E_\ell$ in $[5(\ell - 1), 5\ell]$ for $\ell = 1, 2, 3$ (these machines correspond to the edges in $E$ that are not incident with any vertex in $N_\ell$) in each of the five copies. The interval selector is set to the interval $[5(\ell - 1), 5\ell]$ for each vertex in $N_\ell$, $\ell = 1, 2, 3$, and this setting applies to all its five copies. By König's edge-coloring theorem, Theorem 1.1, a feasible schedule can be obtained in the each of the intervals $[0, 5]$, $[5, 10]$, and $[10, 15]$ on $m$ machines for the job assignment just described. Therefore we obtain a *5-10-15* schedule for the instance.

Suppose that a *5-10-15* schedule exists for the instance. Consider the same copy of the interval selector and the main component. Let $S_\ell$ be the set of vertices with their interval selector selecting $[5(\ell - 1), 5\ell]$ for $\ell = 1, 2, 3$. The sets are pairwise disjoint and their union equals $N$. Thus it remains to prove $|e \cap S_\ell| \leq 1$ for $\ell = 1, 2, 3$ and $e \in E$. To that end, let us consider any vertex $v_i \in S_\ell$ for $\ell = 1, 2, 3$. Exactly one unit-time operation of each job $a_i, b_i, c_i, d_i$, and $f_i$ corresponding to $v_i$ is executed in $[5(\ell - 1), 5\ell]$ on the interval selector machines. Thus by definition of *5-10-15* schedule, each job $a_i, b_i, c_i, d_i$, and $f_i$ corresponding to $v_i$ is *entirely* executed in $[5(\ell - 1), 5\ell]$. Consequently each machine $e_{i_1}, e_{i_2}, e_{i_3}$, and $e_{i_4}$, where each edge $e_{i_1}, e_{i_2}, e_{i_3}$, and $e_{i_4}$ is incident with $v_i$, executes one unit-time operation of each job $a_i, b_i, c_i, d_i$, and $f_i$ in $S$. Now suppose for contradiction that $|e \cap S_\ell| = 2$ for some $e$ and $S_\ell$. Let $e = (v_i, v_j)$. Then the machine $e$ would execute the unit-time operation of each $a_i, b_i, c_i, d_i$, and $f_i$ and $a_j, b_j, c_j, d_j$, and $f_j$, altogether ten unit-time operations, in $[5(\ell - 1), 5\ell]$, which gives a contradiction and proves that $|e \cap S_\ell| \leq 1$ for $\ell = 1, 2, 3$ and $e \in E$ (or that $G$ has 3-coloring).

To complete the proof, it remains to show that if $F = 10$ for $S$, then $S$ is a *5-10-15* schedule. Let $C_1 \le \cdots \le C_m \le \cdots \le C_{2m} \le \cdots \le C_{3m}$ be the completion times of the $3m$ jobs in $S$. Since each job has exactly five unit-time operations, we immediately have $5 \le C_1 \le C_2 \le \cdots \le C_m$; thus, $\sum_{i=1}^{m} C_i \ge 5m$. We now show that $\sum_{i=m+1}^{2m} C_i \ge 10m$ and $\sum_{i=2m+1}^{3m} C_i \ge 15m$. We prove the former first. Consider $C_{m+1} \le \cdots \le C_{2m}$. If $C_{m+1} \ge 10$, then $\sum_{i=m+1}^{2m} C_i \ge 10m$, and the claim holds. Assume $C_{m+1} < 10$. Let $m+1 \le k < 2m$ be the latest job finishing *before* 10 in $S$, i.e., $C_k < 10$ and $C_{k+1} \ge 10$. Such job exists since $C_{2m} \ge 10$ (by König's edge-coloring theorem $2m$ jobs—each having five unit-time operations—cannot complete all their operations before 10 on $m$ processors). Consider total deviation $\sum_{i=m+1}^{k}(10 - C_i)$ from 10 for the jobs completing *before* 10 in $S$, and total deviation $\sum_{i=k+1}^{2m}(C_i - 10)$ from 10 for the jobs completing *at* or *after* 10 in $S$. We now show that $\sum_{i=m+1}^{k}(10 - C_i) \le \sum_{i=k+1}^{2m}(C_i - 10)$, which proves our claim that $\sum_{i=m+1}^{2m} C_i \ge 10m$. Let $I_x = \{\ell : C_{m+1} \le C_\ell \le x\}$ be the set of jobs that complete by $x$, for $x = C_{m+1}, \ldots, 9$, but not before $C_{m+1}$ in $S$. Observe that at least $|I_x|$ unit-time slots are not occupied by jobs $1, \ldots, 2m$ on $m$ processors in $[x, x+1]$ in $S$. We have $\sum_{x=C_{m+1}}^{9} |I_x| = \sum_{i=m+1}^{k}(10 - C_i)$. On the other hand, let $I$ be the number of unit-time slots in $[0, 10]$ not occupied by the jobs $1, \ldots, 2m$. We have $I = \sum_{i=k+1}^{2m}(C_i - 10)$ (by König's edge-coloring theorem $2m$ jobs—each having five unit-time operations—can complete all their operations before 10 on $m$ processors and leaves no time to spare in $[0, 10]$). Clearly, $I \ge \sum_{x=C_{m+1}}^{9} |I_x|$ and thus the claim holds. Similarly we prove $\sum_{i=2m+1}^{3m} C_i \ge 15m$, and we omit details. Thus we just proved that $\sum_{i=1}^{m} C_i \ge 5m$, $\sum_{i=m+1}^{2m} C_i \ge 10m$, and $\sum_{i=2m+1}^{3m} C_i \ge 15m$. If $C_m > 5$, then $\sum_{i=1}^{m} C_i > 5m$ and $\sum_{i=1}^{3m} C_i > 30m$, which contradicts that $S$ has $F = 10$. Thus $C_1 = \cdots = C_m = 5$ and $\sum_{i=1}^{m} C_i = 5m$ in $S$. If $C_{m+1} > 10$, then $\sum_{i=m+1}^{2m} C_i > 10m$ and $\sum_{i=1}^{3m} C_i > 30m$, which again leads to the contradiction. Thus $C_{m+1} = \cdots = C_{2m} = 10$ and $\sum_{i=m+1}^{2m} C_i = 10m$ in $S$. Therefore also $C_{2m+1} = \cdots = C_{3m} = 15$ and $\sum_{i=2m+1}^{3m} C_i = 15m$ in $S$. This proves that $S$ is a *5-10-15* schedule as required and completes the proof of NP-hardness in the strong sense for $O|p_{ij} = 0, 1|\sum C_i$. □

Lushchakova and Kravchenko [62] give a polynomial-time algorithm for $O2|p_{i,j} = 0, 1|\sum w_i C$. Their algorithm runs in $O(n \log n)$ time.

### 3.6.3 Maximum Lateness: Problem $O|p_{ij} = 0, 1, r_i|L_{\max}$; and Number of Tardy Jobs: Problem $O|p_{ij} = 0, 1|\sum U_i$

Kubale [51] shows that the problem $O|p_{ij} = 0, 1, r_i|L_{\max}$ is NP-hard in the strong sense. To close the existing gap, to the author's knowledge the problem has been open thus far, we now show that the minimization of number of tardy jobs is NP-hard in the strong sense.

**Theorem 3.10** *The problem $O|p_{ij} = 0, 1|\sum U_i$ is NP-hard in the strong sense.*

***Proof*** The reduction is from the interval 6-edge coloring of (3, 6) biregular bipartite graphs; see Sect. 9.3 for definitions of biregular bipartite graphs and the problem itself. The problem is shown NP-complete in the strong sense by Asratian and Casselgren [4]; see also Theorem 9.7 in this book. Let $G = (\mathcal{J}, \mathcal{M}, E)$ be a (3, 6) biregular bipartite graph. The graph makes one part of an instance of the problem $O|p_{ij} = 0, 1|\sum U_i$, where $\mathcal{J}$ is a set of $n$ jobs and $\mathcal{M}$ is a set of $m$ machines. Observe that $n = 2m$ in $G$. Each job $J_i$ has three unit-time operations processed on three different machines adjacent to $J_i$ in the bipartite $G$. The operations of $J_i$ on any other machine are missing. The other part, the due dates of jobs, are set to $d_i = 3$ for any job $J_i \in \mathcal{J}$. Finally the threshold for the number of tardy jobs $\sum U_i$ is set to $m$. Since $G$ is a (3, 6) biregular bipartite graph, the instance $O|p_{ij} = 0, 1|\sum U_i$ has each machine workload equal to 6 and each job length equal to 3.

Suppose that there is an interval 6-edge coloring $c$ of $G$. Then one edge incident with each vertex in $\mathcal{M}$ is colored with color 1 in $c$. Thus there is a subset $\mathcal{J}_1 \subset \mathcal{J}$ of $m$ jobs such that each job in $\mathcal{J}_1$ has one edge incident with that job colored with color 1. Since $c$ is an interval coloring, each job in $\mathcal{J}_1$ is colored with colors 1, 2, and 3 and hence can be scheduled to complete by 3. This gives a schedule with $\sum U_i = m$.

Now suppose that $S$ is a schedule for the problem $O|p_{ij} = 0, 1|\sum U_i$ with at least $m$ jobs that are not tardy, i.e., complete by 3. Since each job is of length 3, there are exactly $m$ jobs that are not tardy. Let $\mathcal{J}_0$ be the set of all these jobs. Thus there is an edge coloring $c_0$ of $G$ such that the edges incident with vertices in $\mathcal{J}_0$ are colored with 1, 2, and 3 in $c_0$. Moreover, each vertex in $\mathcal{M}$ has three edges out of six edges incident with the vertices colored with 1, 2, and 3 in $c_0$. Let $G_0 = (\mathcal{J}_0, \mathcal{M}, E_0)$, where $E_0$ is the set of all edges incident with vertices in $\mathcal{J}_0$. The edge coloring $c_0$ limited to $G_0$ is an interval 3-edge coloring of $G_0$. Delete $\mathcal{J}_0$ and all edges incident to the vertices in $\mathcal{J}_0$ from $G$. The resulting graph $G_3 = (\mathcal{J} \setminus \mathcal{J}_0, \mathcal{M}, E \setminus E_0)$ is 3-regular bipartite graph. Hence $G_3$ has a perfect matching; see Bondy and Murty [11], for instance. Thus by Lemma 9.1 $G_3$ has an interval 3-edge coloring $c_1$. Change the colors 1, 2, and 3 in $c_1$ into 4, 5, and 6, respectively, to obtain edge coloring $c'$ of $G_3$. The edge coloring $c_0 \cup c'$ makes an interval 6-edge coloring of $G_0 \cup G_3 = G$. □

## 3.7 Preemptive Open Shop Scheduling

### *3.7.1 An Algorithm for $O|pmtn|C_{max}$*

Gonzalez and Shani [37] give a polynomial-time algorithm for $O|\text{pmtn}|C_{max}$. The algorithm runs in $O(r(\min\{r, m^2\} + m \log n))$ time, where $r$ is the number of operations with positive processing times. The total number of preemptions introduced by the algorithm does not exceed $r(m - 1) + m^2$. Gonzalez [35] later proposes another algorithm that runs in $O(r + \min\{m^4, n^4, r^2\})$ time for the problem. The algorithm introduces $O(\min\{rn, rm, n^3, m^3\})$ preemptions. Vairaktarakis and Sahni [80] introduce further improvements to those algorithms.

We show a polynomial-time algorithm that is based directly on Birkhoff–von Neumann theorem for doubly stochastic matrices for the problem. Let $n \times m$ matrix $\mathbb{P}$ be an instance of $O|\text{pmtn}|C_{max}$ with $n$ jobs and $m$ machines. Let $\Delta(\mathbb{P}) = \{\max_i\{P_i = \sum_h p_{i,h}\}, \max_h\{L_h = \sum_i p_{i,h}\}\}$ be the degree of the instance, where $P_i$ is length of job $J_i$ and $L_h$ is workload of machine $M_h$. Clearly, the degree of the instance $\Delta(\mathbb{P})$ is a lower bound on minimum makespan for the instance $\mathbb{P}$ of $O|\text{pmtn}|C_{max}$. Let $m \times n$ matrix $\mathbb{P}^T$ be an instance of $O|\text{pmtn}|C_{max}$ that is dual to the instance $\mathbb{P}$. Informally speaking, jobs in $\mathbb{P}$ become machines in $\mathbb{P}^T$, and machines in $\mathbb{P}$ become jobs in $\mathbb{P}^T$. Let us define $(n + m) \times (n + m)$ matrix

$$\mathbb{Q} = \begin{bmatrix} \mathbb{P} & \mathbb{A} \\ \mathbb{B} & \mathbb{P}^T \end{bmatrix},$$

where the $n \times n$ diagonal matrix

$$\mathbb{A} = \begin{bmatrix} \Delta(\mathbb{P}) - P_1 & 0 & \dots & 0 \\ 0 & \Delta(\mathbb{P}) - P_2 & \dots & 0 \\ 0 & \dots & & 0 \\ . & . & . & . \\ . & . & . & . \\ . & . & . & . \\ 0 & 0 & \dots & \Delta(\mathbb{P}) - P_n \end{bmatrix}$$

complements the sum of each row of $\mathbb{P}$ and each column of $\mathbb{P}^T$ to $\Delta(\mathbb{P})$, and the $m \times m$ diagonal matrix

$$\mathbb{B} = \begin{bmatrix} \Delta(\mathbb{P}) - L_1 & 0 & \dots & 0 \\ 0 & \Delta(\mathbb{P}) - L_2 & \dots & 0 \\ 0 & \dots & & 0 \\ . & . & . & . \\ . & . & . & . \\ . & . & . & . \\ 0 & 0 & \dots & \Delta(\mathbb{P}) - L_m \end{bmatrix}$$

complements the sum of each column of $\mathbb{P}$ and each row of $\mathbb{P}^T$ to $\Delta(\mathbb{P})$. The matrix $\frac{1}{\Delta(\mathbb{P})}\mathbb{Q}$ is doubly stochastic. By Birkhoff–von Neumann theorem (see Theorem 1.3), there are $(n + m) \times (n + m)$ permutation matrices $\mathbb{P}_1, \dots, \mathbb{P}_q$ and positive real numbers $\lambda_1, \dots, \lambda_q$ such that

$$\frac{1}{\Delta(\mathbb{P})}\mathbb{Q} = \lambda_1 \mathbb{P}_1 + \cdots + \lambda_q \mathbb{P}_q, \tag{3.21}$$

and $\lambda_1 + \cdots + \lambda_q = 1$. Moreover $q \le (n+m)^2 - (n+m) + 1$. Let $\mathbb{S}_i$ be an $n \times m$ matrix obtained by deleting rows $n+1, \ldots, n+m$ and columns $m+1, \ldots, n+m$ from the permutation matrix $\mathbb{P}_i$, $i = 1, \ldots, q$. Thus at most one entry equals 1 in each row and each column of $\mathbb{S}_i$ (and the remaining entries equal 0), for $i = 1, \ldots, q$. We have

$$\frac{1}{\Delta(\mathbb{P})}\mathbb{P} = \lambda_1 \mathbb{S}_1 + \cdots + \lambda_q \mathbb{S}_q. \tag{3.22}$$

A schedule $\mathcal{S}_i$ for $\mathbb{S}_i$ is obtained by scheduling job $J_j$ on machine $M_h$ for $\lambda_i \Delta(\mathbb{P}) S_{j,h}$ units of time, where the entry $S_{j,h}$ of $\mathcal{S}_i$ equals 0 or 1. The schedule is feasible since each job is scheduled on at most one machine in $\mathcal{S}_i$, and each machine processes at most one job in $\mathcal{S}_i$. The makespan of $\mathcal{S}_i$ equals $C_i = \lambda_i \Delta(\mathbb{P}) S_{j,h}$. The schedules $\mathcal{S}_1, \ldots, \mathcal{S}_q$ can be permuted in any way to obtain a complete schedule for $\mathbb{P}$. The concatenation of the schedules for each permutation results in a feasible schedule with makespan $C_1 + \cdots + C_q \le \Delta(\mathbb{P})$. Observe that (3.22) guarantees that each operation $O_{j,h}$ of job $J_j$ is processed for exactly $p_{j,h}$ time units on machine $M_h$, $j = 1, \ldots, n$ and $h = 1, \ldots, m$. Therefore $\mathcal{S}$ is a preemptive schedule for $\mathbb{P}$ with makespan not exceeding $\Delta(\mathbb{P})$. Hence $\mathcal{S}$ is an optimal preemptive schedule for $\mathbb{P}$. The decomposition in (3.21) can be found by the Birkhoff heuristic, for instance (see Brualdi [13] and Dufossé and Uçar [31]) in time $O(r^2\sqrt{n+m})$. Please see Chap. 11 for further discussion of the number of permutation matrices required by Birkhoff–von Neumann theorem. We now illustrate this approach on an instance with $n = 6$ jobs and $m = 3$ machines:

$$\mathbb{P} = \begin{bmatrix} 3 & 2 & 5 \\ 1 & 3 & 1 \\ 2 & 4 & 3 \\ 6 & 1 & 1 \\ 2 & 1 & 0.5 \\ 0.5 & 2 & 3 \end{bmatrix}.$$

We have $\Delta(\mathbb{P}) = 14.5$ for the instance; thus,

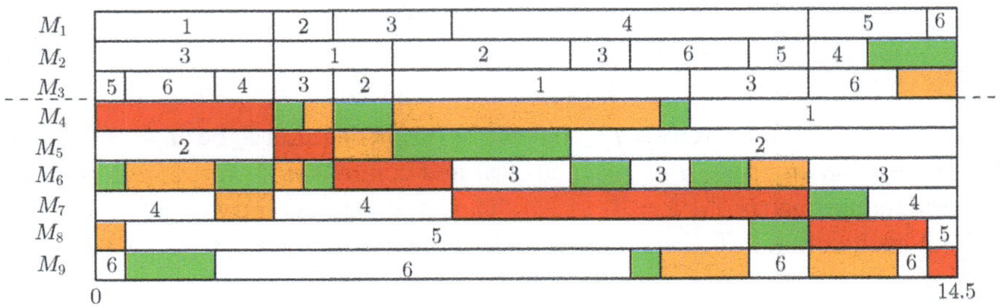


**Fig. 3.12** A schedule obtained for a doubly stochastic matrix $\frac{1}{14.5}\mathbb{Q}$

$$\mathbb{Q} = \begin{bmatrix} 3 & 2 & 5 & 4.5 & 0 & 0 & 0 & 0 & 0 \\ 1 & 3 & 1 & 0 & 9.5 & 0 & 0 & 0 & 0 \\ 2 & 4 & 3 & 0 & 0 & 5.5 & 0 & 0 & 0 \\ 6 & 1 & 1 & 0 & 0 & 0 & 6.5 & 0 & 0 \\ 2 & 1 & 0.5 & 0 & 0 & 0 & 0 & 11 & 0 \\ 0.5 & 2 & 3 & 0 & 0 & 0 & 0 & 0 & 9 \\ 0 & 0 & 0 & 3 & 1 & 2 & 6 & 2 & 0.5 \\ 0 & 1.5 & 0 & 2 & 3 & 4 & 1 & 1 & 2 \\ 0 & 0 & 1 & 5 & 1 & 3 & 1 & 0.5 & 3 \end{bmatrix}.$$

The decomposition of $\frac{1}{14.5}\mathbb{Q}$ into a convex combination of $q = 17$ permutation matrices leads to the schedule in Fig. 3.12. The schedule includes three additional jobs: red, orange, and green; and six additional machines $M_4, \ldots, M_9$ created to obtain a doubly stochastic matrix $\frac{1}{14.5}\mathbb{Q}$.

We have $\lambda_1 = \frac{1}{14.5}$ and $\mathbb{P}_1$ equal to

$$\mathbb{P}_1 = \begin{bmatrix} 1&0&0&0&0&0&0&0&0 \\ 0&0&0&0&1&0&0&0&0 \\ 0&1&0&0&0&0&0&0&0 \\ 0&0&0&0&0&0&1&0&0 \\ 0&0&1&0&0&0&0&0&0 \\ 0&0&0&0&0&0&0&0&1 \\ 0&0&0&1&0&0&0&0&0 \\ 0&0&0&0&0&1&0&0&0 \\ 0&0&0&0&0&0&0&1&0 \end{bmatrix}$$

in the decomposition, and

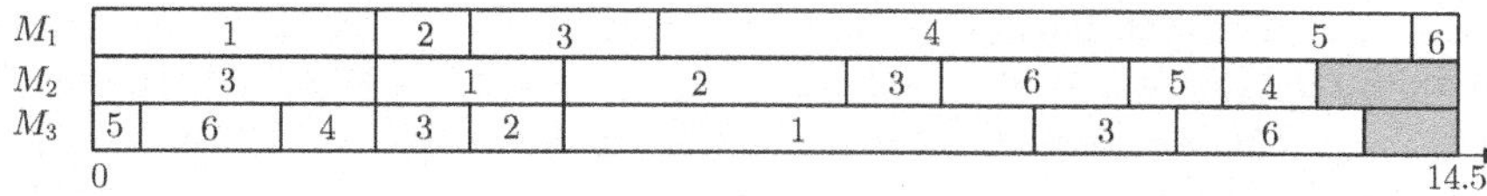


**Fig. 3.13** An optimal schedule for the instance $\mathbb{P}$ with $C_{\max} = 14.5$

$$\mathbb{S}_1 = \begin{bmatrix} 1\,0\,0 \\ 0\,0\,0 \\ 0\,1\,0 \\ 0\,0\,0 \\ 0\,0\,1 \\ 0\,0\,0 \end{bmatrix},$$

which defines the schedule in the interval $[0, 1]$ in Fig. 3.12. The reader is encouraged to write down the complete convex combination for the schedule; see Problem 3.7. The optimal schedule for $\mathbb{Q}$ obtained from the schedule in Fig. 3.12 is shown in Fig. 3.13. There are three preemptions in the schedule: $J_3$ is preempted on $M_2$, and $J_3$ and $J_6$ are preempted on $M_3$.

Shchepin and Vakhania [74] and [75] consider the problem $O|\text{pmtn}|C_{\max}$ by imposing an upper bound on the number of preemptions. They show that the upper bound $m - 3$ renders the problem $NP$-hard. Observe, however, that for $m = 3$ the upper bound equals 0; thus, preemptions are not permitted. Therefore the problem reduces to a non-preemptive problem $O3||C_{\max}$ that is $NP$-hard; see Theorem 7.2.

### 3.7.2 An Algorithm for $O|pmtn, r_j|L_{max}$

Cho and Sahni [26] study a preemptive open shop scheduling with job release dates and deadlines. They begin with a decision problem that asks whether there is a feasible preemptive schedule where each job does not start before its release date yet completes by its deadline, i.e., the problem $O|\text{pmtn}, r_i, \bar{d}_i|-$. A binary search similar to the one presented in Sect. 3.5.2 can turn a polynomial-time algorithm for $O|\text{pmtn}, r_i, \bar{d}_i|-$ into a polynomial-time algorithm for $O|\text{pmtn}, r_j|L_{max}$. Thus we focus on $O|\text{pmtn}, r_i, \bar{d}_i|-$ for the time being. The test proposed by Cho and Sahni [26] for this decision problem is somewhat similar to the test proposed by Horn [46] for the problem $P|\text{pmtn}, r_i, \bar{d}_i|-$; see Sect. 3.5.2. The test first orders the $2n$ values: $r_1, \ldots, r_n, \bar{d}_1, \ldots, \bar{d}_n$ in ascending order to determine all distinct values and their order

$$e_1 < \cdots < e_\ell, \tag{3.23}$$

where $\ell > 1$. For each interval $[e_k, e_{k+1}]$, $k = 1, \ldots, \ell - 1$, let $E_k$ be the set of operations $O_{i,h}$ such that $r_i \le e_k$ and $e_{k+1} \le \bar{d}_i$, i.e., $[e_k, e_{k+1}] \subseteq [r_i, \bar{d}_i]$. Thus any part $x_{i,h,k} \ge 0$ of operation $O_{i,h} \in E_k$ can be scheduled in $[e_k, e_{k+1}]$ provided that it is not longer than $e_{k+1} - e_k$. Moreover, for each $h = 1, \ldots, m$ the total of all parts $x_{i,h,k}$ of operations $O_{i,h} \in E_k$ scheduled in $[e_k, e_{k+1}]$ may not exceed $e_{k+1} - e_k$, which is the processing capacity the machine $M_h$ provides in the interval $[e_k, e_{k+1}]$; and for each $i = 1, \ldots, n$, the total of all parts $x_{i,h,k}$ of operations $O_{i,h} \in E_k$ scheduled in $[e_k, e_{k+1}]$ may not exceed $e_{k+1} - e_k$ to avoid simultaneous processing of job $J_i$ on different machines. The last two constraints guarantee that the solution for the interval $[e_k, e_{k+1}]$, $k = 1, \ldots, \ell - 1$ can be turned into a feasible open shop schedule with makespan $e_{k+1} - e_k$ in the interval $[e_k, e_{k+1}]$. Finally, all parts of operation $O_{i,h}$ need to sum up to the operation's processing time $p_{i,h}$. Observe that by the construction, for each job $J_i$ there are $a(i) \le b(i)$ such that $O_{ih} \in E_k$ for each $k = a(i), \ldots, b(i)$ and $O_{ih} \notin E_k$ for each $k \notin [a(i), b(i)]$. We summarize those constraints in the following system $F$ of linear inequalities:

$$
\begin{aligned}
0 \le x_{i,h,k} &\le e_{k+1} - e_k O_{ih} \in E_k, \; k = 1, \ldots, \ell, \; i = 1, \ldots, n, \; h = 1, \ldots, m; \\
0 &= x_{i,h,k} \qquad O_{ih} \notin E_k, \; k = 1, \ldots, \ell, \; i = 1, \ldots, n, \; h = 1, \ldots, m; \\
\sum_{i: O_{ih} \in E_k} x_{i,h,k} &\le e_{k+1} - e_k \qquad k = 1, \ldots, \ell - 1, \; h = 1, \ldots, m; \\
\sum_{h: O_{ih} \in E_k} x_{i,h,k} &\le e_{k+1} - e_k \qquad k = 1, \ldots, \ell - 1, \; i = 1, \ldots, n; \\
\sum_{k=a(i)}^{b(i)} x_{i,h,k} &= p_{ih} \qquad i = 1, \ldots, n, \; h = 1, \ldots, m.
\end{aligned}
$$

The coefficients of the system $F$ are limited to the values $-1$, 0, or 1; thus, by Tardos [78] the system $F$ can be solved in strongly polynomial time. To obtain a feasible schedule for a solution to $F$, if one exists, we observe that for each interval $[e_k, e_{k+1}]$, $k = 1, \ldots, \ell - 1$ the solution defines an instance of $O|\text{pmtn}|C_{max}$. The processing time of operation $O_{i,h}$ equals $x_{i,h,k}$ , and $\max\{\max_i\{\sum_h x_{i,h,k}\}, \max_h\{\sum_i x_{i,h,k}\}\} \le e_{k+1} - e_k$ in that instance. Therefore we can use any polynomial-time algorithm for $O|\text{pmtn}|C_{max}$, for example, one of those discussed in the previous section, to find a feasible open shop schedule with $C_{max} \le e_{k+1} - e_k$ for the instance in the interval $[e_k, e_{k+1}]$. The algorithm for $O|\text{pmtn}|C_{max}$ needs to be called $\ell$ times to find a complete schedule. This proves that that $O|\text{pmtn}, r_j|L_{max}$ can be solved in polynomial time.

### 3.7.3 *Complexity of Total Completion Time Minimization*

Liu and Bulfin [59] prove that three-machine total completion time minimization problem, $O3|\text{pmtn}|\sum C_i$, is NP-hard in the strong sense. Lenstra [57] shows that weighted total completion, $O2|\text{pmtn}|\sum w_i C_i$, is NP-hard in the strong sense for two machines. Du and Leung [30] show that for equal weights the problem, $O2|\text{pmtn}|\sum C_i$, is NP-hard in the ordinary sense; however, the question whether the two-machine problem is NP-hard in the strong sense remains open. Gladky [34] shows that adding job release dates, $O2|r_i, \text{pmtn}|\sum C_i$, renders the problem NP-hard in the strong sense. Liaw [58] develops a dynamic programing algorithm for $O2|\text{pmtn}|\sum w_i C_i$. The algorithm runs in $O(n2^n P^2)$ time. Queyranne and Sviridenko [68] develop a $(2+\epsilon)$-approximation algorithm for $O|r_i, \text{pmtn}|\sum w_i C_i$. The question whether there is a PTAS for $O|\text{pmtn}|\sum C_i$ remains open.

## 3.8 Exact Algorithms

Brucker et al. [15, 16] apply the ideas of Grabowski et al. [38] and Cariler and Pinson [19] to develop first branch and bound algorithm for $O||C_{\max}$. Guéret et al. [42] and Guéret and Prins [43] introduce improvements and refinements to the branch and bound algorithm. Dorndorf et al. [29] propose another branch and bound algorithm and report that their algorithm performs better than those presented earlier in the literature. Bräsel et al. [12] propose a method to reduce the search space for exact algorithms. Tamura et al. [77] propose a method that reduces constraint satisfaction and optimization problems into the Boolean satisfiability problem and apply their method to solve $O||C_{\max}$. Grimes et al. [41] and Malapert et al. [63] propose constraint programming algorithms. Recently, Ozolins [67] gives a dynamic programming algorithm for the problem. All those algorithms are tested on benchmark instances that typically include those proposed by Taillard [76]. Ahmadian et al. [3] provide a recent survey of exact algorithms.

## Problems

**3.1** Prove that the algorithm for $O|p_{ij} = 1, r_i|L_{max}$ given in Sect. 3.5.2 can be implemented to run in time $O(n^3 lognm)$.

**3.2** Show that the schedule in Fig. 3.9 is optimal.

**3.3** Show that ideal schedules for $O|p_{ij} = 1|\sum C_i$ always exist.

**3.4** Show that the completion time of each unit-time operation in an ideal schedule for $O|p_{ij} = 1|\sum C_i$ can be computed in $O(\log n + \log m)$ time.

**3.5** In the proof of Theorem 3.3 show that a schedule with $C_{\max} \leq 3$ exists if and only if there is 3-edge coloring of $G$ with colors 1, 2, or 3 where each multiedge with multiplicity 2 has one of its edges colored with color 2.

**3.6** Give a polynomial-time algorithm for $O||C_{\max}$ limited to the instances with maximum degree equal to 3.

**3.7** Specify a convex combination of permutation matrices for the doubly stochastic matrix $\frac{1}{14.5}\mathbb{Q}$ using the schedule in Fig. 3.13.

**3.8** Please eliminate any redundant constraints from the system $F$ of linear inequalities in Sect. 3.7.2.

## References


1. J.O. Achugbue, F.Y. Chin, Scheduling the open shop to minimize mean flow time. SIAM J. Comput. **11**, 709–720 (1982)
2. I. Adiri, N. Amit, Openshop and flowshop scheduling to minimize sum of completion times. Comput. Oper. Res. **11**, 275–284 (1984)
3. M.M. Ahmadian, M. Khatami, A. Salehipour, T.C.E. Cheng, Four decades of research on the open-shop scheduling problem to minimize makespan. Eur. J. Oper. Res. (2021). https://doi.org/10.1016/j.ejor.2021.03.026
4. A.S. Asratian, C.J. Casselgren, On interval edge colorings of $(\alpha, \beta)$-biregular bipartite graphs. Discrete Math. **307**, 1951–1956 (2007)
5. A. Atay, P. Calleja, S. Soteras, Open shop scheduling games. Eur. J. Oper. Res. (2021). https://doi.org/10.1016/j.ejor.2021.02.030
6. W. Banaszczyk, The Steinitz constant of the plane. J. Reine Angew. Math. **373**, 218–220 (1987)
7. P. Baptiste, On minimizing the weighted number of late jobs in unit execution time open shops. Eur. J. Oper. Res. **149**, 344–354 (2003)
8. P. Baptiste, P. Brucker, M. Chrobak, C. Dürr, S. A. Kravchenko, F. Sourd, The complexity of mean flow time scheduling problems with release times. J. Sched. **10**, 139–146 (2007)
9. I. Bárány, T. Fiala, Nearly optimum solution of multimachine scheduling problems (in Hungarian). Szigma **15**, 177–191 (1982)
10. J. Błażewicz, K. Ecker, E. Pesch, G. Schmidt, J. Węglarz, *Handbook on Scheduling: From Theory to Applications*. International Handbooks on Information Systems (Springer, Berlin, 2007)
11. J.A. Bondy, U.S.R. Murty, *Graph Theory with Applications* (North-Holland Publishing, Amsterdam, 1976)
12. H. Bräsel, M. Harborth, T. Tautenhahn, P. Willenius, On the set of solutions of the open shop problem. Ann. Oper. Res. **92**, 241–263 (1999)
13. R.A. Brualdi, Notes on the Birkhoff algorithm for doubly stochastic matrices. Can. Math. Bull. **25**, 127–147 (1982)
14. P. Brucker, *Scheduling Algorithms* (Springer, Berlin, 1995)
15. P. Brucker, T. Hilbig, J. Hurink, A branch and bound algorithm for a single-machine scheduling problem with positive and negative time lags. Discrete Appl. Math. **94**, 77–99 (1999)
16. P. Brucker, J. Hurink, B. Jurisch, Wöstmann, A branch and bound algorithm for the open shop problem. Discrete Appl. Math. **76**, 43–49 (1997)
17. P. Brucker, B. Jurisch, M. Jurisch, Open-shop problems with unit processing times. Z. Opns. Res. **37**, 59–73 (1993)

18. P. Brucker, B. Jurisch, T. Tautenhahn, F. Werner, Scheduling unit time open shops to minimize the weighted number of late jobs. Oper. Res. Lett. **14**, 245 –250 (1993)
19. J. Carlier, É. Pinson, An algorithm for solving the job-shop problem. Manag. Sci. **35**, 164–176 (1989)
20. B. Chen, V.A. Strusevich, Approximation algorithms for three machine open shop scheduling. ORSA J. Comput. **5**, 321–328 (1993)
21. B. Chen, W. Yu, How good is a dense shop schedule. Acta Math. Appl. Sin. **17**, 121–128 (2001)
22. R. Chen, *Open Shop Scheduling Problems and Their Dense Schedules*. PhD Thesis, East China University of Science and Technology, 2003
23. R. Chen, W. Huang, Z. Men, G. Tang, Open-shop dense schedules: properties and worst-case performance ratio. J. Sched. **15**, 3–11 (2012)
24. R. Chen, W. Yu, Upper-bound of performance ratio of dense schedules for open-shop (in Chinese). J. East China Univ. Sci. Technol. **26**, 522–526 (2000)
25. R. Chen, W. Yu, Analysis of operation chain's properties of dense schedules for open-shop. J. East China Univ. Sci. Technol. **29**, 522–526 (2003)
26. Y. Cho, S. Sahni, Preemptive scheduling of independent jobs with release and due times on open, flow and job shop. Opns. Res. **29**, 511–522 (1981)
27. E.G. Coffman Jr., D. Dereniowski, W. Kubiak, An efficient algorithm for finding ideal schedules. Acta Inform. **49**, 1–14 (2012)
28. E. Cole, K. Ost, S. Schirra, Edge-coloring Bipartite Multigraphs in $O(E \log D)$. Combinatorica **21**, 5–12 (2001)
29. U. Dorndorf, E. Pesch, T. Phan-Huy, Solving the open shop scheduling problem. J. Sched. **4**, 157–174 (2001)
30. J. Du, J.Y.-T. Leung, Minimizing mean flow time in two-machine open shops and flow shops. J. Algorithms **14**, 24–44 (1993)
31. F. Dufossé, B. Uçar, Notes on Birkhoff-von Neumann decomposition of doubly stochastic matrices. Linear Algebra Appl. **479**, 108–115 (2016)
32. T. Fiala, An algorithm for the open-shop problem. Math. Oper. Res. **8**, 100–109 (1983)
33. M.R. Garey, D.S. Johnson, *Computers and Intractability: A Guide to the Theory of NP-Completeness* (W. H. Freeman, San Francisco, 1979)
34. A.A. Gladky, A two-machine preemptive openshop scheduling problem: an elementary proof of np-completeness. Eur. J. Oper. Res. **103**, 113–116 (1997)
35. T. Gonzalez, A note on open shop preemptive schedules. IEEE Trans. Comput. **C-28**, 782–786 (1979)
36. T. Gonzalez, Unit execution time shop problems. Math. Oper. Res. **7**, 57–66 (1982)
37. T. Gonzalez, S. Sahni, Open shop scheduling to minimize finish time. J. ACM **23**, 665–679 (1976)
38. J. Grabowski, E. Nowicki, S. Zdrzałka, A block approach for single-machine scheduling with release dates and due dates. Eur. J. Oper. Res. **26**, 278 –285 (1986)
39. R.L. Graham, E.L. Lawler, J.K. Lenstra, A.H.G. Rinnooy Kan, Optimization and approximation in deterministic sequencing and scheduling: a survey. Ann. Discr. Math. **5**, 287–326 (1979)
40. A. Grigoriev, *High Multiplicity Scheduling Problems*. PhD Thesis, Maastricht University, 2003
41. D. Grimes, E. Hebrard, A. Malapert, Closing the open shop: Contradicting conventional wisdom, in *International Conference on Principles and Practice of Constraint Programming* (Springer, 2009), pp. 400–408
42. C. Guéret, N. Jussien, C. Prins, Using intelligent backtracking to improve branch-and-bound methods: an application to open-shop problems. Eur. J. Oper. Res. **127**, 344– 354 (2000)
43. C. Guéret, C. Prins, A new lower bound for the open-shop problem. Ann. Oper. Res. **92**, 165–183 (1999)
44. G. Hardy, J. Littlewood, G. Polya, *Inequalities* (Cambridge University Press, Cambridge, 1952)
45. H. Hoogeveen, P. Schuurman, G.J. Woeginger, Nonapproximability results for scheduling problems with minsum criteria. INFORMS J. Comput. **13**, 157–168 (2001)
46. W.A. Horn, Some simple scheduling algorithms. Naval Res. Logist. Q. **21**, 177–185 (1974)

47. J. Józefowska, B. Jurisch, W. Kubiak, Scheduling shops to minimize the weighted number of late jobs. Oper. Res. Lett. **16**, 277–283 (1994)
48. A. Kononov, S. Sevastyanov, M. Sviridenko, A complete 4-parametric complexity classification of short shop scheduling problems. J. Sched. **15**, 427–446 (2012)
49. C. Koulamas, G.J. Kyparisis, Open shop scheduling to minimize the number of late jobs. Naval Res. Logist. **45**, 525–532 (1998)
50. S.A. Kravchenko, On the complexity of minimizing the number of late jobs in unit time open shop. Discrete Appl. Math. **100**, 127–132 (2000)
51. M. Kubale, Open shop problem with zero-one time operations and integer date/deadline intervals. Discrete Appl. Math. **76**, 213–223 (1997)
52. W. Kubiak, A pseudo-polynomial algorithm for a two-machine no-wait job-shop scheduling problem. Eur. J. Oper. Res. **43**, 267–270 (1989)
53. W. Kubiak, C. Sriskandarajah, K. Zaras, A note on the complexity of open shop scheduling problems. INFOR **29** (1991)
54. G.J. Kyparisis, C. Koulamas, Open shop scheduling with makespan and total completion time criteria. Comput. Oper. Res. **27**, 15–27 (2000)
55. E.L. Lawler, J.K. Lenstra, A.H.G. Rinnooy Kan, Minimizing maximum lateness in a two-machine open shop. Math. Oper. Res. **6**, 153–158 (1981)
56. E.L. Lawler, J.K. Lenstra, A.H.G. Rinnooy Kan, Minimizing maximum lateness in a two-machine open shop (erratum). Math. Oper. Res. **7**, 635 (1982)
57. J.K. Lenstra, Unpublished.
58. C-F. Liaw, Scheduling two-machine preemptive open shops to minimize total completion times. Comput. Oper. Res. **31**, 1349–1363 (2004)
59. C.Y. Liu, R.L. Bulfin, On the complexity of preemptive open-shop scheduling problems. Oper. Res. Lett. **4**, 71–74 (1985)
60. C.Y. Liu, R.L. Bulfin, Scheduling open shops with unit execution times to minimize functions of due dates. Oper. Res. **36**, 553–559 (1988)
61. L. Lu, M.E. Posner, An NP-hard open shop scheduling problem with polynomial average time complexity. Math. Oper. Res. **18**, 12–38 (1993)
62. I.N. Lushchakova, S.A. Kravchenko, Two-machine shop scheduling with zero and unit processing times. Eur. J. Oper. Res. **107**, 378–388 (1998)
63. A. Malapert, H. Cambazard, C. Guéret, N. Jussien, A. Langevin, L.M. Rousseau, An optimal constraint programming approach to the open-shop problem. INFORMS J. Comput. **24**, 228–244 (2012)
64. V.M. Malhotra, M.P. Kumar, S.N. Maheshwari, An $O(|V|^3)$ Algorithm for finding maximum flows in networks. Info. Process. Lett. **7**, 277–278 (1978)
65. T. Masuda, H. Ishii, Two machine open shop scheduling problem with bi-criteria. Discrete Appl. Math. **52**, 253–259 (1994)
66. R. McNaughton, Scheduling with deadlines and loss functions. Manag. Sci. **6**, 1–12 (1959)
67. A. Ozolins, Dynamic programming approach for solving the open shop problem. Central Eur. J. Oper. Res. **29**, 291–306 (2021)
68. M. Queyranne, M. Sviridenko, A (2+$\epsilon$)-approximation algorithm for the generalized preemptive open shop problem with minsum objective. J. Algorithms **45**, 202–212 (2002)
69. S.V. Sevastianow, A polynomially solvable case of the open shop problem with arbitrary number of machines (In Russian). Kibernetika i Systemnii Analiz **6**, 135–154 (1992)
70. S.V. Sevastianow, On some geometric methods in scheduling theory: a survey. Discrete Appl. Math. **55**, 59–82 (1994)
71. S.V. Sevastianow, G.J. Woeginger, Makespan minimization in open shops: a polynomial time approximation scheme. Math. Program. **82**, 191–198 (1998)
72. S.V. Sevast'janow, Vector summation in Banach Space and polynomial algorithms for flow shops and open shops. Math. Oper. Res **20**, 90–103 (1995)
73. S.V. Sevast'janow, W. Banaszczyk, To the Steinitz lemma in coordinate form. Discrete Math. **169**, 145–152 (1997)

74. E.V. Shchepin, N. Vakhania, On the geometry, preemptions and complexity of multiprocessor and shop scheduling. Ann. Oper. Res. **159**, 183–213 (2008)
75. E.V. Shchepin, N. Vakhania, A note on the proof of the complexity of the little preemptive open-shop problem. Ann. Oper. Res. **191**, 251–253 (2011)
76. E. Taillard, Benchmarks for basic scheduling problems. Eur. J. Oper. Res. **64**, 278–285 (1993)
77. N. Tamura, A. Taga, S. Kitagawa, M. Banbara, Compiling finite linear CSP into SAT. Constraints **14**, 254–272 (2009)
78. É. Tardos, A strongly polynomial algorithm to solve combinatorial linear programs. Oper. Res. **34**, 250–256 (1986)
79. V.G. Timkovsky, Identical parallel machines vs. unit-time shops and preemptions vs. chains in scheduling complexity. Eur. J. Oper. Res. **149**, 355–376 (2003)
80. G. Vairaktarakis, S. Sahni, Dual criteria preemptive open shop problems with minimum finish time. Naval Res. Logist. **42**, 103–121 (1995)
81. D.P. Williamson, L.A. Hall, H. Hoogeveen, C.A.J. Hurkens, J.K. Lenstra, S.V. Sevast'janow, D.B. Shmoys, Short shop schedules. Oper. Res. **45**, 288–294 (1997)
82. G.J. Woeginger, The open shop scheduling problem, in *35th Symposium on Theoretical Aspects of Computer Science (STACS 2018)*, ed. by R. Niedermeier, B. Vallée. Leibniz International Proceedings in Informatics (Dagstuhl Publishing, 2018), pp. 4:1–4:12

# Chapter 4
# Multiprocessor Operations

## 4.1 Introduction

In the open shop scheduling with multiprocessor operations a set of jobs $\mathcal{J} = \{J_1, \ldots, J_n\}$ is scheduled on machines $\mathcal{M} = \{M_1, \ldots, M_m\}$. The set of machines is partitioned into $p$ *disjoint* groups $\mathcal{G}_\ell$, $\ell = 1, \ldots, p$. Each job consists of *single-processor* and *multiprocessor* operations. A single-processor operation $O_{j,h}$ of job $J_j$ requires a single machine $M_h \in \mathcal{M}$, and a multiprocessor operation $\hat{O}_{j,\ell}$ requires *all* machines from the group, $\mathcal{G}_\ell$, $\ell = 1, 2, \ldots, p$ simultaneously. The processing time of $O_{j,h}$ equals $b_{jh} \geq 0$, and the processing time of $\hat{O}_{j,\ell}$ equals $a_{j\ell} \geq 0$. In this chapter we depart from the notation $p_{j,h}$ (we use $b_{jh}$ instead) introduced in Chap. 1 for processing time of operation $O_{j,h}$. This is to further emphasize the presence of individual and group operations in the open shop with multiprocessor operations. All processing times are integers for the time being. The processing time $b_{jh}$ equals 0 means that $J_j$ is missing on $M_h$, similarly the processing time $a_{j\ell}$ equals 0 means that $J_j$ is missing on $\mathcal{G}_\ell$. In a feasible schedule each machine can process at most one operation at a time, and no two operations of the same job can be processed simultaneously. Any operation can be preempted at any moment and resumed at any moment later at no cost. The makespan is to be minimized.

An instance of the open shop scheduling with multiprocessor operations naturally decomposes into two instances of open shops. One referred to as the *group* open shop consists of $p$ group machines $\mathcal{G}_1, \ldots, \mathcal{G}_p$, and $n$ jobs in $\hat{\mathcal{J}}$, where the processing time of job $\hat{J}_j \in \hat{\mathcal{J}}$ on a group machine $\mathcal{G}_\ell$, $\ell = 1, \ldots, p$ equals $a_{j\ell}$. The other referred to as *individual* open shop consists of $m$ machines $\mathcal{M} = \{M_1, \ldots, M_m\}$, and $n$ jobs in $\mathcal{J}$, where the processing time of job $J_j \in \mathcal{J}$ on an individual machine $M_h \in \mathcal{M}$ equals $b_{j,h}$. For the group open shop machine $\mathcal{G}_\ell$ workload equals $\Delta(\mathcal{G}_\ell) = \sum_j a_{j\ell}$ for $\ell = 1, \ldots, p$, and job length equals $\Delta(\hat{J}_j) = \sum_\ell a_{j\ell}$ for $\hat{J}_j \in \hat{\mathcal{J}}$. Thus by König's edge-coloring theorem there is an

 
W. Kubiak, *A Book of Open Shop Scheduling*, International Series in Operations Research & Management Science 325,
https://doi.org/10.1007/978-3-030-91025-9_4

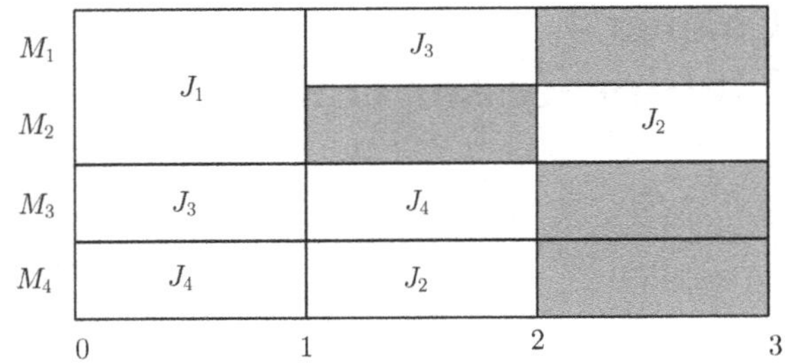


**Fig. 4.1** An instance with $\Delta = 2$ and optimal schedule with $C_{\max} = 3$

optimal schedule $S_{\mathcal{G}}$ with makespan $\Delta(\mathcal{G}) = \max\{\max_\ell \Delta(\mathcal{G}_\ell), \max_j \Delta(\hat{J}_j)\}$ for the group open shop. For the individual open shop machine $M_h \in \mathcal{M}$ workload equals $\Delta(M_h) = \sum_j b_{jh}$, and job length equals $\Delta(J_j) = \sum_h b_{jh}$ for $J_j \in \mathcal{J}$. Thus again by König's edge-coloring theorem there is an optimal schedule $S_{\mathcal{M}}$ with makespan $\Delta(\mathcal{M}) = \max\{\max_h \Delta(M_h), \max_j \Delta(J_j)\}$ for the individual open shop. Both schedules, $S_{\mathcal{G}}$ and $S_{\mathcal{M}}$, respectively, can be obtained in polynomial time, see Gabow and Kariv [11] or Cole et al. [6]. Either schedule permits preemptions at integer points only and so does their concatenation $S_{\mathcal{G}}S_{\mathcal{M}}$. The makespan of the concatenation equals $\Delta(\mathcal{G}) + \Delta(\mathcal{M})$.

Now instead of looking at the two instances of the decomposition one at a time let us consider the original instance. The machine $M_h$ workload equals $L_h = \sum_j a_{j\ell} + \sum_j b_{jh} = \Delta(\mathcal{G}_\ell) + \Delta(M_h)$, where $M_h \in \mathcal{G}_\ell$, and job $J_j$ length $P_j = \sum_\ell a_{j\ell} + \sum_h b_{jh} = \Delta(\hat{J}_j) + \Delta(J_j)$. Therefore, $\Delta = \max\{\max_j P_j, \max_h L_h\}$ is a lower bound on the makespan of an optimal schedule. Since $\Delta \geq \Delta(\mathcal{G})$ and $\Delta \geq \Delta(\mathcal{M})$, the algorithm that gives the concatenation $S_{\mathcal{G}}S_{\mathcal{M}}$ is a 2-approximation algorithm for the makespan minimization of the open shop scheduling problem with multiprocessor operations. To illustrate consider the instance in Fig. 4.1, we have $p = 2$, $\Delta(\mathcal{G}_1) = 1$, $\Delta(\mathcal{G}_2) = 0$, $\Delta(M_1) = \Delta(M_2) = 1$, $\Delta(M_3) = \Delta(M_4) = 2$, and $\Delta(J_2) = \Delta(J_3) = \Delta(J_4) = 2$, $\Delta(J_1) = 0$, $\Delta(\hat{J}_1) = 1$, $\Delta(\hat{J}_2) = \Delta(\hat{J}_3) = \Delta(\hat{J}_4) = 0$. Thus $\Delta(\mathcal{G}) = 1$ and $\Delta(\mathcal{M}) = 2$ and the schedule $S_{\mathcal{G}}S_{\mathcal{M}}$ has makespan 3. On the other hand $\Delta = 2$. Observe that a schedule with $C_{\max} = 2$ does not exist for this instance. Such a schedule would need individual operations of four different jobs to be scheduled in parallel on individual machines. This however contradicts the fact that only three jobs have individual operations in the instance.

Observe also that allowing preemptions at any point, not necessarily at integer points, may reduce schedule makespan. The schedule in Fig. 4.2 by allowing preemptions at any point reduces the makespan from $C_{\max} = 3$ to $C_{\max} = \frac{7}{3}$ for the instance in Fig. 4.2. Sections 4.2–4.9 focus on schedules with preemptions allowed at integer points only. Those schedules are solutions to the University timetabling problem. Section 4.10 considers preemptive schedules which solve preemptive open shop scheduling problem with multiprocessor operations. Those schedules allow preemptions at any points thus they do not necessarily solve the University timetabling problem; however, they become a good point of departure for approximate solutions, see Sect. 4.11.

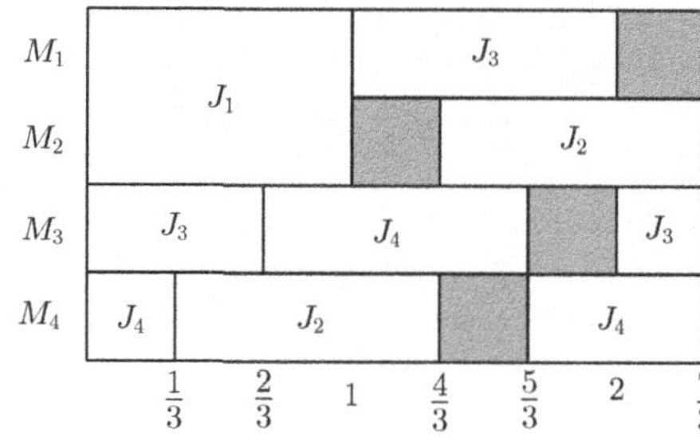


**Fig. 4.2** An optimal schedule with preemptions allowed at any point for the instance in Fig. 4.1

## 4.2 Complexity of Short Schedules with Preemptions at Integer Points

Asratian and de Werra [1] prove the following.

**Theorem 4.1** *The problem to determine if there is a schedule with* $C_{\max} \le 3$ *for an open shop with multiprocessor operations and preemptions allowed at integer points only is NP-complete in the strong sense even for* $p \le 4$.

***Proof*** The proof is by reduction from the following edge-coloring problem with pre-assigned colors. Let $G = (X, Y, E)$ be a bipartite graph with $\Delta(G) = 3$, where each vertex $v \in X$ is of degree 2 or 3. Moreover each vertex $v \in X$ has a set $C(v) \subseteq \{1, 2, 3\}$ of colors pre-assigned, and $|C(v)| = \deg_G(v)$. Can the edges of $G$ be colored with colors 1, 2, and 3 so that the edges incident with $v \in X$ are colored with colors in $C(v)$? The problem is shown NP-complete in the strong sense in Even at al. [10], see also Asratian and Kamalian [2]. In the corresponding open shop instance we have $m = |X|$ machines, $\mathcal{M} = X$, partitioned into four disjoint groups $\mathcal{G}_1 = \{v \in X : C(x) = \{1, 3\}\}$, $\mathcal{G}_2 = \{v \in X : C(x) = \{2, 3\}\}$, $\mathcal{G}_3 = \{v \in X : C(x) = \{1, 2\}\}$, and $\mathcal{G}_4 = \{v \in X : C(x) = \{1, 2, 3\}\}$. The jobs in $Y$ are processed on machines in $X$ so that the operations of job $u \in Y$ are of unit processing time each, and processed on machines $v \in X$ adjacent with $u$ in $G$. Moreover, there is one more job, the job $J$, with three group operations on $\mathcal{G}_1$, $\mathcal{G}_2$, and $\mathcal{G}_3$, no individual operations, and no group operation on $\mathcal{G}_4$ in the open shop instance. The three group operations of the job $J$ have unit processing time each. Thus $\mathcal{J} = Y \cup \{J\}$, and $C_{\max} = 3$.

Suppose there is an edge-coloring of $G$ with three colors 1, 2, and 3 so that the edges incident with $v \in X$ are colored with the colors in $C(x)$. Then a schedule can be readily obtained where each individual machine in $\mathcal{G}_1$ is occupied in [0, 1] and [2, 3], each individual machine in $\mathcal{G}_2$ is occupied in [1, 3], each individual machine in $\mathcal{G}_3$ is occupied in [0, 2], and each individual machine in $\mathcal{G}_4$ is occupied in [0, 3]. This allows to schedule $J$ on $\mathcal{G}_1$ in [1, 2], on $\mathcal{G}_2$ in [0, 1], and on $\mathcal{G}_3$ in [2, 3] to get a schedule with $C_{\max} = 3$, see the schedule in Fig. 4.3.

Now suppose $\mathcal{S}$ is a schedule with $C_{\max} = 3$. Thus job $J$ is processed at any time in the interval [0, 3]. Without loss of generality we can assume that group operation of $J$ on $\mathcal{G}_2$ is in [0, 1], on $\mathcal{G}_1$ is in [1, 2], and on $\mathcal{G}_3$ is in [2, 3] in $\mathcal{S}$. To see this

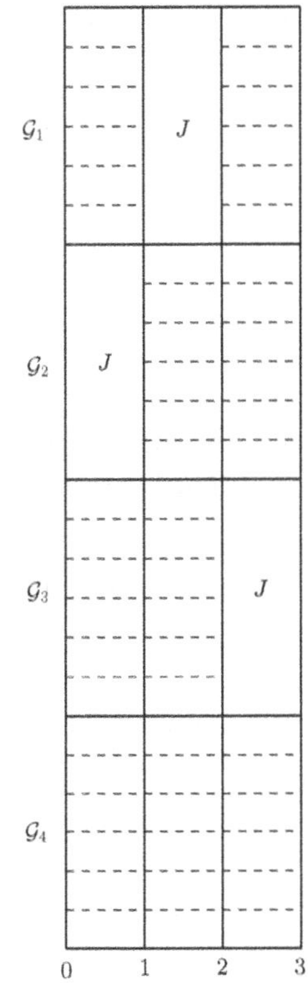


**Fig. 4.3** Scheduling job $J$ and individual operations of jobs in $\mathcal{J}$

suppose that $J$ is processed in the interval $[i-1, i]$ on $\mathcal{G}_1$, in $[j-1, j]$ on $\mathcal{G}_2$, and in $[k-1, k]$ on $\mathcal{G}_3$ in $\mathcal{S}$. We have $\{i, j, k\} = \{1, 2, 3\}$. Let $O_i$, $O_j$, and $O_k$ be the sets of all unit-time operations, group or individual, processed in the unit-time intervals $[i-1, i]$, $[j-1, j]$, and in $[k-1, k]$, respectively, in $\mathcal{S}$. Schedule each operation from $O_i$ in $[1, 2]$, each operation from $O_j$ in $[0, 1]$, and each operation for $O_k$ in $[2, 3]$. This permutation of the three unit-time intervals gives a feasible schedule with $C_{\max} = 3$, and the required order of processing for the operations of job $J$.

Thus an individual machine $v \in \mathcal{G}_1$ processes individual operations in $[0, 1]$ and $[2, 3]$. Those operations belong to jobs $u_1, u_2 \in Y$. Thus the edges $(v, u_1), (v, u_2) \in E$ incident with $v$ will be colored with colors 1 and 3 which makes precisely the set $C(v) = \{1, 3\}$ required for vertex $v$. Similar argument works for any individual machine $v \in \mathcal{G}_2$, and any individual machine $v \in \mathcal{G}_3$. Thus the edges incident with $v \in \mathcal{G}_2$ and $v \in \mathcal{G}_3$ will be colored with colors 2 and 3, and 1 and 2 respectively. Therefore we obtain the required edge-coloring of $G$. □

de Werra et al [8] further strengthen Theorem 4.1 by proving it for three groups, $p = 3$. We will omit the proof and leave it as an exercise, see Problem 4.1. However the schedules, if any exist, with $C_{max} \leq 2$ and for an arbitrary number of groups $p$ can be obtained in polynomial time. We have the following theorem.

**Theorem 4.2** *The problem to determine if there is a schedule with* $C_{\max} \leq 2$ *for an open shop with multiprocessor operations and preemptions allowed at integer points only is polynomial.*

***Proof*** Without loss of generality we can assume that each operation $o$, individual or group, has processing time 0, 1, or 2. We assume an arbitrary number of groups $p$. Split each operation $o$ with processing time 2 into two unit-time operations $o'$ and $o''$. The two belong to the same job and require the same machines, individual, or group, for processing as does $o$. For each unit-time operation $o$, define $\alpha_o = \mathcal{G}_\ell$ and $\beta_o = \{j\}$ if $o = \hat{O}_{j,\ell}$ is a group operation, and $\alpha_o = \{M_h\}$ and $\beta_o = \{j\}$ if $o = O_{j,h}$ is an individual operation. Let $G = (O, E)$ be a simple graph where $O$ is the set of all unit-time operations, and $E$ is a set of edges $(o, o')$ such that the operations $o$ and $o'$ either share a machine, i.e., $\alpha_o \cap \alpha_{o'} \neq \emptyset$, or a job, i.e., $\beta_o \cap \beta_{o'} \neq \emptyset$. We claim that there is a schedule with $C_{\max} \leq 2$ if and only if the *vertices* of $G$ can be colored with at most two colors so that any two vertices connected by and edge in $E$ are colored with different colors. That is $G$ is 2-colorable, Bondy and Murty [3]. Suppose $G$ is 2-colorable with colors 1 and 2. Schedule each operation $o \in O$ on machines in $\alpha_o$ in the interval $[0, 1]$ if the vertex $o$ is colored with color 1, and in the interval $[1, 2]$ if the vertex $o$ is colored with color 2. The schedule is feasible since $\alpha_o \cap \alpha_{o'} = \emptyset$ for any two operations $o$ and $o'$ both scheduled in the same time interval $[0, 1]$ or $[1, 2]$, i.e., no two such operations share a machine (each machine processes at most one operation at a time). Moreover, $\beta_o \cap \beta_{o'} = \emptyset$ for any two operations $o$ and $o'$ both scheduled in the same time $[0, 1]$ or $[1, 2]$, i.e., no two such operations belong to the same job (each job is processed by at most one machine, individual, or group, at a time). Therefore, there is a feasible schedule with $C_{\max} \leq 2$. Now suppose that there is a feasible schedule $\mathcal{S}$ with $C_{\max} \leq 2$. Without loss of generality we may assume that each operation, group, or individual completes at 1 or 2 in $\mathcal{S}$. Color each $o$ that completes at 1 with color 1, and each $o$ that completes at 2 with 2. Suppose for contradiction that there are operations $o$ and $o'$ connected by an edge $(o, o') \in E$ and colored with the same color $i = 1$ or 2 by the coloring. Thus both are scheduled in the same time interval $[i - 1, i]$ in $\mathcal{S}$, and since the schedule is feasible they must belong to different jobs and must not share a machine. Therefore $(o, o') \notin E$ which gives a contradiction.

Any simple graph is 2-colorable if and only if it is bipartite, Bondy and Murty [3]. Therefore there is a schedule with $C_{max} \leq 2$ if and only if the graph $G$ is bipartite. The test whether $G$ is bipartite or not can be done in $O(|O| + |E|)$ time. Therefore we just obtained a linear-time algorithm to test if there is a schedule with $C_{\max} \leq 2$. □

## 4.3 University Timetabling. A Polynomial-Time Algorithm and Conjecture for Two Groups

The University timetabling studied in this chapter was first introduced by Asratian and de Werra in [1]. The University timetabling is a generalization of the well-known, see Gotlieb [13], de Werra [7], and Bondy and Murty [3], class–teacher timetabling model. In the generalization, in addition to the lectures given by a single teacher to a single class, there are some lectures given by a single teacher to a group of classes simultaneously. We look for a minimum number of periods (period is a unit of time allocated to a lecture and it cannot be fractional in a solution to the timetabling problem) in which to complete all lectures without conflicts. The University timetabling model is motivated by the situation where various study programs share some courses which are common to all programs (classes). Asratian and de Werra [1] point out that such situation arises at Luleå University of Technology in Sweden and Ecole Polytechnique Fédérale de Lausanne (EPFL) in Switzerland. At EPFL for instance groups of three or four classes are created for courses of mathematics or physics which correspond to group-lectures. Besides those group-lectures there are individual lectures for courses given to one class (program) only, [1]. de Werra et al. [8] describe a similar situation at some French autonomous universities. Later on de Werra et al. [9], and Kis et al. [15] recast the problem as an equivalent open shop scheduling with multiprocessor operations and preemptions allowed at integer points only.

For two groups, $p = 2$, de Werra et al. [9] and Kis et al. [15] observe that a feasible schedule can be partitioned in the following four parts: part (a) consists of multiprocessor operations on $\mathcal{G}_1$, and single-processor operations or idle time on the machines in $\mathcal{G}_2$; part (b) consists of multiprocessor operations on both groups $\mathcal{G}_1$ and $\mathcal{G}_2$; part (c) consists of multiprocessor operations on $\mathcal{G}_2$, and single-processor operations or idle time on the machines in $\mathcal{G}_1$; and part (d) consists of single-processor operations or idle time on all machines, see Fig. 4.4. The parts (a), (b), (c), and (d) have sizes $\Delta(\mathcal{G}_1) - r$, $r$, $\Delta(\mathcal{G}_2) - r$, and $w$ respectively for some $r$ and $w$, where $\Delta(\mathcal{G}_\ell) = \sum_{j \in \mathcal{J}} a_{j,\ell}$ for $\ell = 1, 2$. Therefore the total of $\Delta(\mathcal{G}_1) + \Delta(\mathcal{G}_2) - r + w$ equals the schedule makespan, and the minimization of makespan reduces to the minimization of $w - r$. To simplify the notation we will often use $h$ instead of $M_h$ when referring to machine $M_h \in \mathcal{M}$, and $j$ instead of $J_j$ when referring to job $J_j \in \mathcal{J}$ in the remainder of this chapter.

The following integer linear program $\mathcal{ILP}$ with variables $r$, $w$, and $y_{jh}, x_{j\ell}$, for $j \in \mathcal{J}$, $h \in \mathcal{M}$, and $\ell = 1, 2$ was given in de Werra et al. [9] and Kis et al. [15] to minimize the makespan for $p = 2$:

$$ILP = \min(w - r). \tag{4.1}$$

Subject to

$$\sum_j b_{jh} - (\Delta(\mathcal{G}_2) - r) \le \sum_j y_{jh} \le w \qquad h \in \mathcal{G}_1 \tag{4.2}$$

$$\sum_j b_{jh} - (\Delta(\mathcal{G}_1) - r) \le \sum_j y_{jh} \le w \qquad h \in \mathcal{G}_2 \tag{4.3}$$

$$\sum_h y_{jh} \le w \qquad j \in \mathcal{J} \tag{4.4}$$

$$0 \le y_{jh} \le b_{jh} \qquad h \in \mathcal{M} \qquad j \in \mathcal{J} \tag{4.5}$$

$$\sum_j x_{j1} = r \tag{4.6}$$

$$\sum_j x_{j2} = r \tag{4.7}$$

$$x_{j1} + x_{j2} \le r \qquad j \in \mathcal{J} \tag{4.8}$$

$$0 \le x_{j\ell} \le a_{j\ell} \qquad j \in \mathcal{J} \qquad \ell = 1, 2 \tag{4.9}$$

$$\sum_{h \in \mathcal{G}_1} (b_{jh} - y_{jh}) + a_{j2} - x_{j2} \le \Delta(\mathcal{G}_2) - r \qquad j \in \mathcal{J} \tag{4.10}$$

$$\sum_{h \in \mathcal{G}_2} (b_{jh} - y_{jh}) + a_{j1} - x_{j1} \le \Delta(\mathcal{G}_1) - r \qquad j \in \mathcal{J} \tag{4.11}$$

all variables $r$, $w$, and $y_{jh}$, $x_{j\ell}$, for $j \in \mathcal{J}$, $h \in \mathcal{M}$, and $\ell = 1, 2$ are integers. (4.12)

The variable $y_{jh}$ represents the amount of $j \in \mathcal{J}$ on $h \in \mathcal{M}$ in part (d). The variable $x_{j\ell}$ represents the amount of $j \in \mathcal{J}$ on $\mathcal{G}_\ell$, $\ell = 1, 2$, in part (b). The variable $w$ is the size of (d), and the variable $r$ is the size of (b). The constraints (4.2)–(4.5) guarantee that the size of part (d) does not exceed $w$. The constraints (4.6)–(4.9) guarantee that the size of part (b) equals $r$. The constraints (4.10)–(4.11) along with the left hand side inequalities in (4.2) and (4.3) guarantee that the size of part (a) does not exceed $\Delta(\mathcal{G}_1) - r$ and that the size of part (c) does not exceed $\Delta(\mathcal{G}_2) - r$.

Kis et al. [15] show how to solve the $\mathcal{ILP}$ in polynomial time. They further show that $\lceil LP \rceil \le ILP \le \lceil LP \rceil + 1$, where $LP$ is the value of an optimal solution to the $LP$-relaxation of $\mathcal{ILP}$, and conjecture that:

*Conjecture 4.1* $ILP = \lceil LP \rceil$.

We prove this conjecture in this chapter. We follow the proof given in Kubiak [16]. Observe that $\mathcal{G}_1 = \emptyset$ or $\mathcal{G}_2 = \emptyset$ results in integral solutions with makespan $\Delta(\mathcal{G}_2) + \max\{\max_j\{\sum_h b_{jh}\}, \max_h\{\sum_j b_{jh}\}\}$ or $\Delta(\mathcal{G}_1)+\max\{\max_j\{\sum_h b_{jh}\}, \max_h\{\sum_j b_{jh}\}\}$, respectively. Thus the conjecture holds in this case and we assume non-empty $\mathcal{G}_1$ and non-empty $\mathcal{G}_2$ from now on in Sects. 4.3–4.9. We begin in the next section by focusing on those solutions to the $LP$-relaxation with the value of objective function $\lceil LP \rceil$ that minimize $r$. The goal will be to show that the minimum $r$ must be integer. This will be shown in Sects. 4.3–4.9. A detailed outline of the proof will be given in Sect. 4.3.2 once necessary notation and preliminary concepts are introduced there.

### 4.3.1 *LP* Relaxation with Minimum *r*

Let $(\mathbf{y}^*, \mathbf{x}^*, , r^*, w^*)$ be an optimal solution to the $LP$-relaxation of $\mathcal{ILP}$. Let $w^* = \lfloor w^* \rfloor + \lambda_{w^*}$ and $r^* = \lfloor r^* \rfloor + \lambda_{r^*}$, where $0 \leq \lambda_{w^*} < 1$ and $0 \leq \lambda_{r^*} < 1$. Consider the following linear program $\ell p$:

$$lp = \min r.$$

Subject to

$$w - r = \lceil w^* - r^* \rceil \tag{4.13}$$

$$\lfloor r^* \rfloor \leq r \tag{4.14}$$

$$\sum_j b_{jh} - (\Delta(\mathcal{G}_2) - r) \leq \sum_j y_{jh} \leq w \qquad h \in \mathcal{G}_1 \tag{4.15}$$

$$\sum_j b_{jh} - (\Delta(\mathcal{G}_1) - r) \leq \sum_j y_{jh} \leq w \qquad h \in \mathcal{G}_2 \tag{4.16}$$

$$\sum_h y_{jh} \leq w \qquad j \in \mathcal{J} \tag{4.17}$$

$$0 \leq y_{jh} \leq b_{jh} \qquad h \in \mathcal{M} \qquad j \in \mathcal{J} \tag{4.18}$$

$$\sum_j x_{j1} = r \tag{4.19}$$

$$\sum_j x_{j2} = r \tag{4.20}$$

$$x_{j1} + x_{j2} \leq r \qquad j \in \mathcal{J} \tag{4.21}$$

$$0 \le x_{j\ell} \le a_{j\ell} \qquad j \in \mathcal{J} \qquad \ell = 1, 2 \tag{4.22}$$

$$\sum_{h \in \mathcal{G}_1} (b_{jh} - y_{jh}) + a_{j2} - x_{j2} \le \Delta(\mathcal{G}_2) - r \qquad j \in \mathcal{J} \tag{4.23}$$

$$\sum_{h \in \mathcal{G}_2} (b_{jh} - y_{jh}) + a_{j1} - x_{j1} \le \Delta(\mathcal{G}_1) - r \qquad j \in \mathcal{J}. \tag{4.24}$$

All entries in the constraint matrix of $\ell p$ are 0, +1, or −1, thus $\ell p$ can be solved by a strongly polynomial algorithm given in Tardos [19]. Let $(\mathbf{y}, \mathbf{x}, r, w)$ be an optimal solution to $\ell p$. The solution exists since $(\mathbf{y}^*, \mathbf{x}^*, r^*, \lfloor w^* \rfloor + \lambda_{r^*})$ is feasible for $\ell p$ if $\lambda_{w^*} \le \lambda_{r^*}$, and $(\mathbf{y}^*, \mathbf{x}^*, r^*, \lceil w^* \rceil + \lambda_{r^*})$ is feasible for $\ell p$ if $\lambda_{w^*} > \lambda_{r^*}$, thus $\ell p$ is feasible and clearly it is also bounded. Observe that $\lfloor w^* \rfloor + \lambda_{r^*} - r^* = \lfloor w^* \rfloor - \lfloor r^* \rfloor = \lceil w^* - r^* \rceil$ for $\lambda_{w^*} \le \lambda_{r^*}$, and $\lceil w^* \rceil + \lambda_{r^*} - r^* = \lceil w^* \rceil - \lfloor r^* \rfloor = \lceil w^* - r^* \rceil$ for $\lambda_{w^*} > \lambda_{r^*}$.

We assume without loss of generality that the solution meets the machine *saturation condition*, i.e., the upper and lower bounds in (4.15) and (4.16) are equal. If the machine saturation is not met by the solution for some machine $h$, then a job $j(h)$ with $b_{j(h)h} = w - \sum_j b_{jh} + (\Delta(\mathcal{G}_2) - r)$, $a_{j(h)1} = a_{j(h)2} = 0$ should be added to the instance for each such machine to make the solution meet the saturation condition. Observe that by (4.13) $b_{j(h)h}$ is integral so the extended instance is a valid instance of the open shop with multiprocessor operations problem. We take $y_{j(h)h} = w - \sum_j y_{jh}$ in the extended solution. Observe that $n = |\mathcal{J}| \ge |\mathcal{G}_1| + |\mathcal{G}_2|$ for the solutions that meet the saturation condition.

An integral solution $(\mathbf{y}, \mathbf{x}, r, w)$ to $\ell p$ is feasible for $\mathcal{ILP}$, and $w - r = \lceil w^* - r^* \rceil = \lceil LP \rceil$. Moreover this solution is optimal for $\mathcal{ILP}$ since by definition of $LP$-relaxation we have $LP \le ILP$ for any feasible solution to $\mathcal{ILP}$. This proves Conjecture 4.1. Therefore it suffices to prove that there is an integral solution to $\ell p$. To that end, we prove the following theorem in Sects. 4.3–4.9.

**Theorem 4.3** *The $r$ in an optimal solution to $\ell p$ is integral. Moreover, there is optimal solution to $\ell p$ that is integral.*

***Proof*** Let $\mathbf{s} = (\mathbf{y}, \mathbf{x}, r, w)$ be an optimal solution to $\ell p$. Suppose for a contradiction that the $r$ in $\mathbf{s}$ equals

$$r = \lfloor r \rfloor + \epsilon,$$

where $0 < \epsilon < 1$. Thus by (4.13)

$$w = \lfloor w \rfloor + \epsilon\ .$$

In Sects. 4.3–4.9 we show that such $\mathbf{s}$ cannot be optimal which leads to a contradiction and proves the first part of the theorem. We then show that an optimal solution that is integral can be found in polynomial time. An outline of the proof will be given at the end of the next section after we first introduce the necessary notations and definitions. □

### 4.3.2 Preliminaries

Consider the solution $\mathbf{s} = (\mathbf{y}, \mathbf{x}, r, w)$. Let $B_1$ be the set of all jobs $j$ with fractional $x_{j1}$, and let $B_2$ be the set of all jobs $j$ with fractional $x_{j2}$. Clearly both sets are non-empty because $\epsilon > 0$. By (4.19) and (4.20) the fractions in $B_\ell$ sum up to $i_\ell + \epsilon$ ($\sum_{j \in B_\ell} \varepsilon_j = i_\ell + \epsilon$), where $i_\ell$ is a non-negative integer, for $\ell = 1, 2$.

A job $j$ is *d-tight* if

$$\sum_h y_{jh} = w.$$

Denote by $D$ the set of all $d$-tight jobs.

A job $j$ is *a-tight* if

$$\sum_{h \in \mathcal{G}_2} (b_{jh} - y_{jh}) + a_{j1} - x_{j1} = \Delta(\mathcal{G}_1) - r.$$

A job $j$ is *c-tight* if

$$\sum_{h \in \mathcal{G}_1} (b_{jh} - y_{jh}) + a_{j2} - x_{j2} = \Delta(\mathcal{G}_2) - r.$$

For jobs $g$ and $k$ such that $x_{g1} > 0$ and $x_{k2} > 0$ define

$$\varepsilon_r(g, k) = \begin{cases} \min_{j \in (B_1 \cup B_2) \setminus \{g,k\}} \{r - (x_{j1} + x_{j2}), \epsilon\} & \text{if } (B_1 \cup B_2) \setminus \{g, k\} \neq \emptyset; \\ \epsilon & \text{if } B_1 \cup B_2 \subseteq \{g, k\}. \end{cases}$$

Observe that jobs $g$ and $k$ with $\varepsilon_r(g, k) > 0$ can potentially be used to obtain a solution to $lp$ with smaller $r$ since the reduction of both $x_{g1}$ and $x_{k2}$ by some small enough $\varepsilon > 0$ will leave the resulting constraint (4.21) satisfied. Moreover define

$$\varepsilon_c(k) = \sum_{h \in \mathcal{G}_1} y_{kh} - \left( \sum_{h \in \mathcal{G}_1} b_{kh} + a_{k2} - x_{k2} - \Delta(\mathcal{G}_2) + r \right),$$

$$\varepsilon_a(g) = \sum_{h \in \mathcal{G}_2} y_{gh} - \left( \sum_{h \in \mathcal{G}_2} b_{gh} + a_{g1} - x_{g1} - \Delta(\mathcal{G}_1) + r \right).$$

Let $G$ be a job–machine bipartite graph such that there is an edge between machine $h \in \mathcal{M}$ and job $j \in \mathcal{J}$ if and only if $y_{jh} > 0$. The edge has multiplicity $y_{jh}$ (the multiplicity may be fractional). A *column* $I = (M_I, \varepsilon_I)$ consists of a matching $M_I$ in $G$ that matches all $m$ machines in $\mathcal{M}$ with a subset of exactly $m$ jobs in $\mathcal{J}$ that are scheduled simultaneously in the solution $\mathbf{s}$, and its multiplicity $\varepsilon_I > 0$ (the multiplicity may be fractional). That is the jobs matched in the column $I$ are processed simultaneously for $\varepsilon_I$ time units. Let $\mathcal{J}_I$ be the set of all jobs matched in $M_I$, i.e., $\mathcal{J}_I = \{j \in \mathcal{J} : (j, h) \in M_I \quad \text{for some} \quad h \in \mathcal{M}\}$. By definition of $D$ we require that $D \subseteq \mathcal{J}_I$ for a column in $\mathbf{s}$. By Gonzalez and Sahni [12], see also Gabow and Kariv [11] and Sect. 3.7.1 (Birkhoff–von Neumann theorem), part (d) can be represented by a set of columns $d(\mathbf{y}, w) = \{I_1, \ldots, I_p\}$. In the spirit of Birkhoff–von Neumann theorem, we can recast $d(\mathbf{y}, w)$ as follows. Let $\mathbb{Y}$ be an $n \times m$ matrix where the entry in row $i$ and column $h$ equals $y_{ih}$, and let $\mathbb{P}_I$ be an $n \times m$, 0-1 matrix corresponding to column $I = (M_I, \varepsilon_I) \in d(\mathbf{y}, w)$. The entry in row $i$ and column $h$ of $\mathbb{P}_I$ equals 1 if and only if job $i$ is matched with machine $h$ in $M_I$. We then can decompose $\mathbb{Y}$ as follows:

$$\mathbb{Y} = \varepsilon_{I_1}\mathbb{P}_{I_1} + \cdots + \varepsilon_{I_p}\mathbb{P}_{I_p}.$$

For a set $X$ of columns let $l(X)$ denote the total multiplicity of all columns in $X$. We have $l(d(\mathbf{y}, w)) = w$ and $l(X_j) = \sum_h y_{jh} \leq w$ where $X_j$ is the set of all columns that match job $j \in \mathcal{J}$. Let $I_1 = (M_{I_1}, \varepsilon_{I_1}), \ldots, I_q = (M_{I_q}, \varepsilon_{I_q})$ be a subset of $q \geq 1$ columns from $d(\mathbf{y}, w)$, the set of columns $Y = \{(M_{I_1}, \lambda_1), \ldots, (M_{I_q}, \lambda_q)\}$, where $0 \leq \lambda_1 \leq \varepsilon_{I_1}, \ldots, 0 \leq \lambda_q \leq \varepsilon_{I_q}$ and $\lambda_1 + \ldots + \lambda_q = \lambda$ is called the *interval* of length $\lambda$ in $d(\mathbf{y}, w)$. Let $d(\mathbf{y}, w) \setminus Y$ be the set of all columns in $d(\mathbf{y}, w)$ with columns in $Y$ removed. For each $j \in \mathcal{J}$ we have $l(Z_j) \leq l(d(\mathbf{y}, w) \setminus Y) = w - \lambda$ where $Z_j$ is the set of all columns that match $j$ in $d(\mathbf{y}, w) \setminus Y$.

Let $u_1, \ldots, u_p$ and $l_1, \ldots, l_q$ be different jobs from $\mathcal{J}$, and $I$ be a column. We say that $I$ is of type

$$\begin{pmatrix} *, u_1, \ldots, u_p \\ *, l_1, \ldots, l_q \end{pmatrix}$$

if $\{(u_1, h_1), \ldots, (u_p, h_p)\} \subseteq M_I$ for some machines $h_1, \ldots, h_p$ in $\mathcal{G}_1$, and $\{(l_1, H_1), \ldots, (l_q, H_q)\} \subseteq M_I$ for some machines $H_1, \ldots, H_q$ in $\mathcal{G}_2$. The asterisk denotes any matching for other jobs. For convenience, we sometimes use the following notation:

$$\begin{pmatrix} *, U \\ *, L \end{pmatrix},$$

where $U = \{u_1, \ldots, u_p\}$ and $L = \{l_1, \ldots, l_q\}$ instead. By definition if $p = 0$ or $q = 0$, then the asterisk alone denotes any matching on $\mathcal{G}_1$ or $\mathcal{G}_2$, respectively. We extend this notation for convenience as follows. Let $u$ and $l$ be different jobs from $\mathcal{J}$, and $I$ be a column. We say that $I$ is of type

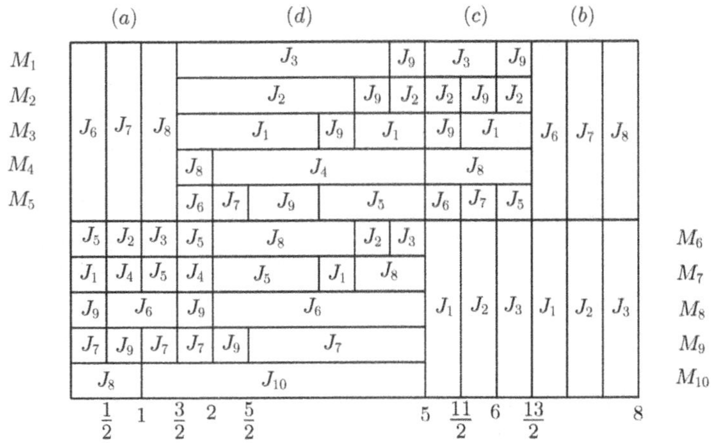


**Fig. 4.4** An example of solution $\mathbf{s} = (\mathbf{y}, \mathbf{x}, \frac{3}{2}, \frac{7}{2})$ and its corresponding schedule $S$ (parts (a), (d), and (c))

$$\begin{pmatrix} *, \overline{u} \\ *, \overline{l} \end{pmatrix}$$

if $(u, h) \notin M_I$ for any machine $h \in \mathcal{G}_1$, and $(l, H) \notin M_I$ for any machine $H \in \mathcal{G}_2$.

The concepts just introduced are illustrated in Fig. 4.4. The makespan of $S$ equals 8, and the schedule $S$ is clearly the shortest possible. The instance itself consists of $n = 10$ jobs and $m = 10$ machines, $\mathcal{G}_1 = \{M_1, M_2, M_3, M_4, M_5\}$ and $\mathcal{G}_2 = \{M_6, M_7, M_8, M_9, M_{10}\}$. The processing times of operations can easily be obtained from $S$, for example, for job $J_1$ we have $b_{13} = 4$, $b_{17} = 1$, $a_{12} = 1$ and all remaining operations have processing time 0, and for job $J_9$ we have $b_{91} = b_{92} = b_{93} = b_{95} = b_{98} = b_{99} = 1$ and all remaining operations have processing time 0. The solution $\mathbf{s}$ can also be easily obtained from $S$, for example, for job $J_1$ we have $y_{13} = 3$, $y_{17} = \frac{1}{2}$, $x_{12} = \frac{1}{2}$ and all remaining variables are set to 0, and for job $J_9$ we have $y_{91} = y_{92} = y_{93} = y_{98} = y_{99} = \frac{1}{2}$ and $y_{95} = 1$ all remaining variables are set 0. In $S$: $w = \frac{7}{2}$, $r = \frac{3}{2}$, $\epsilon = \frac{1}{2}$, $i_1 = i_2 = 1$, $B_1 = \{J_1, J_2, J_3\}$, $B_2 = \{J_6, J_7, J_8\}$, and $\varepsilon_r(g, k) = \frac{1}{2}$ for each pair $g \in B_1$ and $k \in B_2$. All jobs are $d$-tight; jobs $J_1$, $J_2$, $J_3$, $J_8$, and $J_9$ are $c$-tight; jobs $J_6$, $J_7$, and $J_8$ are $a$-tight. The matching $M = \{(M_1, J_3), (M_2, J_2), (M_3, J_1), (M_4, J_8), (M_5, J_6), (M_6, J_5), (M_7, J_4), (M_8, J_9), (M_9, J_7), (M_{10}, J_{10})\}$, and the multiplicity $\frac{1}{2}$ make up a column $(M, \frac{1}{2})$ which is the schedule $S$ in the interval $[\frac{3}{2}, 2]$. All other details of $\mathbf{s}$ and $S$ should now be clear from Fig. 4.4. We show later in Fig. 4.5 that $\mathbf{s}$ is not optimal for $\ell p$ since $\ell p$ admits solution with $r = 1$ and the same makespan 8.

### 4.3.3 Outline of the Proof

We now give a high level informal overview of the proof of the conjecture before moving to its details in the remaining sections. The proof is by contradiction. The solution **s** defines four open shops, one for each part $(a)$, $(b)$, $(c)$, and $(d)$. The bipartite graph $G$ with the edge multiplicities $y_{jh}$ obtained from the solution **s** defines an $m$-machine open shop with operation processing times equal $y_{jh}$ for part $(d)$, we call this part $d$-open shop. The groups $\mathcal{G}_1$ and $\mathcal{G}_2$ define a two-machine open shop with operation processing times $x_{j1}$ and $x_{j_2}$ for part $(b)$, we call this part $b$-open shop. The group $\mathcal{G}_1$ and the individual machines in $\mathcal{G}_2$ make up a $(|\mathcal{G}_2| + 1)$-machine open shop with operation processing times $b_{jh} - y_{jh}$ on the individual machines in $\mathcal{G}_2$ and $a_{j1} - x_{j1}$ on the group $\mathcal{G}_1$ for part $(a)$. Similarly, the group $\mathcal{G}_2$ and individual machines in $\mathcal{G}_1$ make up a $(|\mathcal{G}_1| + 1)$-machine open shop with operation processing times $b_{jh} - y_{jh}$ on the individual machines in $\mathcal{G}_1$ and $a_{j2} - x_{j2}$ on the group $\mathcal{G}_2$ for part $(c)$. We call these two $a$-open shop and $c$-open shop, respectively. All four open shops are interrelated since they share jobs, individual machines, or groups, thus a local change to one affects the other open shops as well. Notice that all open shops are defined by the solution **s** rather than directly by the problem instance which normally is the case for open shops.The open shops for $(a)$, $(d)$, and $(c)$ are shown in Fig. 4.4 for illustration. The makespan of each open shop is fractional, both $r$ and $w$ are fractional in **s**; however, the total makespan is integral since $w - r$ is integral in **s**.

Sections 4.3.4 and 4.4 give a matching-based approach to characterize those columns in $d$-open shop that cannot occur in **s** with $\epsilon > 0$ since their presence would contradict the optimality of $r$. Namely, those columns, if occurred in **s**, could be used along with the $x_{j1}, x_{j2}$ to find another feasible solution with parts $(b)$ and $(d)$ shorter by $\varepsilon$, $0 < \varepsilon \leq \epsilon$, each, and parts $(a)$ and $(c)$ longer by $\varepsilon$, $0 < \varepsilon \leq \epsilon$, each so that the total makespan does not change. More precisely the approach uses the column matchings in $d$-open shop on $\mathcal{G}_1$ and $\mathcal{G}_2$ separately; this structure is reflected in the notation for the column type, in order to match the former with some $x_{j2}$ and the latter with some $x_{j1}$ so that we get a feasible solution with the same makespan yet $d$-open shop shorter by $\varepsilon$. The matching-based approach leads to the characterizations of the $d$-open shop given in Sects. 4.4.1 and 4.5, and the $b$-open shop in Sect. 4.4.2; however, it is insufficient to prove the conjecture. Nevertheless both characterizations are key for the subsequent sections.

Therefore we introduce a network flow-based approach to shorten the $d$ open shop makespan from $w$ to $\lfloor w \rfloor$ and the $b$ open shop makespan from $r$ to $\lfloor r \rfloor$ in order to obtain a feasible solution with the same total makespan. We show that this approach works by constructing two network flow problems for $d$-open shop, one for the case with $\sum_{j \in B_1} \varepsilon_j = \epsilon$ or $\sum_{j \in B_2} \varepsilon_j = \epsilon$ in Sect. 4.6, and the other for the case with $\sum_{j \in B_1} \varepsilon_j = i_1 + \epsilon$ and $\sum_{j \in B_2} \varepsilon_j = i_2 + \epsilon$ for some positive integers $i_1$ and $i_2$ in Sect. 4.7. The network flow problems have integral lower and upper bounds on the arc flows which means they admit integral flows provided that feasible flows *exist* at all. To that end we show how to construct a feasible flow for each network from the solution $\mathbf{s}$ in Sects. 4.6 and 4.8. The construction relies on the characteristics of $d$-open shop given in Sects. 4.4.1 and 4.5. The characteristics naturally focus on the sets $B_1$ and $B_2$ in $\mathbf{s}$ since any change to the four open shops needs to involve the changes to the jobs in $B_1 \cup B_2$. That is not all since the integral solutions to the network flow problems give integral solutions to the $d$-open shop only. Those solutions need to be subsequently extended to the other three open shops while preserving the whole solution feasibility and the total makespan. This is also done in Sects. 4.6 and 4.8. The extension relies on characteristics of the $b$-open shop proved in Sect. 4.4.2 where we prove that $B_1 \cap B_2 = \emptyset$ in $\mathbf{s}$, i.e., $x_{j1}$ and $x_{j2}$ cannot be both fractional, and Sect. 4.5 where we prove that the product $x_{j1}x_{j2} = 0$ for each job $j \in \mathcal{J}$ in $\mathbf{s}$ except for the case where $B_1 = \{j\}$ or $B_2 = \{j\}$. The characteristics make it possible to find integral feasible solutions for the $b$-, $c$-, and $a$-open shops consistent with the network-flow solutions to the $d$-open shop. Finally we show in Sect. 4.9 that the network-flow based approach leads to contradiction since it shortens $r$ assumed to be the shortest possible. This proves the conjecture.

### 4.3.4 Columns Absent from $d(\mathbf{y}, w)$ in s

In this section we show that for two different jobs $g$ and $k$ such that $x_{g1} > 0$ and $x_{k2} > 0$ certain columns or subsets of columns must be missing from $d(\mathbf{y}, w)$ if $\epsilon > 0$. Though these results are contingent on $\varepsilon_r(g, k) > 0$, we show that this condition often holds, for instance in Sect. 4.4.2 we show that this inequality holds for each pair $g \in B_1$ and $k \in B_2$.

Let $g$ and $k$ be two different jobs such that $x_{g1} > 0$ and $x_{k2} > 0$. A $(g, k)$-*feasible semi-matching* in $G$ is a set of edges $E = E_1 \cup E_2$ of $G$ of cardinality $m = |\mathcal{G}_1| + |\mathcal{G}_2|$ such that

1. $E_1 = \{(j, h) \in E : h \in \mathcal{G}_1\}$ and $E_2 = \{(j, h) \in E : h \in \mathcal{G}_2\}$ are matchings.
2. There are $h \in \mathcal{M}$ and $(j, h) \in E$ for each $j \in D$.
3. If $\varepsilon_a(g) = 0$, then $(g, h) \notin E_2$ for any $h \in \mathcal{G}_2$.
4. If $\varepsilon_c(k) = 0$, then $(k, h) \notin E_1$ for any $h \in \mathcal{G}_1$.

If $E$ is a matching, then a $(g, k)$-feasible semi-matching in $G$ is called a $(g, k)$-*feasible matching* in $G$.

We define solution $(\mathbf{y}(E), \mathbf{x}(g, k), r(g, k), w(g, k), \varepsilon)$ for jobs $g, k$, and a $(g, k)$-feasible semi-matching $E$, where

$$\varepsilon = \begin{cases} \varepsilon' & \text{if } \varepsilon_a(g) = 0 \text{ and } \varepsilon_c(k) = 0\,; \\ \min\{\varepsilon', \varepsilon_a(g)\} & \text{if } \varepsilon_a(g) > 0 \text{ and } \varepsilon_c(k) = 0\,; \\ \min\{\varepsilon', \varepsilon_c(k)\} & \text{if } \varepsilon_a(g) = 0 \text{ and } \varepsilon_c(k) > 0\,; \\ \min\{\varepsilon', \varepsilon_a(g), \varepsilon_c(k)\} & \text{if } \varepsilon_a(g) > 0 \text{ and } \varepsilon_c(k) > 0\,; \end{cases} \tag{4.25}$$

and

$$\varepsilon' = \min\{\varepsilon_r(g,k), x_{g1}, x_{k2}, \min_{(j,h)\in E}\{y_{jh}\}, \min_{j\in\mathcal{J}\setminus D}\{w - \sum_h y_{jh}\}\}, \tag{4.26}$$

as follows:

$$y_{jh}(E) = \begin{cases} y_{jh} - \varepsilon & \text{if } (j,h) \in E \text{ ;} \\ y_{jh} & \text{otherwise ;} \end{cases} \tag{4.27}$$

$$x_{j1}(g,k) = \begin{cases} x_{g1} - \varepsilon & \text{if } j = g \text{ ;} \\ x_{j1} & \text{if } j \neq g; \end{cases} \tag{4.28}$$

$$x_{j2}(g,k) = \begin{cases} x_{k2} - \varepsilon & \text{if } j = k \text{ ;} \\ x_{j2} & \text{if } j \neq k; \end{cases} \tag{4.29}$$

$$r(g,k) = r - \varepsilon; \tag{4.30}$$

$$w(g,k) = w - \varepsilon \,. \tag{4.31}$$

We have the following lemma.

**Lemma 4.1** *Let $g$ and $k$ be two different jobs such that $x_{g1} > 0$ and $x_{k2} > 0$. If $\varepsilon_r(g,k) > 0$, then no $(g,k)$-feasible semi-matching $E$ in $G$ exists.*

***Proof*** Details can be found in Kubiak [16]. □

**Lemma 4.2** *Let $g$ and $k$ be two different jobs such that $x_{g1} > 0$ and $x_{k2} > 0$. If $\varepsilon_r(g,k) > 0$, then no column of type $\binom{*,\overline{k}}{*,\overline{g}}$ exists in $d(\mathbf{y}, w)$.*

***Proof*** If such a column $I = (M_I, \varepsilon_I)$ exists, then $M_I$ is $(g,k)$-feasible semi-matching $E$ in $G$ which contradicts Lemma 4.1. □

We now consider another forbidden configuration of columns in $d(\mathbf{y}, w)$. Let $I_1 = (M_{I_1}, \varepsilon_{I_1})$ and $I_2 = (M_{I_2}, \varepsilon_{I_2})$ be two columns. Let $g$, $k$, $a$, and $b$ be four different jobs such that $x_{g1} > 0$, $x_{k2} > 0$, $x_{a1} > 0$, and $x_{b2} > 0$. Define solution $(\mathbf{y}(I_1, I_2), \mathbf{x}', r', w', \varepsilon)$, where

$$\varepsilon = \min\{\varepsilon_r(g,k), \varepsilon_r(a,b), x_{g1}, x_{a1}, x_{b2}, x_{k2}, \varepsilon_{I_1}, \varepsilon_{I_2}, \min_{j\in\mathcal{J}\setminus D}\{w - \sum_h y_{jh}\}\} \tag{4.32}$$

as follows:

$$y_{jh}(I_1, I_2) = \begin{cases} y_{jh} - \varepsilon & \text{if } (j,h) \in M_{I_1} \text{ and } (j,h) \in M_{I_2} \text{ ;} \\ y_{jh} - \varepsilon/2 & \text{if } (j,h) \in M_{I_1} \text{ and } (j,h) \notin M_{I_2} \text{ ;} \\ y_{jh} - \varepsilon/2 & \text{if } (j,h) \notin M_{I_1} \text{ and } (j,h) \in M_{I_2} \text{ ;} \\ y_{jh} & \text{otherwise ;} \end{cases} \tag{4.33}$$

$$x'_{j1} = \begin{cases} x_{j1} - \varepsilon/2 & \text{if } j = g \text{ or } j = a \text{ ;} \\ x_{j1} & \text{otherwise ;} \end{cases} \tag{4.34}$$

$$x'_{j2} = \begin{cases} x_{j2} - \varepsilon/2 & \text{if } j = k \text{ or } j = b \text{ ;} \\ x_{j2} & \text{otherwise ;} \end{cases} \tag{4.35}$$

$$r' = r - \varepsilon; \tag{4.36}$$

$$w' = w - \varepsilon \ . \tag{4.37}$$

We have the following lemma

**Lemma 4.3** *Let $g$, $k$, $a$, and $b$ be four different jobs such that $x_{g1} > 0$, $x_{k2} > 0$, $x_{a1} > 0$, and $x_{b2} > 0$. If $\varepsilon_r(g, k) > 0$ and $\varepsilon_r(a, b) > 0$, then a column of type $\left(\begin{smallmatrix} *,a,b,g,k \\ * \end{smallmatrix}\right)$ does not exist in $d(\mathbf{y}, w)$ or a column of type $\left(\begin{smallmatrix} * \\ *,a,b,k,g \end{smallmatrix}\right)$ does not exist in $d(\mathbf{y}, w)$.*

***Proof*** Details can be found in Kubiak [16]. □

The following two corollaries follow immediately from the proof of Lemma 4.3.

**Corollary 4.1** *Let $g$, $k$, and $a$ be three different jobs such that $x_{g1} > 0$, $x_{k2} > 0$, and $x_{a1}x_{a2} > 0$. If $\varepsilon_r(g, a) > 0$ and $\varepsilon_r(a, k) > 0$, then a column of type $\left(\begin{smallmatrix} *,a,g,k \\ * \end{smallmatrix}\right)$ does not exist in $d(\mathbf{y}, w)$ or a column of type $\left(\begin{smallmatrix} * \\ *,a,k,g \end{smallmatrix}\right)$ does not exist in $d(\mathbf{y}, w)$.*

**Corollary 4.2** *Let $g$ and $k$ be two different jobs such that $x_{g1}x_{g2} > 0$, and $x_{k1}x_{k2} > 0$. If $\varepsilon_r(g, k) > 0$, then a column of type $\left(\begin{smallmatrix} *,g,k \\ * \end{smallmatrix}\right)$ does not exist in $d(\mathbf{y}, w)$ or a column of type $\left(\begin{smallmatrix} * \\ *,k,g \end{smallmatrix}\right)$ does not exist in $d(\mathbf{y}, w)$.*

## 4.4 Pairs of Columns Absent from $d(\mathbf{y}, w)$ in s

Let $g$ and $k$ be two different jobs such that $x_{g1} > 0$, $x_{k2} > 0$. Let $I_k = (M_{I_k}, \varepsilon_{I_k})$ be a column of type $\left(\begin{smallmatrix} *,\bar{k} \\ * \end{smallmatrix}\right)$, and $I_g = (M_{I_g}, \varepsilon_{I_g})$ a column of type $\left(\begin{smallmatrix} * \\ *,\bar{g} \end{smallmatrix}\right)$. Without loss of generality we assume $\varepsilon_{I_k} = \varepsilon_{I_g} = \varepsilon$. Let $G(I_g, I_k) = (M_{I_g} \cup M_{I_k})$ be a job–machine bipartite multigraph, where an edge connects a machine $h$ and a job $j$ if and only if $(j, h) \in M_{I_g} \cup M_{I_k}$. The degree of each machine-vertex in $G(I_g, I_k)$ is exactly 2 and the degree of each job-vertex in $G(I_g, I_k)$ is either 1 or 2. Thus, $G(I_g, I_k)$ is a collection of connected components each of which is either a job–machine path or a job–machine cycle. We have the following lemma for $I_k$ and $I_g$.

**Lemma 4.4** *If $I_k, I_g \in d(\mathbf{y}, w)$, and $\varepsilon_r(g, k) > 0$, then $I_k$ is of type $\left(\begin{smallmatrix} * \\ *,k,g \end{smallmatrix}\right)$ and $I_g$ is of type $\left(\begin{smallmatrix} *,k,g \\ * \end{smallmatrix}\right)$ and both $k$ and $g$ belong to the same connected component of $G(I_g, I_k)$.*

***Proof*** Column $I_k$ either has no job $k$ on any machine (we say $I_k$ is *k-free*) or it is of type $\binom{*}{*,k}$. In the former case $k$ is either missing from $G(I_g, I_k)$ or it is of degree 1 in $G(I_g, I_k)$. In the latter case $I_k$ is either of type $\binom{*}{*,k,g}$ or of type $\binom{*,g}{*,k}$ or it is $g$-free. Since $\varepsilon_r(g, k) > 0$, by Lemma 4.2 $I_k$ cannot be of type $\binom{*,g}{*,k}$ nor can $I_k$ be $g$-free. Thus $I_k$ is of type $\binom{*}{*,k,g}$ or $I_k$ is $k$-free. In the latter case, by Lemma 4.2, $I_g$ must be of type $\binom{*,k}{*,\overline{g}}$. A similar argument shows that $I_g$ is of type $\binom{*,k,g}{*}$ or $I_g$ is $g$-free. In the latter case, by Lemma 4.2, $I_k$ must be of type $\binom{*,\overline{k}}{*,g}$. Thus we end up with the following four cases:

1. $I_k$ is of type $\binom{*}{*,k,g}$, and $I_g$ is of type $\binom{*,k,g}{*}$;
2. $I_k$ is of type $\binom{*}{*,k,g}$, and $I_g$ is $g$-free. By Lemma 4.2, $I_g$ cannot be $k$-free. Hence $k$ is of degree 2 and $g$ is of degree 1 in $G(I_g, I_k)$;
3. $I_k$ is $k$-free, and $I_g$ is of type $\binom{*,k,g}{*}$. By Lemma 4.2, $I_k$ cannot be $g$-free. Hence $g$ is of degree 2 and $k$ is of degree 1 in $G(I_g, I_k)$;
4. $I_k$ is $k$-free and is of type $\binom{*}{*,g}$, and $I_g$ is $g$-free and is of type $\binom{*,k}{*}$. Hence both $g$ and $k$ are of degree 1 in $G(I_g, I_k)$.

In Case (2), let $g$ and $k$ be in the same connected component $P$ of $G(I_g, I_k)$. Then $P$ is a job–machine path

$$g, h_1, j_1, \ldots h_i, k, h_{i+1}, j_{i+1}, \ldots, h_\ell, j_\ell,$$

where $h_1 \in \mathcal{G}_2$ and $\{h_i, h_{i+1}\} \cap \mathcal{G}_2 \neq \emptyset$. If $h_i \in \mathcal{G}_2$, then match the jobs with the machines as follows:

$$M = \{(h_1, j_1), \ldots, (h_{i-1}, j_{i-1}), (h_i, k), (h_{i+1}, j_{i+1}), \ldots, (h_\ell, j_\ell)\}$$

in the component $P$. If $h_i \in \mathcal{G}_1$, then there is a job $j_{i^*} \in \{j_1, \ldots, j_{i-1}\}$ such that $h_{i^*} \in \mathcal{G}_2$ and $h_{i^*+1} \in \mathcal{G}_1$. Then match the jobs with the machines as follows:

$$M = \{(h_1, j_1), \ldots, (h_{i^*-1}, j_{i^*-1}), (h_{i^*}, j_{i^*}), (h_{i^*+1}, j_{i^*}), \ldots, (h_i, j_{i-1}), (h_{i+1}, k), \ldots, (h_\ell, j_{\ell-1})\}$$

in the component $P$. Thus each machine in $P$ is matched exactly once, each job of degree 2 in $P$ is matched at least once (actually each such job is matched exactly once except job $j_{i^*}$ that is matched exactly twice: with $h_{i^*} \in \mathcal{G}_2$ and $h_{i^*+1} \in \mathcal{G}_1$), $g$ is omitted from the matching, and $k$ is matched with a machine in $\mathcal{G}_2$. The matching can easily be extended by adding matchings from the remaining connected components of $G(I_g, I_k)$. The result is a $(g, k)$-feasible semi-matching in $G(I_g, I_k)$. We proceed in a similar fashion in Case (3) if $k$ and $g$ are in the same component $P$ of $G(I_g, I_k)$. In Case (4) if $g$ and $k$ are in the same connected component $P$ of $G(I_g, I_k)$, then $P$ is a job–machine path

$$g, h_1, j_1, \dots h_i, k,$$

with $h_1 \in \mathcal{G}_2$ and $h_i \in \mathcal{G}_1$. Then there is job $j_{i^*} \in \{j_1, \dots, j_{i-1}\}$ such that $h_{i^*} \in \mathcal{G}_2$ and $h_{i^*+1} \in \mathcal{G}_1$. Match the jobs with the machines as follows:

$$M = \{(h_1, j_1), \dots, (h_{i^*-1}, j_{i^*-1}), (h_{i^*}, j_{i^*}), (h_{i^*+1}, j_{i^*}), \dots, (h_i, j_{i-1})\}$$

in the component $P$. Thus each machine in $P$ is matched exactly once, each job of degree 2 in $P$ is matched at least once (actually each such job is matched exactly once except job $j_{i^*}$ that is matched exactly twice: with $h_{i^*} \in \mathcal{G}_2$ and $h_{i^*+1} \in \mathcal{G}_1$), $g$ and $k$ are omitted from the matching. The matching can easily be extended by adding matchings from the remaining connected components of $G(I_g, I_k)$. The result is a $(g, k)$-feasible semi-matching in $G(I_g, I_k)$.

Let us now assume that $k$ is in connected component $C_k$ and $g$ is in a connected component $C_g$ and $C_k \neq C_g$. We have

1. In Case (1), $k$ is of degree 2 and both on a machine in $\mathcal{G}_1$ and on a machine in $\mathcal{G}_2$ in $C_k$, and $g$ is of degree 2 and both on a machine in $\mathcal{G}_1$ and on a machine in $\mathcal{G}_2$ in $C_g$.
2. In Case (2), $k$ is of degree 2 and on $h \in \mathcal{G}_2$ in $C_k$, and $g$ is of degree 1 in $C_g$.
3. In Case (3), $g$ is of degree 2 and on $h \in \mathcal{G}_1$ in $C_g$, and $k$ is of degree 1 in $C_k$.
4. In Case (4), $g$ is of degree 1 in $C_g$, and $k$ is of degree 1 in $C_k$.

A matching for $C_k$ is selected so that $k$ is matched with the machine in $\mathcal{G}_2$, if $k$ is of degree 2, or omitted from the matching, if $k$ is of degree 1. Similarly a matching for $C_g$ is selected so that $g$ is matched with the machine in $\mathcal{G}_1$, if $g$ is of degree 2, or omitted from the matching if $g$ is of degree 1. The matching can easily be extended by adding matchings from the remaining connected components of $G(I_g, I_k)$. The result is a $(g, k)$-feasible semi-matching in $G(I_g, I_k)$. Thus in all cases, except Case (1) with both $k$ and $g$ being in the same connected component of $G(I_g, I_k)$, we showed how to obtain $(g, k)$-feasible semi-matching $M$ in $G(I_g, I_k)$. This however contradicts Lemma 4.1 since $I_k, I_g$ in $d(\mathbf{y}, w)$ can be replaced by columns $I' = (M, \varepsilon)$ and $I'' = ((M_{I_g} \cup M_{I_k}) \setminus M, \varepsilon)$ resulting into another feasible solution to $\ell p$ with the same value $r$ of objective function but with a $(g, k)$-feasible semi-matching $M$. □

### 4.4.1 The a-, c-, and d-Tightness in s

We show that each job in $B_1$ is both $a$-tight and $d$-tight, and each job in $B_2$ is both $c$-tight and $d$-tight. We begin by showing the $a$- and $c$- tightness.

**Lemma 4.5** *Each job $g \in B_1$ is a-tight and*

$$\sum_{h \in \mathcal{G}_2} y_{gh} < w, \tag{4.38}$$

*and each job $k \in B_2$ is c-tight and*

$$\sum_{h \in \mathcal{G}_1} y_{kh} < w. \tag{4.39}$$

***Proof*** Details can be found in Kubiak [16]. □

We now prove $d$-tightness for each job in $B_1 \cup B_2$.

**Theorem 4.4** *Each job in $B_1 \cup B_2$ is d-tight.*

***Proof*** By (4.38) in Lemma 4.5, there is a column $I_g$ of type $\binom{*}{*,\overline{g}}$ in $d(\mathbf{y}, w)$ for each $g \in B_1$. By (4.39) in Lemma 4.5, there is a column $I_k$ of type $\binom{*,\overline{k}}{*}$ in $d(\mathbf{y}, w)$ for each $k \in B_2$.

Consider job $g$ with the largest $x_{i1} + x_{i2}$ among the jobs $i \in B_1 \cup B_2$. Suppose $g \in B_1$. If $g \in B_2 \backslash B_1$, then the proof proceeds in a similar way and thus it will be omitted. Take any $k \in B_2 \setminus \{g\}$ or $k = g$ if $B_2 = \{g\}$. Observe that by our choice of $g$, if $x_{i1} + x_{i2} = r$ for some $i \in (B_1 \cup B_2) \setminus \{g, k\}$, then $x_{g1} + x_{g2} = r$. Therefore $\{k, i, g\} \subseteq B_1 \cup B_2$ which leads to a contradiction by (4.19) and (4.20) if $k \neq g$. Otherwise, if $k = g$, then by (4.20) $B_1 \cup B_2 = \{i, g\}$ and $g \in B_1 \cap B_2$. Thus $i \in B_1 \cap B_2$ and we get contradiction since $i \notin B_2$. Thus $\varepsilon_r(g, k) > 0$.

If $k$ is not $d$-tight, then there is a column $I$ of type $\binom{*,\overline{k}}{*,\overline{k}}$ in $d(\mathbf{y}, w)$. Thus, if $I \neq I_g$, then we get a contradiction with Lemma 4.4 applied to $I$ and $I_g$. Otherwise, if $I = I_g$, then $I$ is of type $\binom{*,\overline{k}}{*,\overline{g}}$ which contradicts Lemma 4.2. Similarly, if $g$ is not $d$-tight, then there is a column $I$ of type $\binom{*,\overline{g}}{*,\overline{g}}$ in $d(\mathbf{y}, w)$. Thus, if $I \neq I_k$, then we get a contradiction with Lemma 4.4 applied to $I_k$ and $I$. Otherwise, if $I = I_k$, then $I$ is of type $\binom{*,\overline{k}}{*,\overline{g}}$ which contradicts Lemma 4.2. Therefore the theorem holds for each job in $\{g\} \cup B_2$. Moreover, there is a column $I'_g$ of type $\binom{*,\overline{g}}{*}$. Otherwise all columns in $d(\mathbf{y}, w)$ are of type $\binom{*,g}{*}$ and thus $I_k$ is of type $\binom{*,\overline{k}}{*,\overline{g}}$ for any $k \in B_2$ which contradicts Lemma 4.2.

It remains to prove the theorem for each $a \in B_1 \setminus \{g\}$ whenever $B_1 \setminus \{g\} \neq \emptyset$. Observe that if $x_{g1} + x_{g2} = r$, then $x_{g2} > 0$. Otherwise $B_1 = \{g\}$ and we get a contradiction. Take a job $k = g$, if $x_{g1} + x_{g2} = r$, or any job $k \in B_2$, if $x_{g1} + x_{g2} < r$. W have $\varepsilon_r(a, k) > 0$. This holds since there is no $i \in (B_1 \cup B_2) \setminus \{a, k\}$ that meets $x_{i1} + x_{i2} = r$. Suppose for a contradiction that $x_{i1} + x_{i2} = r$ for some $i \in (B_1 \cup B_2) \setminus \{a, k\}$. Then $x_{k1} + x_{k2} = r$. Since $a \neq k$, we have $\{k, i, a\} \subseteq B_1 \cup B_2$ which leads to a contradiction by (4.19) and (4.20).

Thus if $a$ is not $d$-tight, then there is a column $I$ of type $\binom{*,\overline{a}}{*,\overline{a}}$ in $d(\mathbf{y}, w)$. Then, if $\varepsilon_r(a, k) > 0$ for $k \in B_2$, we have either $I \neq I_k$ which leads a contradiction with Lemma 4.4 applied to $I_k$ and $I$ or $I = I_k$ which implies that $I$ is of type $\binom{*,\overline{k}}{*,\overline{a}}$ which contradicts Lemma 4.2. If $\varepsilon_r(a, k) > 0$ for $k \notin B_2$, then $k = g$. Thus, if $I \neq I'_g$, then we get a contradiction with Lemma 4.4 applied to $I$ and $I'_g$. Otherwise, if $I = I'_g$, then $I$ is of type $\binom{*,\overline{g}}{*,\overline{a}}$ which contradicts Lemma 4.2. □

For $j \in B_1 \cup B_2$ define

$$\alpha_j = \sum_{h \in \mathcal{G}_1} y_{jh} \text{ and } \beta_j = \sum_{h \in \mathcal{G}_2} y_{jh}.$$

The following two lemmas relate the fractions of $x_{j1}$, $x_{j2}$, $\alpha_j$, and $\beta_j$ for $j \in B_1 \cup B_2$. The lemmas follow from Lemmas 4.5 and Theorem 4.4 and will prove useful in the remainder of the proof.

**Lemma 4.6** *For $g \in B_1$, let*

$$x_{g1} = \lfloor x_{g1} \rfloor + \varepsilon_g \ , \ \beta_g = \lfloor \beta_g \rfloor + \lambda_g, \ \textit{and} \ \alpha_g = \lfloor \alpha_g \rfloor + \omega_g,$$

*where $0 \leq \lambda_g, \omega_g < 1$, $0 < \varepsilon_g < 1$ for $g \in B_1$. Then, $\omega_g = \varepsilon_g$, and $\lambda_g = \epsilon - \varepsilon_g$ for $\epsilon \geq \varepsilon_g$, and $\lambda_g = 1 - (\varepsilon_g - \epsilon)$ for $\epsilon < \varepsilon_g$.*

***Proof*** Details can be found in Kubiak [16]. □

**Lemma 4.7** *For $k \in B_2$, let*

$$x_{k2} = \lfloor x_{k2} \rfloor + \varepsilon_k \ \textit{and} \ \beta_k = \lfloor \beta_k \rfloor + \lambda_k \ \textit{and} \ \alpha_k = \lfloor \alpha_k \rfloor + \omega_k,$$

*where $0 \leq \lambda_k, \omega_k < 1$, $0 < \varepsilon_k < 1$ for a job $k \in B_2$. Then, $\lambda_k = \varepsilon_k$, and $\omega_k = \epsilon - \varepsilon_k$ for $\epsilon \geq \varepsilon_k$, and $\lambda_k = 1 - (\varepsilon_k - \epsilon)$ for $\epsilon < \varepsilon_k$.*

***Proof*** The proof is similar to the proof of Lemma 4.6 and will be omitted. □

### 4.4.2 *The Absence of Crossing Jobs in* s

Each job $k \in B_1 \cap B_2$ is called *crossing*. We call a job $a \in B_1 \cup B_2$ an *e-crossing job*, if it meets the following conditions:

- $0 < x_{a2}$ and $0 < x_{a1}$.
- Both $B_1 \setminus \{a\}$ and $B_2 \setminus \{a\}$ are not empty.

We have the following.

**Theorem 4.5** *Each crossing job is e-crossing.*

***Proof*** Suppose for a contradiction that job $a$ is crossing but not $e$-crossing. By Theorem 4.4 job $a$ is $d$-tight and thus

$$\sum_{h \in \mathcal{G}_2} y_{ah} + \sum_{h \in \mathcal{G}_1} y_{ah} = w. \tag{4.40}$$

By Lemma 4.5 job $a$ is both $a$-tight and $c$-tight, thus

$$a_{a1} - x_{a1} + \sum_{h \in \mathcal{G}_2} (b_{ah} - y_{ah}) = \Delta(\mathcal{G}_1) - r \tag{4.41}$$

and

$$a_{a2} - x_{a2} + \sum_{h \in \mathcal{G}_1} (b_{ah} - y_{ah}) = \Delta(\mathcal{G}_2) - r. \tag{4.42}$$

By summing up (4.40), (4.41), and (4.42) side by side we obtain

$$a_{a1} + a_{a2} + \sum_{h} b_{ah} - \Delta(\mathcal{G}_1) - \Delta(\mathcal{G}_2) + r - w = -r + x_{a1} + x_{a2}\,. \tag{4.43}$$

Since $a$ is not $e$-crossing, $B_1 \setminus \{a\} = \emptyset$ or $B_2 \setminus \{a\} = \emptyset$. Thus, $x_{a1} = \lfloor x_{a1} \rfloor + \epsilon$ or $x_{a2} = \lfloor x_{a2} \rfloor + \epsilon$. Therefore, the left hand side of (4.43) is integral but its right hand side is fractional since both $x_{a1}$ and $x_{a2}$ are fractional. This leads to contradiction and thus the theorem holds. □

**Theorem 4.6** *For each e-crossing job a we have* $x_{a1} + x_{a2} < r$.

***Proof*** By contradiction. Let $a$ be $e$-crossing with $x_{a1} + x_{a2} = r$. Let $g \in B_1 \setminus \{a\}$ and $k \in B_2 \setminus \{a\}$. By Theorem 4.4 and Lemma 4.5 there are columns $I_k$ of type $\binom{*}{*,k}$ and $I_g$ of type $\binom{*,g}{*}$ in $d(\mathbf{y}, w)$. By Theorem 4.4, $I_k$ is either of type $\binom{*}{*,k,g}$ or of type $\binom{*,g}{*,k}$, and $I_g$ is either of type $\binom{*,g,k}{*}$ or of type $\binom{*,g}{*,k}$. Suppose that $I_k$ or $I_g$ is of type $\binom{*,g}{*,k}$, then $g \neq k$. Since $a$ is $e$-crossing, by Theorem 4.4 this column, say $I$, is either of type $\binom{*,a,g}{*,k}$ or of type $\binom{*,g}{*,a,k}$. The former is of type $\binom{*,\bar{k}}{*,\bar{a}}$ and the latter of type $\binom{*,\bar{a}}{*,\bar{g}}$. Since $g \neq k$, $a$ is the only job $i$ with $x_{i1} + x_{i2} = r$. Thus $\varepsilon_r(a, k) > 0$ and $\varepsilon_r(g, a) > 0$. Therefore we get a contradiction with Lemma 4.2 which implies that $I_g$ is of type $\binom{*,g,k}{*}$ and $I_k$ is of type $\binom{*}{*,k,g}$ (observe that we may now have $g = k$). Since $a$ is $e$-crossing, by Theorem 4.4 we have $I_g$ of type $\binom{*,a,g,k}{*}$ or of type $\binom{*,g,k}{*,a}$, and $I_k$ is of type $\binom{*}{*,a,k,g}$ or of type $\binom{*,a}{*,k,g}$. The $I_g$ of type $\binom{*,g,k}{*,a}$ is of type $\binom{*,\bar{a}}{*,\bar{g}}$, and the $I_k$ of type $\binom{*,a}{*,k,g}$ is of type $\binom{*,\bar{k}}{*,\bar{a}}$. Moreover, if $g \neq k$, then $a$ is the only job $i$ with $x_{i1} + x_{i2} = r$, and if $k = g$, then either $x_{k1} + x_{k2} = r$ or $a$ is the only job $i$ with $x_{i1} + x_{i2} = r$. Thus $\varepsilon_r(a, k) > 0$ and $\varepsilon_r(g, a) > 0$. Therefore, $I_g$ being of type $\binom{*,g,k}{*,a}$ or $I_k$ being of type $\binom{*,a}{*,k,g}$ contradicts Lemma 4.2. Thus it remains to consider $I_g$ of type $\binom{*,a,g,k}{*}$ and $I_k$ is of type $\binom{*}{*,a,k,g}$. This leads to a contradiction by Corollaries 4.1 and 4.2 since $\varepsilon_r(g, a) >$ and $\varepsilon_r(a, k) > 0$. The last two inequalities clearly hold if $a$ is the only job $i$ with $x_{i1} + x_{i2} = r$, otherwise $g = k$ and $k$ is the other job $i$ with $x_{i1} + x_{i2} = r$. □

The following corollary follows immediately from the proof of Theorem 4.6 since the assumption $x_{i1} + x_{i2} < r$ for each $i \in B_1 \cup B_2$ implies $\varepsilon_r(g, k) > 0$ for each $g \in B_1$ and $k \in B_2$.

**Corollary 4.3** *If $x_{i1} + x_{i2} < r$ for each $i \in B_1 \cup B_2$, then no job is e-crossing.*

We are now ready to prove two main results of this section.

**Theorem 4.7** *No crossing job exists.*

***Proof*** By contradiction. Suppose $a$ is a crossing job. Take a crossing job with the largest $x_{a1}+x_{a2}$. By Theorem 4.5 $a$ is $e$-crossing, and by Theorem 4.6 $x_{a1}+x_{a2} < r$. By Corollary 4.3 $x_{i1}+x_{i2} = r$ for some $i \in B_1 \cup B_2$. Thus $i \neq a$. By Theorem 4.6 $i$ is not $e$-crossing. Thus ($x_{i1} = 0$ or $x_{i2} = 0$) which implies ( $B_1 = \{i\}$ or $B_2 = \{i\}$). This leads to contradiction since $a \in B_1 \cap B_2$ and $a \neq i$. □

**Theorem 4.8** *For each $g \in B_1$ and $k \in B_2$, $\varepsilon_r(g, k) > 0$.*

***Proof*** Suppose for a contradiction that $\varepsilon_r(g, k) = 0$ for some $g \in B_1$ and $k \in B_2$. By Theorem 4.7, $g \neq k$. Then $r = x_{j1} + x_{j2}$ for some $j \in (B_1 \cup B_2) \setminus \{g, k\}$. By Theorem 4.7, $j$ is not crossing thus $\{j, g\} \subseteq B_1$ and $j \notin B_2$, or $\{j, k\} \subseteq B_2$ and $j \notin B_1$. Suppose the former, the proof for the latter is similar and thus will be omitted. We have $x_{j2}$ integral. However, by Theorem 4.6 $j$ is not $e$-crossing. Hence $x_{j2} = 0$. Thus $r = x_{j1}$ and $B_1 = \{j\}$ which gives a contradiction. □

## 4.5 Characterization of $d(\mathbf{y}, w)$ in s

We give a characterization of $d(\mathbf{y}, w)$ that will be used in the remainder of the proof.

**Lemma 4.8** *For each $g \in B_1$ and $k \in B_2$ , any column $I$ in $d(\mathbf{y}, w)$ is either of type $\binom{*,k}{*,g}$ or of type $\binom{*}{*,k,g}$ or of type $\binom{*,k,g}{*}$. Moreover, for each $g \in B_1$ and $k \in B_2$ there is $I_k$ of type $\binom{*}{*,k,g}$, and there is $I_g$ of type $\binom{*,k,g}{*}$ in $d(\mathbf{y}, w)$. Finally, if there is $i \in B_1 \cup B_2$ such that $x_{i1} + x_{i2} = r$, then either $B_1 = \{i\}$ or $B_2 = \{i\}$.*

***Proof*** Let $g \in B_1$ and $k \in B_2$. By Lemma 4.5 and Theorem 4.4 each column $I$ in $d(\mathbf{y}, w)$ is either of type $\binom{*,k}{*}$ or of type $\binom{*}{k,*}$. By Theorem 4.4 $I$ is either of type $\binom{*,k,g}{*}$ or of type $\binom{*,k}{*,g}$, or of type $\binom{*}{*,g,k}$ or of type $\binom{*,g}{*,k}$. By Theorem 4.8 we have $\varepsilon_r(g, k) > 0$ and thus by Lemma 4.2 $I$ is *not* of type $\binom{*,g}{*,k}$. This proves the first part of the lemma. Again, by Lemma 4.5 and Theorem 4.4 there are columns $I_k$ of type $\binom{*}{*,k}$ and $I_g$ of type $\binom{*,g}{*}$ in $d(\mathbf{y}, w)$. By Theorem 4.4 $I_k$ is either of type $\binom{*}{*,k,g}$ or of type $\binom{*,g}{*,k}$, and $I_g$ is either of type $\binom{*,g,k}{*}$ or of type $\binom{*,g}{*,k}$. By Theorem 4.8 we have $\varepsilon_r(g, k) > 0$ and thus by Lemma 4.2 neither $I_k$ nor $I_g$ is of type $\binom{*,g}{*,k}$. This proves the second part of the lemma.

If there is $i \in B_1 \cup B_2$ such that $x_{i1} + x_{x2} = r$. By Theorem 4.6 $i$ is not $e$-crossing thus $x_{i1} = 0$ or $x_{i2} = 0$ or $B_1 \setminus \{i\} = \emptyset$ or $B_2 \setminus \{i\} = \emptyset$. In all the cases, either $B_1 = \{i\}$ or $B_2 = \{i\}$. This completes the proof. □

**Theorem 4.9** *If there is a job $j$ such that $x_{j1}x_{j2} > 0$, then $B_1 = \{j\}$ or $B_2 = \{j\}$.*

***Proof*** Let $x_{j1}x_{j2} > 0$ for a job $j$. Without loss of generality let $j$ be a job with the largest value of $x_{j1} + x_{j2}$ among jobs with $x_{j1}x_{j2} > 0$. Suppose for a contradiction that $B_1 \setminus \{j\} \neq \emptyset$ and $B_2 \setminus \{j\} \neq \emptyset$. Thus if $j \in B_1 \cup B_2$, then $j$ is $e$-crossing. By Theorem 4.6, $x_{j1} + x_{j2} < r$. Take $g \in B_1 \setminus \{j\}$ and $k \in B_2 \setminus \{j\}$. If $j \notin B_1 \cup B_2$, then both $x_{j1}$ and $x_{j2}$ are integral. Thus $x_{j1} + x_{j2} < r$. Take $g \in B_1$ and $k \in B_2$. Thus we can pick three jobs $g \in B_1$, $k \in B_2$, and $j$ such that $x_{j1} + x_{j2} < r$ and $g \neq j$ and $k \neq j$. Moreover, by Theorem 4.7 we have $g \neq k$. We now show that $\varepsilon_r(g, j) > 0$ and $\varepsilon_r(j, k) > 0$. To prove the former inequality we observe that by our choice of job $j$ for any job $i \in B_1 \cup B_2$ different from $g$ and $j$, and such that $x_{i1} + x_{i2} = r$ must be either $r = x_{i1}$ or $r = x_{i2}$. Otherwise $x_{i1}x_{i2} > 0$, thus $i$ would have been chosen instead of $j$. The proof of the latter inequality follows by a similar argument. Thus by Corollary 4.1 a column of type $\binom{*,j,g,k}{*}$ does not exist in $d(\mathbf{y}, w)$ or a column of type $\binom{*}{*,j,k,g}$ does not exist in $d(\mathbf{y}, w)$. Suppose the former holds, then by Lemma 4.8 a column of type $\binom{*,g,k}{*}$ exists in $d(\mathbf{y}, w)$. This column is either of type $\binom{*,g,k}{*,j}$ or of type $\binom{*,\bar{j},g,k}{*,\bar{j}}$ which implies that the column is of type $\binom{*,\bar{j}}{*,\bar{g}}$. This however contradicts Lemma 4.2. For the latter, we prove in a similar fashion that a column of type $\binom{*,\bar{k}}{*,\bar{j}}$ exists in $d(\mathbf{y}, w)$ which contradicts Lemma 4.2. Therefore we get a contradiction which proves the theorem. □

### *4.5.1 The Overlap*

An overlap of $B_1$ is a column $I = (M_I, \varepsilon) \in d(\mathbf{y}, w)$ that matches at least two different jobs from $B_1$ with machines in $\mathcal{G}_1$. Similarly, an overlap of $B_2$ is a column $I = (M_I, \varepsilon) \in d(\mathbf{y}, w)$ that matches at least two different jobs from $B_2$ with machines in $\mathcal{G}_2$.

**Lemma 4.9** *An overlap of $B_1$ and an overlap of $B_2$ do not occur simultaneously.*

***Proof*** Suppose for contradiction that both overlaps occur simultaneously. Then there are different jobs $a$ and $g$ both from $B_1$ done on $\mathcal{G}_1$ in a column $I_{a,g} \in d(\mathbf{y}, w)$ of type $\binom{*,a,g}{*}$, and different jobs $b$ and $k$ both from $B_2$ done on $\mathcal{G}_2$ in a column $I_{b,k} \in d(\mathbf{y}, w)$ of type $\binom{*}{*,b,k}$. By Theorem 4.7 there are no crossing jobs thus all four jobs $a$, $g$, $b$, and $k$ are different. On the other hand for $g \in B_1$ and $k \in B_2$, by Lemma 4.8, any column $I$ in $d(\mathbf{y}, w)$ is either of type $\binom{*,k}{*,g}$ or of type $\binom{*}{*,k,g}$ or of type $\binom{*,k,g}{*}$ . Thus $I_{a,g}$ must be of type $\binom{*,a,k,g}{*}$. For $a \in B_1$ and $b \in B_2$, again by Lemma 4.8, any column $I$ in $d(\mathbf{y}, w)$ is either of type $\binom{*,b}{*,a}$ or of type $\binom{*}{*,b,a}$ or of type $\binom{*,b,a}{*}$. Thus $I_{a,g}$ must be of type $\binom{*,a,b,g}{*}$. Therefore $I_{a,g}$ is of type $\binom{*,a,b,k,g}{*}$. We show similarly that $I_{b,k}$ is of type $\binom{*}{*,a,b,k,g}$. This, by Theorem 4.8, contradicts Lemma 4.3 and proves the lemma. □

## 4.6 Integral Optimal Solution to $\ell p$ for $\sum_{j\in B_1}\varepsilon_j = \epsilon$ or $\sum_{j\in B_2}\varepsilon_j = \epsilon$

In this section we prove that an integral optimal solution for $\ell p$ exists if $\epsilon > 0$ and $\sum_{j\in B_1}\varepsilon_j = \epsilon$ or $\sum_{j\in B_2}\varepsilon_j = \epsilon$. We first prove this assuming $\sum_{j\in B_2}\varepsilon_j = \epsilon$ in **s** throughout this section. The proof for $\sum_{j\in B_1}\varepsilon_j = \epsilon$ proceeds in a similar fashion and thus will be omitted.

Consider the following network-flow problem $\mathcal{F}$ with variables $t_{jh}$ for $j$ and $h \in \mathcal{G}_2$, and $z_{jh}$ for $j$ and $h \in \mathcal{G}_1$. The $r$, $w$, and $x_{j\ell}$ for $j \in \mathcal{J}$ and $\ell = 1, 2$ in $\mathcal{F}$ are constants obtained from the solution $\mathbf{s} = (\mathbf{y}, \mathbf{x}, r, w)$.

$$F = \max \sum_{j\in B_1}\sum_{h\in\mathcal{G}_2} t_{jh}.$$

Subject to

$$\sum_j t_{jh} = \lfloor w \rfloor \quad h \in \mathcal{G}_2 \tag{4.44}$$

$$\sum_{h\in\mathcal{G}_2} b_{jh} + a_{j1} - \Delta(\mathcal{G}_1) + \lfloor r\rfloor - x_{j1} \le \sum_{h\in\mathcal{G}_2} t_{jh} \quad j \in \mathcal{J}\setminus B_1 \tag{4.45}$$

$$\sum_{h\in\mathcal{G}_2} b_{jh} + a_{j1} - \Delta(\mathcal{G}_1) + \lfloor r\rfloor - \lceil x_{j1}\rceil \le \sum_{h\in\mathcal{G}_2} t_{jh}$$

$$\le \sum_{h\in\mathcal{G}_2} b_{jh} + a_{j1} - \Delta(\mathcal{G}_1) + \lfloor r\rfloor - \lfloor x_{j1}\rfloor \quad j \in B_1 \tag{4.46}$$

$$\sum_j z_{jh} = \lfloor w\rfloor \quad h \in \mathcal{G}_1 \tag{4.47}$$

$$\sum_{h\in\mathcal{G}_1} b_{jh} + a_{j2} - \Delta(\mathcal{G}_2) + \lfloor r\rfloor - \lfloor x_{j2}\rfloor \le \sum_{h\in\mathcal{G}_1} z_{jh} \quad j \in \mathcal{J} \tag{4.48}$$

$$0 \le t_{jh} \le b_{jh} \qquad h \in \mathcal{M} \qquad j \in \mathcal{J} \tag{4.49}$$

$$0 \le z_{jh} \le b_{jh} \qquad h \in \mathcal{M} \qquad j \in \mathcal{J} \tag{4.50}$$

$$\sum_{h\in\mathcal{G}_1} z_{jh}+\sum_{h\in\mathcal{G}_2} t_{jh}\leq\lfloor w\rfloor \quad j\in\mathcal{J}. \tag{4.51}$$

**Lemma 4.10** *There is a feasible solution to $\mathcal{F}$ with value*

$$\sum_{j\in B_1}\sum_{h\in\mathcal{G}_2} b_{jh}-\sum_{j\in\mathcal{J}\setminus B_1}(a_{j1}-x_{j1})-(|B_1|-1)(\Delta(\mathcal{G}_1)-\lfloor r\rfloor)-\epsilon. \tag{4.52}$$

***Proof*** For **s**, consider the set $Y_j$ of all columns of type $\binom{*}{*,j}$ in $d(\mathbf{y},w)$ for $j\in B_2$. By Lemma 4.7, $l(Y_j)=\beta_j=\lfloor\beta_j\rfloor+\varepsilon_j$. If there is no overlap of $B_2$ or $\sum_{j\in B_2}\lfloor\beta_j\rfloor>0$, then take an interval $Y\subseteq\bigcup_{j\in B_2}Y_j$ such that $l(Y)=\epsilon$, $l(Y\cap Y_j)\geq\varepsilon_j$ for $j\in B_2$. Otherwise, if there is overlap of $B_2$ and $\sum_{j\in B_2}\lfloor\beta_j\rfloor=0$, then take an interval $Y\subseteq(\bigcup_{j\in B_2}Y_j)\cup Z$ such that $l(Y)=\epsilon$, $l(Y\cap Y_j)\geq\varepsilon_j$ for $j\in B_2$. Here the $Z$ is the set of all columns of type $\binom{*,B_2}{*,B_1}$ in $d(\mathbf{y},w)$. In order for such $Y$ to exist we show that $l((\bigcup_{j\in B_2}Y_j)\cup Z)\geq 1$. By Lemma 4.9 there is no overlap of $B_1$, thus $l(\bigcup_{j\in B_1}W_j)=\sum_{j\in B_1}l(W_j)=\sum_{j\in B_1}\alpha_j$ where $W_j$ is the set of all columns of type $\binom{*,j}{*}$ for $j\in B_1$ in $d(\mathbf{y},w)$. Hence by Lemma 4.6, $l(\bigcup_{j\in B_1}W_j)=\sum_{j\in B_1}\lfloor\alpha_j\rfloor+\sum_{j\in B_1}\varepsilon_j$. By definition $\sum_{j\in B_1}\varepsilon_j=i_1+\epsilon$ for some integer $i_1\geq 0$. Therefore $l(\bigcup_{j\in B_1}W_j)=i+\epsilon$ for some integer $i\geq 0$. Thus $l(d(\mathbf{y},w)\setminus\bigcup_{j\in B_1}W_j)$ is integral since $l(d(\mathbf{y},w))=w$, and positive. However $d(\mathbf{y},w)\setminus\bigcup_{j\in B_1}W_j=(\bigcup_{j\in B_2}Y_j)\cup Z$ by Theorem 4.4 and Lemma 4.8. This proves $l((\bigcup_{j\in B_2}Y_j)\cup Z)\geq 1$, and the required $Y$ exists.

Let $Y_{jh}$ be the set of columns $I\in Y$ such that $(j,h)\in M_I$, set $\gamma_{jh}:=l(Y_{jh})$. Informally, $\gamma_{jh}$ is the amount of $j\in\mathcal{J}$ done on $h\in\mathcal{M}$ in the interval $Y$. We define a truncated solution as follows: $z^*_{jh}:=y_{jh}-\gamma_{jh}$ for $h\in\mathcal{G}_1$, and $t^*_{jh}:=y_{jh}-\gamma_{jh}$ for $h\in\mathcal{G}_2$. By Theorem 4.4 each $j\in B_2$ is $d$-tight thus

$$\sum_{h\in\mathcal{G}_1}\gamma_{jh}+\sum_{h\in\mathcal{G}_2}\gamma_{jh}=\epsilon \quad j\in B_2 \tag{4.53}$$

and

$$\sum_{h\in\mathcal{G}_2}\gamma_{jh}=\eta_j\geq\varepsilon_j \quad j\in B_2. \tag{4.54}$$

We prove that this truncated solution is feasible for $\mathcal{F}$ and meets (4.52). □

We first prove the following lemma.

**Lemma 4.11** *If $\sum_{j\in B_2}\varepsilon_j=\epsilon$, then truncated solution meets (4.48).* □

***Proof*** We have the following for the truncated solution:

$$\sum_{h \in \mathcal{G}_1} z^*_{jh} = \sum_{h \in \mathcal{G}_1} y_{jh} - (\epsilon - \eta_j) \quad j \in B_2. \tag{4.55}$$

By Lemma 4.5 each $j \in B_2$ is $c$-tight. Thus we get

$$\sum_{h \in \mathcal{G}_1} y_{jh} = \sum_{h \in \mathcal{G}_1} b_{jh} + a_{j2} - \Delta(\mathcal{G}_2) - \lfloor x_{j2} \rfloor + \lfloor r \rfloor + \epsilon - \varepsilon_j \quad j \in B_2. \tag{4.56}$$

Therefore by (4.55) and (4.56) we get

$$\sum_{h \in \mathcal{G}_1} z^*_{jh} + (\varepsilon_j - \eta_j) = \sum_{h \in \mathcal{G}_1} b_{jh} + a_{j2} - \Delta(\mathcal{G}_2) - \lfloor x_{j2} \rfloor + \lfloor r \rfloor \quad j \in B_2,$$

and by (4.54)

$$\sum_{h \in \mathcal{G}_1} z^*_{jh} \geq \sum_{h \in \mathcal{G}_1} b_{jh} + a_{j2} - \Delta(\mathcal{G}_2) - \lfloor x_{j2} \rfloor + \lfloor r \rfloor \quad j \in B_2,$$

which proves (4.48) holds for $j \in B_2$ in the truncated solution $\mathbf{t}^*$ and $\mathbf{z}^*$. For $j \in \mathcal{J} \setminus B_2$, $x_{j2}$ is integral thus

$$\sum_{h \in \mathcal{G}_1} z^*_{jh} \geq \sum_{h \in \mathcal{G}_1} b_{jh} + a_{j2} - \Delta(\mathcal{G}_2) - \lfloor x_{j2} \rfloor + \lfloor r \rfloor + \epsilon - \sum_{h \in \mathcal{G}_1} \gamma_{jh} \quad j \in \mathcal{J} \setminus B_2,$$

since $\epsilon - \sum_{h \in \mathcal{G}_1} \gamma_{jh} \geq 0$ for $j \in \mathcal{J}$ we get

$$\sum_{h \in \mathcal{G}_1} z^*_{jh} \geq \sum_{h \in \mathcal{G}_1} b_{jh} + a_{j2} - \Delta(\mathcal{G}_2) - \lfloor x_{j2} \rfloor + \lfloor r \rfloor \quad j \in \mathcal{J} \setminus B_2.$$

Thus (4.48) holds for $j \in \mathcal{J}$. □

Let $\mathbf{t}^*$ and $\mathbf{z}^*$ be a solution of Lemma 4.11. The $\mathbf{t}^*$ and $\mathbf{z}^*$ clearly meet (4.44), (4.47), (4.49), (4.50), (4.51). By Lemma 4.11 (4.48) holds. Then (4.45) also holds for $\mathbf{t}^*$ and $\mathbf{z}^*$. To show that we observe that by feasibility of $\mathbf{s} = (\mathbf{y}, \mathbf{x}, r, w)$ we have

$$\sum_{h \in \mathcal{G}_2} b_{jh} + a_{j1} - x_{j1} - \Delta(\mathcal{G}_1) + r \leq \sum_{h \in \mathcal{G}_2} (y_{jh} - t^*_{jh}) + \sum_{h \in \mathcal{G}_2} t^*_{jh} \quad j \in \mathcal{J} \setminus B_1.$$

Since for $\mathbf{t}^*$ we have

$$0 \leq \sum_{h \in \mathcal{G}_2} (y_{jh} - t^*_{jh}) \leq \epsilon \quad j \in \mathcal{J},$$

and $x_{j1}$ is integral for $\mathcal{J} \setminus B_1$, the $\mathbf{t}^*$ satisfies the (4.45).

To prove (4.46) we observe that by Lemma 4.5 each $j\in B_1$ is $a$-tight and thus

$$\sum_{h\in\mathcal{G}_2} b_{jh}+a_{j1}-x_{j1}-\Delta(\mathcal{G}_1)+r=\sum_{h\in\mathcal{G}_2}(y_{jh}-t^*_{jh})+\sum_{h\in\mathcal{G}_2}t^*_{jh}\quad j\in B_1. \tag{4.57}$$

By Theorem 4.4 $j\in B_1$ is $d$-tight. Thus by Lemma 4.2 and definition of truncated solution we have

$$\epsilon=\sum_{h\in\mathcal{G}_2}(y_{jh}-t^*_{jh}), \tag{4.58}$$

for $j\in B_1$.

Thus by (4.57) and (4.58)

$$\sum_{h\in\mathcal{G}_2} b_{jh}+a_{j1}-\Delta(\mathcal{G}_1)+\lfloor r\rfloor-\lfloor x_{j1}\rfloor+\epsilon-\epsilon-\varepsilon_j=\sum_{h\in\mathcal{G}_2}t^*_{jh}\quad j\in B_1.$$

Hence (4.46) is met by the truncated solution $\mathbf{t}^*$ and $\mathbf{z}^*$. Therefore the truncated solution $\mathbf{t}^*$ and $\mathbf{z}^*$ is feasible for $\mathcal{F}$.

To prove the lower bound on the value of objective function we observe that by (4.57) and (4.58)

$$\sum_{h\in\mathcal{G}_2} b_{jh}+a_{j1}-x_{j1}-\Delta(\mathcal{G}_1)+\lfloor r\rfloor+\epsilon-\epsilon=\sum_{h\in\mathcal{G}_2}t^*_{jh}\quad j\in B_1. \tag{4.59}$$

Summing up (4.59) side by side over all $j\in B_1$ we get by (4.19) for $(\mathbf{y},\mathbf{x},r,w)$

$$\sum_{j\in B_1}\Big(\sum_{h\in\mathcal{G}_2} b_{jh}+a_{j1}\Big)-(r-c)-|B_1|(\Delta(\mathcal{G}_1)-\lfloor r\rfloor)=\sum_{j\in B_1}\sum_{h\in\mathcal{G}_2}t^*_{jh},$$

where $c=\sum_{j\in\mathcal{J}\setminus B_1}x_{j1}$ is integral by definition of $B_1$. Thus

$$\sum_{j\in B_1}\sum_{h\in\mathcal{G}_2} b_{jh}+\Delta(\mathcal{G}_1)-\lfloor r\rfloor-\sum_{j\in\mathcal{J}\setminus B_1}(a_{j1}-x_{j1})-|B_1|(\Delta(\mathcal{G}_1)-\lfloor r\rfloor)-\epsilon=\sum_{j\in B_1}\sum_{h\in\mathcal{G}_2}t^*_{jh}$$

and

$$\sum_{j\in B_1}\sum_{h\in\mathcal{G}_2} b_{jh}-\sum_{j\in\mathcal{J}\setminus B_1}(a_{j1}-x_{j1})-(|B_1-1)|(\Delta(\mathcal{G}_1)-\lfloor r\rfloor)-\epsilon=\sum_{j\in B_1}\sum_{h\in\mathcal{G}_2}t^*_{jh}$$

as required.

**Lemma 4.12** *If* $\sum_{j\in B_1}\varepsilon_j=\epsilon$, *then*

$$F = \sum_{j \in B_1} \sum_{h \in \mathcal{G}_2} b_{jh} + \sum_{j \in B_1} (a_{j1} - \lfloor x_{j1} \rfloor) - |B_1|(\Delta(\mathcal{G}_1) - \lfloor r \rfloor) \tag{4.60}$$

*and*

$$\sum_{h \in \mathcal{G}_2} t_{jh} = \sum_{h \in \mathcal{G}_2} b_{jh} + a_{j1} - \lfloor x_{j1} \rfloor - \Delta(\mathcal{G}_1) + \lfloor r \rfloor \quad j \in B_1. \tag{4.61}$$

***Proof*** By (4.59)

$$\sum_{h \in \mathcal{G}_2} b_{jh} + a_{j1} - \Delta(\mathcal{G}_1) + \lfloor r \rfloor - \lfloor x_{j1} \rfloor - \varepsilon_j = \sum_{h \in \mathcal{G}_2} t^*_{jh} \quad j \in B_1, \tag{4.62}$$

summing up side by side for $j \in B_1$ and taking $\sum_{j \in B_1} \varepsilon_j = \epsilon$ we get

$$\sum_{j \in B_1} \sum_{h \in \mathcal{G}_2} b_{jh} + \sum_{j \in B_1} (a_{j1} - \lfloor x_{j1} \rfloor) - |B_1|(\Delta(\mathcal{G}_1) - \lfloor r \rfloor) - \epsilon = \sum_{j \in B_1} \sum_{h \in \mathcal{G}_2} t^*_{jh}, \tag{4.63}$$

for the truncated solution $\mathbf{t}^*$ and $\mathbf{z}^*$, which by Lemma 4.10 is feasible for $\mathcal{F}$. Let $\mathbf{t}$ and $\mathbf{z}$ be an optimal solution for $\mathcal{F}$. Since all upper and lower bounds in $\mathcal{F}$ are integral, we may assume both $\mathbf{t}$ and $\mathbf{z}$ integral by the Integral Circulation Theorem, see Lawler [17]. Thus by (4.63)

$$\sum_{j \in B_1} \sum_{h \in \mathcal{G}_2} b_{jh} + \sum_{j \in B_1} (a_{j1} - \lfloor x_{j1} \rfloor) - |B_1|(\Delta(\mathcal{G}_1) - \lfloor r \rfloor) \le \sum_{j \in B_1} \sum_{h \in \mathcal{G}_2} t_{jh}, \tag{4.64}$$

and the upper bounds in (4.46) give

$$\sum_{j \in B_1} \sum_{h \in \mathcal{G}_2} b_{jh} + \sum_{j \in B_1} (a_{j1} - \lfloor x_{j1} \rfloor) - |B_1|(\Delta(\mathcal{G}_1) - \lfloor r \rfloor) \ge \sum_{j \in B_1} \sum_{h \in \mathcal{G}_2} t_{jh}. \tag{4.65}$$

Hence by (4.64) and (4.65) we get

$$\sum_{j \in B_1} \sum_{h \in \mathcal{G}_2} b_{jh} + \sum_{j \in B_1} (a_{j1} - \lfloor x_{j1} \rfloor) - |B_1|(\Delta(\mathcal{G}_1) - \lfloor r \rfloor) = \sum_{j \in B_1} \sum_{h \in \mathcal{G}_2} t_{jh} = F,$$

which proves (4.60) in the lemma. Finally, in order to reach this optimal value all upper bounds in (4.46) must be reached, which proves (4.61). □

**Theorem 4.10** *For $\sum_{j \in B_2} \varepsilon_j = \epsilon$, an optimal solution to $\mathcal{F}$ can be extended to an integral feasible solution to $\ell p$ with $lp = \lfloor r \rfloor < r$.*

***Proof*** Let $\mathbf{t}$ and $\mathbf{z}$ be an optimal solution to $\mathcal{F}$. This solution exists since by Lemma 4.10 there is a feasible solution to $\mathcal{F}$. Since all upper and lower bounds in $\mathcal{F}$ are integral, we may assume both $\mathbf{t}$ and $\mathbf{z}$ integral by the Integral Circulation

Theorem, see Lawler [17]. Thus by Lemma 4.10

$$\sum_{j\in B_1}\sum_{h\in\mathcal{G}_2} t_{jh} \geq \sum_{j\in B_1}\sum_{h\in\mathcal{G}_2} b_{jh} - \sum_{j\in\mathcal{J}\setminus B_1}(a_{j1}-x_{j1}) - (|B_1|-1)(\Delta(\mathcal{G}_1)-\lfloor r\rfloor). \tag{4.66}$$

For the partial solution $((\mathbf{t},\mathbf{z}), r'=\lfloor r\rfloor, w'=\lfloor w\rfloor)$ we have (4.51) implies (4.17), (4.49) and (4.50) imply (4.18), (4.44) implies (4.16), and (4.47) implies (4.15). To prove the last two implications we observe that

$$\sum_j b_{jh} - \Delta(\mathcal{G}_2) + \lfloor r\rfloor \leq \lfloor w\rfloor \qquad h\in\mathcal{G}_1$$

and

$$\sum_j b_{jh} - \Delta(\mathcal{G}_1) + \lfloor r\rfloor \leq \lfloor w\rfloor \qquad h\in\mathcal{G}_2$$

for **s**. The (4.44) guarantees

$$\sum_j t_{jh} = \lfloor w\rfloor \quad h\in\mathcal{G}_2,$$

and (4.47) guarantees

$$\sum_j z_{jh} = \lfloor w\rfloor \quad h\in\mathcal{G}_1.$$

Therefore (4.16) and (4.15) are satisfied by the partial solution $((\mathbf{t},\mathbf{z}), r' = \lfloor r\rfloor, w'=\lfloor w\rfloor)$.

Let us now extend the solution $((\mathbf{t},\mathbf{z}), r'=\lfloor r\rfloor, w'=\lfloor w\rfloor)$ by setting $x^*_{j2}:=\lfloor x_{j2}\rfloor$, for $j\in B_2$ and $x^*_{j2}:=x_{j2}$ for $j\in\mathcal{J}\setminus B_2$. Since $\sum_{j\in B_2}\varepsilon_j=\epsilon$, (4.20) is met by this extension. Clearly (4.22) is also met for $\ell=2$. By (4.48) we have

$$\sum_{h\in\mathcal{G}_1} b_{jh} + a_{j2} - \lfloor x_{j2}\rfloor - \Delta(\mathcal{G}_2) + \lfloor r\rfloor \leq \sum_{h\in\mathcal{G}_1} z_{jh} \quad j\in\mathcal{J}.$$

Thus (4.23) is met for the extended solution $((\mathbf{t},\mathbf{z}), r'=\lfloor r\rfloor, w'=\lfloor w\rfloor)$, and $x^*_{j2}$ for $j\in\mathcal{J}$.

We now extend this solution further by setting

$$x^*_{j1} := \sum_{h\in\mathcal{G}_2} b_{jh} + a_{j1} - \Delta(\mathcal{G}_1) + \lfloor r\rfloor - \sum_{h\in\mathcal{G}_2} t_{jh} \tag{4.67}$$

for $j \in B_1$ and $x^*_{j1} := x_{j1}$ for $j \in \mathcal{J} \setminus B_1$. To prove that (4.24) is met for the extended solution $((\mathbf{t}, \mathbf{z}), r' = \lfloor r \rfloor, w' = \lfloor w \rfloor)$, and $x^*_{j2}, x^*_{j1}$ for $j \in \mathcal{J}$ we need to show that

$$\sum_{h \in \mathcal{G}_2} b_{jh} + a_{j1} - x^*_{j1} - \Delta(\mathcal{G}_1) + \lfloor r \rfloor \leq \sum_{h \in \mathcal{G}_2} t_{jh} \tag{4.68}$$

for each $j \in \mathcal{J}$. By the definition (4.67) this holds for $j \in B_1$. For $j \in \mathcal{J} \setminus B_1$ we have $x_{j1}$ integral and thus (4.68) holds since (4.45) holds. Thus (4.24) is met for the extended solution $((\mathbf{t}, \mathbf{z}), r' = \lfloor r \rfloor, w' = \lfloor w \rfloor)$, and $x^*_{j2}, x^*_{j1}$ for $j \in \mathcal{J}$. Moreover $a_{j1} \geq x^*_{j1} \geq 0$ for each $j \in \mathcal{J} \setminus B_1$ and thus (4.22) holds for $\ell = 1$ in this extended solution. It suffices to prove this for $j \in B_1$.

Then, since $\lfloor r \rfloor \geq \lfloor r \rfloor - \lfloor x_{j1} \rfloor$, $x^*_{j1} \geq 0$ by (4.67) and the right hand side inequality of (4.46). Moreover, $a_{j1} \geq \lceil x_{j1} \rceil$. Thus by the left hand side inequality of (4.46)

$$\sum_{h \in \mathcal{G}_2} b_{jh} - \Delta(\mathcal{G}_1) + \lfloor r \rfloor \leq \sum_{h \in \mathcal{G}_2} t_{jh}$$

and by (4.67)

$$x^*_{j1} = \sum_{h \in \mathcal{G}_2} b_{jh} - \Delta(\mathcal{G}_1) + \lfloor r \rfloor - \sum_{h \in \mathcal{G}_2} t_{jh} + a_{j1} \leq a_{j1}.$$

Therefore (4.22) holds for $\ell = 1$ for $j \in B_1$. For $j \in \mathcal{J} \setminus B_1$ the (4.22) for $\ell = 1$ in the extended solution $((\mathbf{t}, \mathbf{z}), r' = \lfloor r \rfloor, w' = \lfloor w \rfloor)$, and $x^*_{j2}, x^*_{j1}$ follows from (4.22) for $\ell = 1$ in the solution $(\mathbf{y}, \mathbf{x}, r, w)$.

By definition of the extended solution $((\mathbf{t}, \mathbf{z}), r' = \lfloor r \rfloor, w' = \lfloor w \rfloor)$, and $x^*_{j2}, x^*_{j1}$ for $j \in \mathcal{J}$, and since by Theorem 4.7 there are no crossing jobs we have

$$x^*_{j1} + x^*_{j2} \leq \lfloor r \rfloor \tag{4.69}$$

for $j \in \mathcal{J} \setminus B_1$. We now need to show this inequality for $j \in B_1$. For these jobs by the left hand side inequality of (4.46), and by (4.67) we get $x^*_{j1} - \lfloor r \rfloor + \lfloor r \rfloor - \lceil x_{j1} \rceil \leq 0$. Thus $x^*_{j1} \leq \lceil x_{j1} \rceil$ for each job $j \in B_1$. This unfortunately does not guarantee (4.69) for $j \in B_1$. However, we either have $\lceil x_{j1} \rceil + x_{j2} \leq \lfloor r \rfloor$ for each $j \in B_1$, in which case (4.69) holds for $j \in B_1$, or $\lceil x_{k1} \rceil + x_{k2} > \lfloor r \rfloor$ for some $k \in B_1$. The latter implies $\sum_{j \in B_1} \varepsilon_j = \epsilon$, which by Lemma 4.12, implies

$$\sum_{h \in \mathcal{G}_2} t_{jh} = \sum_{h \in \mathcal{G}_2} b_{jh} + a_{j1} - \Delta(\mathcal{G}_1) + \lfloor r \rfloor - \lfloor x_{j1} \rfloor \quad j \in B_1$$

in the optimal solution $\mathbf{t}$ and $\mathbf{z}$ to $\mathcal{F}$. Thus by definition (4.67), $x^*_{j1} = \lfloor x_{j1} \rfloor$ for $j \in B_1$. Since by Theorem 4.7 there are no crossing jobs the (4.69) is satisfied. Hence it remains to prove that if $\lceil x_{k1}\rceil + x_{k2} > \lfloor r\rfloor$ for some $k\in B_1$, then $\sum_{j\in B_1}\varepsilon_j=\epsilon$. For contradiction assume $\lceil x_{k1}\rceil + x_{k2} > \lfloor r\rfloor$ for some $k\in B_1$ and $\sum_{j\in B_1}\varepsilon_j>\epsilon$. If $x_{j1}x_{j2}=0$ for each $j\in\mathcal{J}$, then $x_{k2}=0$. Thus $\lceil x_{k1}\rceil > \lfloor r\rfloor$ which implies $\sum_{j\in B_1}\varepsilon_j=\epsilon$ and gives contradiction. Otherwise, if $x_{i1}x_{i2}>0$ for some $i\in\mathcal{J}$, then by Theorem 4.9 we have $B_1=\{i\}$ or $B_2=\{i\}$. If $B_1=\{i\}$, then $\sum_{j\in B_1}\varepsilon_j=\epsilon$ which gives contradiction. Hence $B_2=\{i\}$ and $x_{j2}=0$ for each $j\in B_1$. Since by Theorem 4.7 there are no crossing jobs and $x_{i1}$ is integral and positive. Thus $x_{i1}\geq 1$, and $i\neq k$. By (4.19) $\sum_j x_{j1}=\sum_{j\neq i}x_{j1}+x_{i1}=r$. Hence $\sum_{j\neq i}x_{j1}\leq r-1$ which gives $x_{k1}\leq r-1$. Since $x_{k2}=0$, we get $x_{k1}+1+x_{k2}\leq r$. Thus $\lceil x_{k1}\rceil+x_{k2}\leq\lfloor r\rfloor$ which again gives contradiction. This proves that if $\lceil x_{k1}\rceil+x_{k2}>\lfloor r\rfloor$ for some $k\in B_1$, then $\sum_{j\in B_1}\varepsilon_j=\epsilon$ as required. Hence (4.21) holds for the extended solution $((\mathbf{t},\mathbf{z}), r'=\lfloor r\rfloor, w'=\lfloor w\rfloor)$, and $x^*_{j2}, x^*_{j1}$.

Finally we need to prove that (4.19) holds for an extended solution. By (4.67) and (4.66)

$$\sum_j x^*_{j1}\leq\lfloor r\rfloor \tag{4.70}$$

for the extended solution $((\mathbf{t},\mathbf{z},\lfloor r\rfloor,\lfloor w\rfloor))$, and $x^*_{j2}, x^*_{j1}$ for $j\in\mathcal{J}$. This solution satisfies all constraints (4.15)–(4.18) and (4.20)–(4.24) of $\ell p$. To complete the proof it suffices to modify the extension $x^*_{j1}$ for $j\in\mathcal{J}$ in order to ensure the equality in (4.70) to satisfy (4.19), and to keep other constraints (4.15)–(4.18) and (4.20)–(4.24) of $\ell p$ satisfied.

If $\sum_j x^*_{j1}<\lfloor r\rfloor$, then take a $j\in B_1$ with a positive $d_j=\min\{\lceil x_{j1}\rceil-x^*_{j1}, \lfloor r\rfloor - x^*_{j1}-x_{j2}\}$. Recall that by Theorem 4.7, $x_{j2}$ is integral for each $j\in B_1$. Such $j$ exists. To prove this existence define $X=\{j\in B_1:\lceil x_{j1}\rceil=x^*_{j1}\}$ and $Y=\{j\in B_1: x^*_{j1}=\lfloor x_{j1}\rfloor\}$. By definition (4.67) and (4.46) we have $B_1=X\cup Y$, and since

$$\sum_j x^*_{j1}<\lfloor r\rfloor<\sum_j\lceil x_{j1}\rceil \tag{4.71}$$

we have $Y\neq\emptyset$. Suppose for a contradiction that for each job $j\in Y$ we have $\lfloor r\rfloor=x^*_{j1}+x_{j2}$. Thus we have

$$\sum_j x^*_{j1}=\sum_{j\in\mathcal{J}\setminus B_1}x_{j1}+\sum_{j\in X}\lceil x_{j1}\rceil+\sum_{j\in Y}\lfloor x_{j1}\rfloor<\lfloor r\rfloor. \tag{4.72}$$

Since for each job $j\in Y$ we have $\lfloor r\rfloor=\lfloor x_{j1}\rfloor+x_{j2}$, we obtain

$$\sum_{j\in\mathcal{J}\setminus B_1}x_{j1}+\sum_{j\in X}\lceil x_{j1}\rceil+|Y|\lfloor r\rfloor-\sum_{j\in Y}x_{j2}<\lfloor r\rfloor,$$

and by (4.71) the set $Y$ is not empty. Since $\sum_{j\in Y} x_{j2} \le \lfloor r \rfloor$ by (4.20) we get

$$\sum_{j\in\mathcal{J}\setminus B_1} x_{j1} + \sum_{j\in X} \lceil x_{j1} \rceil + |Y| \lfloor r \rfloor < 2\lfloor r \rfloor,$$

and thus $|Y| \le 1$, and since $Y$ is not empty we have $|Y| = 1$. However

$$\lfloor r \rfloor = \lfloor \sum_j x_{j1} \rfloor = \sum_j \lfloor x_{j1} \rfloor + \lfloor \sum_{j\in B_1} \varepsilon_j \rfloor,$$

where

$$\lfloor \sum_{j\in B_1} \varepsilon_j \rfloor \le |B_1| - 1.$$

Thus

$$\lfloor r \rfloor = \lfloor \sum_j x_{j1} \rfloor \le \sum_{j\in\mathcal{J}\setminus B_1} x_{j1} + \sum_{j\in B_1} \lfloor x_{j1} \rfloor + |B_1| - 1 = \sum_{j\in\mathcal{J}\setminus B_1} x_{j1} + \sum_{j\in X} \lceil x_{j1} \rceil + \sum_{j\in Y} \lfloor x_{j1} \rfloor$$

since $|Y| = 1$ which contradicts (4.72) and proves that $j \in Y$ with $d_j = 1$ exists. Set $d := \min\{\min_{j,d_j>0}\{d_j\}, \lfloor r \rfloor - \sum_j x^*_{j1}\} = 1$. Then, set $x^*_{j1} := x^*_{j1} + 1$ for some $j \in Y$ with $d_j = 1$. We have $x^*_{j1} \le \min\{\lceil x_{j1} \rceil, \lfloor r \rfloor - x_{j2}\}$ and $\sum_j x^*_{j1} \le \lfloor r \rfloor$ for the new extended solution, which ensures that all constraints (4.15)–(4.18) and (4.20)–(4.24) of $\ell p$ are met in the new extended solution. Since $d = 1$ the $\sum_j x^*_{j1}$ gets closer to but does not exceed $\lfloor r \rfloor$. Therefore by (4.71) we finally reach an extended solution $\mathbf{t}$, $\mathbf{z}$, and $x^*_{j2}, x^*_{j1}$ for $j \in \mathcal{J}$ that meets all (4.15)–(4.24) of $\ell p$. The solution is integral with $w' = \lfloor w \rfloor$, and $r' = \lfloor r \rfloor$ which proves the lemma. □

## 4.7 The Projection

Consider the following system $S$ that defines the set of feasible solutions to the $LP$-relaxation of $\mathcal{ILP}$,

$$\sum_j b_{jh} - (\Delta(\mathcal{G}_2) - r) \le \sum_j y_{jh} \le w \qquad h \in \mathcal{G}_1 \tag{4.73}$$

$$\sum_j b_{jh} - (\Delta(\mathcal{G}_1) - r) \le \sum_j y_{jh} \le w \qquad h \in \mathcal{G}_2 \tag{4.74}$$

$$\sum_h y_{jh} \le w \qquad j \in \mathcal{J} \tag{4.75}$$

$$0 \le y_{jh} \le b_{jh} \qquad h \in \mathcal{M} \qquad j \in \mathcal{J} \tag{4.76}$$

$$\sum_j x_{j1} = r \tag{4.77}$$

$$\sum_j x_{j2} = r \tag{4.78}$$

$$x_{j1} + x_{j2} \le r \qquad j \in \mathcal{J} \tag{4.79}$$

$$0 \le x_{j\ell} \le a_{j\ell} \qquad j \in \mathcal{J} \qquad \ell = 1, 2 \tag{4.80}$$

$$\sum_{h \in \mathcal{G}_1} (b_{jh} - y_{jh}) + a_{j2} - x_{j2} \le \Delta(\mathcal{G}_2) - r \qquad j \in \mathcal{J} \tag{4.81}$$

$$\sum_{h \in \mathcal{G}_2} (b_{jh} - y_{jh}) + a_{j1} - x_{j1} \le \Delta(\mathcal{G}_1) - r \qquad j \in \mathcal{J}. \tag{4.82}$$

Now consider the system $S_r$ obtained from $S$ by dropping (4.77) and (4.78) and adding the constraints (4.91), (4.92), and (4.93). We use $\alpha_{j1} = \sum_{h \in \mathcal{G}_2} (b_{jh} - y_{jh}) + a_{j1} - \Delta(\mathcal{G}_1)$ and $\alpha_{j2} = \sum_{h \in \mathcal{G}_1} (b_{jh} - y_{jh}) + a_{j2} - \Delta(\mathcal{G}_2)$ for $j \in \mathcal{J}$ for convenience.

$$\sum_j b_{jh} - (\Delta(\mathcal{G}_2) - r) \le \sum_j y_{jh} \le w \qquad h \in \mathcal{G}_1 \tag{4.83}$$

$$\sum_j b_{jh} - (\Delta(\mathcal{G}_1) - r) \le \sum_j y_{jh} \le w \qquad h \in \mathcal{G}_2 \tag{4.84}$$

$$\sum_h y_{jh} \le w \qquad j \in \mathcal{J} \tag{4.85}$$

$$0 \le y_{jh} \le b_{jh} \qquad h \in \mathcal{M} \qquad j \in \mathcal{J} \tag{4.86}$$

$$x_{j1} + x_{j2} \le r \qquad j \in \mathcal{J} \tag{4.87}$$

$$0 \le x_{j\ell} \le a_{j\ell} \qquad j \in \mathcal{J} \qquad \ell = 1, 2 \tag{4.88}$$

$$\sum_{h \in \mathcal{G}_1} (b_{jh} - y_{jh}) + a_{j2} - x_{j2} \leq \Delta(\mathcal{G}_2) - r \qquad j \in \mathcal{J} \tag{4.89}$$

$$\sum_{h \in \mathcal{G}_2} (b_{jh} - y_{jh}) + a_{j1} - x_{j1} \leq \Delta(\mathcal{G}_1) - r \qquad j \in \mathcal{J} \tag{4.90}$$

$$\sum_{j} \alpha_{j1} + (n-1)r \leq 0 \tag{4.91}$$

$$\sum_{j} \alpha_{j2} + (n-1)r \leq 0 \tag{4.92}$$

$$0 \leq r \leq \min\{\Delta(\mathcal{G}_1), \Delta(\mathcal{G}_2)\}. \tag{4.93}$$

Finally consider the following projection on $\mathbf{y}, w, r$.

**Lemma 4.13** *Let $\mathcal{P}$ be the polyhedron that consists of feasible solutions to $S_r$. Then the projection of $\mathcal{P}$ on $\mathbf{y}, w, r$, denoted by $\mathcal{Q}$, is the set of solutions to the following system of inequalities $Q$:*

$$\sum_{j} b_{jh} - (\Delta(\mathcal{G}_2) - r) \leq \sum_{j} y_{jh} \leq w \qquad h \in \mathcal{G}_1 \tag{4.94}$$

$$\sum_{j} b_{jh} - (\Delta(\mathcal{G}_1) - r) \leq \sum_{j} y_{jh} \leq w \qquad h \in \mathcal{G}_2 \tag{4.95}$$

$$\sum_{h} y_{jh} \leq w \qquad j \in \mathcal{J} \tag{4.96}$$

$$0 \leq y_{jh} \leq b_{jh} \qquad h \in \mathcal{M} \qquad j \in \mathcal{J} \tag{4.97}$$

$$\sum_{h \in \mathcal{G}_2} (b_{jh} - y_{jh}) + a_{j1} - \Delta(\mathcal{G}_1) \leq 0 \qquad j \in \mathcal{J} \tag{4.98}$$

$$\sum_{h \in \mathcal{G}_1} (b_{jh} - y_{jh}) + a_{j2} - \Delta(\mathcal{G}_2) \leq 0 \qquad j \in \mathcal{J} \tag{4.99}$$

$$\sum_{h \in \mathcal{G}_2} (b_{jh} - y_{jh}) + r - \Delta(\mathcal{G}_1) \leq 0 \qquad j \in \mathcal{J} \tag{4.100}$$

$$\sum_{h \in \mathcal{G}_1} (b_{jh} - y_{jh}) + r - \Delta(\mathcal{G}_2) \leq 0 \qquad j \in \mathcal{J} \tag{4.101}$$

$$\sum_h (b_{jh} - y_{jh}) + a_{j1} + a_{j2} - \Delta(\mathcal{G}_1) - \Delta(\mathcal{G}_2) + r \le 0 \quad j \in \mathcal{J} \tag{4.102}$$

$$\sum_j \alpha_{j1} + (n-1)r \le 0 \tag{4.103}$$

$$\sum_j \alpha_{j2} + (n-1)r \le 0 \tag{4.104}$$

$$0 \le r \le \min\{\Delta(\mathcal{G}_1), \Delta(\mathcal{G}_2)\}. \tag{4.105}$$

***Proof*** The lemma follows by the Fourier-Motzkin elimination, see Schrijver [18], of variables $x_{j\ell}$ from the system $S_r$. □

We summarize the results of this section in the following theorem and lemma.

**Theorem 4.11** *Let* $(\mathbf{y}, r, w)$ *be feasible for* $Q$. *There exists* $\mathbf{x}$ *such that the solution* $(\mathbf{y}, \mathbf{x}, w, r)$ *is feasible for* $S$.

***Proof*** Let $s = (\mathbf{y}, r, w)$ be a feasible solution for $Q$. By Lemma 4.13 there exist $\mathbf{x} = (x_{j\ell})$, where $j \in \mathcal{J}$ and $\ell = 1, 2$, such that $s = (\mathbf{y}, \mathbf{x}, w, r)$ is feasible for $S_r$. Let $X$ be the set of all such $\mathbf{x}$. Take $\mathbf{x} \in X$ with minimum distance $d = |r - \sum_j x_{j1}| + |r - \sum_j x_{j2}|$. We show that $d = 0$ for $\mathbf{x}$. Suppose that $r < \sum_j x_{j1}$ or $r < \sum_j x_{j2}$. Let $r < \sum_j x_{j1}$. If there is $k$ such that $\alpha_{k1} + r < x_{k1}$, then set $x_{k1} := x_{k1} - \lambda$ where $\lambda = \min\{x_{k1} - (\alpha_k + r), \sum_j x_{j1} - r\}$. The new solution is in $X$ and reduces $d$ which gives a contradiction. Thus we have $\alpha_{j1} + r = x_{j1}$ for each $j$. Therefore $\sum_j \alpha_{j1} + nr = \sum_j x_{j1} \le r$ by the constraint (4.103) which contradicts this case assumption. The proof for $r < \sum_j x_{j2}$ is similar. Therefore we have $r \ge \sum_j x_{j1}$ and $r \ge \sum_j x_{j2}$ for the $\mathbf{x}$. Suppose that $r > \sum_j x_{j1}$ or $r > \sum_j x_{j2}$. If there is $k$ such that $x_{k1} + x_{k2} < r$ and ($x_{k1} < a_{k1}$ or $x_{k2} < a_{k2}$), then set $x_{k1} + \lambda$, where $\lambda = \min\{r - (x_{k1} + x_{k2}), a_{k1} - x_{k1}, d\}$ provided $x_{k1} < a_{k1}$. Otherwise, if $x_{k1} = a_{k1}$ and $x_{k2} < a_{k2}$, set $x_{k2} + \lambda$, where $\lambda = \min\{r - (x_{k1} + x_{k2}), a_{k2} - x_{k2}, d\}$. The new solution is in $X$ but has smaller $d$ which gives a contradiction. Thus we have $x_{j1} + x_{j2} = r$ or ($x_{j1} = a_{j1}$ and $x_{j2} = a_{j2}$) for each $j$. We have at least one $j$ with $x_{j1} + x_{j2} = r$. Otherwise, $r > \min\{\Delta(\mathcal{G}_1), \Delta(\mathcal{G}_2)\}$ which contradicts (4.105). On the other hand, we can have at most one $j$ with $x_{j1} + x_{j2} = r$. Otherwise $\sum_j (x_{j1} + x_{j2}) \ge 2r$ and since $r \ge \sum_j x_{j1}$ and $r \ge \sum_j x_{j2}$ for the $\mathbf{x}$ we get $r = \sum_j x_{j1}$ and $r = \sum_j x_{j2}$ which contradicts the assumption. Therefore there is exactly one $j$ such that $x_{j1} + x_{j2} = r$, and $x_{k1} = a_{k1}$, and $x_{k2} = a_{k2}$ for $k \in \mathcal{J} \setminus \{j\}$. Hence $\Delta(\mathcal{G}_1) - a_{j1} + x_{j1} < r$ or $\Delta(\mathcal{G}_2) - a_{j2} + x_{j2} < r$. Since $\Delta(\mathcal{G}_1) - a_{j1} + x_{j1} \le r$ and $\Delta(\mathcal{G}_2) - a_{j2} + x_{j2} \le r$, we have $\Delta(\mathcal{G}_1) + \Delta(\mathcal{G}_2) - a_{j2} + x_{j2} - a_{j1} + x_{j1} < 2r$. Hence $\Delta(\mathcal{G}_1) + \Delta(\mathcal{G}_2) - a_{j2} - a_{j1} < r$ since $x_{j1} + x_{j2} = r$. However by (4.102) and (4.97) we have $a_{j1} + a_{j2} + r \le \Delta(\mathcal{G}_1) + \Delta(\mathcal{G}_2)$ which gives a contradiction. Thus we have $d = 0$ and the solution is feasible for $S$. □

We have the following lemma.

**Lemma 4.14** *If* $(\mathbf{y}, \mathbf{x}, r, w)$ *is feasible for* $S$*, then* $(\mathbf{y}, r, w)$ *is feasible for* $Q$.

***Proof*** If $(\mathbf{y}, \mathbf{x}, r, w)$ is feasible for $S$, then it is also feasible for $S_r$. Observe that (4.77), (4.78), and (4.80) in $S$ imply (4.93) in $S_r$, the (4.77) and (4.82) in $S$ imply (4.91) in $S_r$, and the (4.78) and (4.81) in $S$ imply (4.92) in $S_r$. Finally, by Lemma 4.13 the $(\mathbf{y}, r, w)$ is feasible for $Q$. □

The system $Q$ is a network-flow model with lower and upper bounds on the arcs for fixed $w$ and $r$.

## 4.8 Integral Optimal Solution to $\ell p$ for $\sum_{j\in B_i} \varepsilon_j > \epsilon$, $i = 1, 2$

Consider **s** with $\sum_{j\in B_i} \varepsilon_j > \epsilon$ for $i = 1, 2$. By Lemma 4.9 overlap of $B_1$ and of $B_2$ do not occur simultaneously. Without loss of generality let us assume no overlap of $B_2$.

Consider the set $Y_j$ of all columns of type $\binom{*}{*,j}$ in $d(\mathbf{y}, w)$ for $j \in B_2$. By Lemma 4.7, $l(Y_j) = \beta_j = \lfloor \beta_j \rfloor + \varepsilon_j$. Take an interval $Y \subseteq \bigcup_{j\in B_2} Y_j$ such that $l(Y) = \epsilon$. Such $Y$ exists since there is no overlap of $B_2$ and $\sum_{j\in B_2} \varepsilon_j > \epsilon$. Let $Y_{jh}$ be the set of columns $I \in Y$ such that $(j, h) \in M_I$, set $\gamma_{jh} := l(Y_{jh})$. Informally, $\gamma_{jh}$ is the amount of $j \in \mathcal{J}$ done on $h \in \mathcal{M}$ in the interval $Y$. We define a truncated solution as follows $z^*_{jh} := y_{jh} - \gamma_{jh}$ for $h \in \mathcal{G}_1$, and $t^*_{jh} := y_{jh} - \gamma_{jh}$ for $h \in \mathcal{G}_2$, and $\lfloor r \rfloor$, $\lfloor w \rfloor$. Thus

$$\sum_{h\in\mathcal{G}_1} \gamma_{jh} + \sum_{h\in\mathcal{G}_2} \gamma_{jh} \le \epsilon \qquad j \in \mathcal{J}.$$

**Theorem 4.12** *For* $\sum_{j\in B_i} \varepsilon_j > \epsilon$, $i = 1, 2$, *there is a feasible integral solution to* $\ell p$ *with* $lp = \lfloor r \rfloor < r$.

***Proof*** We begin by proving that the truncated solution $(y^* = (z^*, t^*), \lfloor r \rfloor, \lfloor w \rfloor)$ is feasible for $Q$.

The constraints (4.98) and (4.99): For **s** we have

$$\sum_{h\in\mathcal{G}_1} b_{jh} + a_{j2} - \Delta(\mathcal{G}_2) + r - x_{j2} \le \sum_{h\in\mathcal{G}_1} y_{jh} \qquad j \in \mathcal{J}$$

$$\sum_{h\in\mathcal{G}_2} b_{jh} + a_{j1} - \Delta(\mathcal{G}_1) + r - x_{j1} \le \sum_{h\in\mathcal{G}_2} y_{jh} \qquad j \in \mathcal{J}.$$

If $r - x_{j1} \ge \epsilon$ and $r - x_{j2} \ge \epsilon$ for each $j \in \mathcal{J}$, then $\sum_{h\in\mathcal{G}_1} y_{jh} - (r - x_{j2}) \le \sum_{h\in\mathcal{G}_1} z^*_{jh}$ and $\sum_{h\in\mathcal{G}_2} y_{jh} - (r - x_{j1}) \le \sum_{h\in\mathcal{G}_2} t^*_{jh}$ for each $j$. Hence (4.98) and (4.99) hold for the truncated solution. Otherwise, if $r - x_{j1} < \epsilon$ or $r - x_{j2} < \epsilon$ for some $j \in \mathcal{J}$, then $\lfloor r \rfloor \le x_{j1}$ or $\lfloor r \rfloor \le x_{j2}$ for some $j$. This implies $\sum_{j\in B_1} \varepsilon_j = \epsilon$ or $\sum_{j\in B_2} \varepsilon_j = \epsilon$ which contradicts the theorem's assumption.

The constraints (4.100) and (4.101): For **s** we have

$$\sum_{h\in\mathcal{G}_2} b_{jh} + r - \Delta(\mathcal{G}_1) + a_{j1} - x_{j1} \le \sum_{h\in\mathcal{G}_2} y_{jh} \quad j \in \mathcal{J},$$

and

$$\sum_{h\in\mathcal{G}_1} b_{jh} + r - \Delta(\mathcal{G}_2) + a_{j2} - x_{j2} \le \sum_{h\in\mathcal{G}_1} y_{jh} \quad j \in \mathcal{J}.$$

By constraint (4.22) and definition of the truncated solution

$$\sum_{h\in\mathcal{G}_2} b_{jh} + \lfloor r \rfloor - \Delta(\mathcal{G}_1 \le \sum_{h\in\mathcal{G}_2} y_{jh} - \epsilon \le \sum_{h\in\mathcal{G}_2} t^*_{jh} \quad j \in \mathcal{J},$$

and

$$\sum_{h\in\mathcal{G}_1} b_{jh} + \lfloor r \rfloor - \Delta(\mathcal{G}_2 \le \sum_{h\in\mathcal{G}_1} y_{jh} - \epsilon \le \sum_{h\in\mathcal{G}_1} z^*_{jh} \quad j \in \mathcal{J}.$$

Hence (4.100) and (4.101) hold.

The constraints (4.102): For **s** by (4.23) and (4.24) we have

$$\sum_{h\in\mathcal{G}_1} (b_{jh} - y_{jh}) + a_{j2} - x_{j2} \le \Delta(\mathcal{G}_2) - r$$

and

$$\sum_{h\in\mathcal{G}_2} (b_{jh} - y_{jh}) + a_{j1} - x_{j1} \le \Delta(\mathcal{G}_1) - r,$$

by summing up the two side by side we get

$$\sum_h (b_{jh} - y_{jh}) + a_{j1} + a_{j2} - x_{j1} - x_{j2} \le \Delta(\mathcal{G}_1) + \Delta(\mathcal{G}_2) - 2r$$

or

$$\sum_h b_{jh} + a_{j1} + a_{j2} - \Delta(\mathcal{G}_1) - \Delta(\mathcal{G}_2) + \lfloor r \rfloor \le \sum_h y_{jh} - r + x_{j1} + x_{j2} - \epsilon.$$

Since $\sum_h y_{jh} - \epsilon \le \sum_{h\in\mathcal{G}_1} z^*_{jh} + \sum_{h\in\mathcal{G}_2} t^*_{jh}$, we have

$$\sum_h y_{jh} - r + x_{j1} + x_{j2} - \epsilon \le \sum_{h \in \mathcal{G}_1} z^*_{jh} + \sum_{h \in \mathcal{G}_2} t^*_{jh} - r + x_{j1} + x_{j2}.$$

But $-r + x_{j1} + x_{j2} \le 0$ by the constraint (4.21) and thus we get

$$\sum_h b_{jh} + a_{j1} + a_{j2} - \Delta(\mathcal{G}_1) - \Delta(\mathcal{G}_2) + \lfloor r \rfloor \le \sum_{h \in \mathcal{G}_1} z^*_{jh} + \sum_{h \in \mathcal{G}_2} t^*_{jh}$$

which proves that (4.102) holds for $y^* = (z^*, t^*)$.

The constraints (4.94)–(4.95): For **s** by (4.15), and (4.16) we have

$$\sum_j b_{jh} - \Delta(\mathcal{G}_2) + \lfloor r \rfloor \le \sum_j y_{jh} - \epsilon \le \lfloor w \rfloor \qquad h \in \mathcal{G}_1,$$

and

$$\sum_j b_{jh} - \Delta(\mathcal{G}_1) + \lfloor r \rfloor \le \sum_j y_{jh} - \epsilon \le \lfloor w \rfloor \qquad h \in \mathcal{G}_2.$$

For the truncated solution we have $\sum_j y_{jh} = \sum_j z^*_{jh} + \sum_j \gamma_{jh}$ for $h \in \mathcal{G}_1$, and $\sum_j y_{jh} = \sum_j t^*_{jh} + \sum_j \gamma_{jh}$ for $h \in \mathcal{G}_2$. Because of the machine saturation $\sum_j \gamma_{jh} = \epsilon$ for $h \in \mathcal{G}_1 \cup \mathcal{G}_2$. Thus

$$\sum_j b_{jh} - \Delta(\mathcal{G}_2) + \lfloor r \rfloor \le \sum_j z^*_{jh} \le \lfloor w \rfloor \qquad h \in \mathcal{G}_1,$$

$$\sum_j b_{jh} - \Delta(\mathcal{G}_1) + \lfloor r \rfloor \le \sum_j t^*_{jh} \le \lfloor w \rfloor \qquad h \in \mathcal{G}_2,$$

and (4.94) and (4.95) are satisfied by the truncated solution. By the machine saturation we have $\sum_j z^*_{jh} = \lfloor w \rfloor$ for $h \in \mathcal{G}_1$, and $\sum_j t^*_{jh} = \lfloor w \rfloor$ for $h \in \mathcal{G}_2$.

The constraint (4.96): For **s** by (4.17) we have $l(X_j) \le l(d(\mathbf{y}, w)) = w$ where $X_j$ is the set of all columns in $d(\mathbf{y}, w)$ that match $j \in \mathcal{J}$. Since $l(Y) = \epsilon$, we get $l(Z_j) \le l(d(\mathbf{y}, w) \setminus Y) = \lfloor w \rfloor$ where $Z_j$ is the set of all columns in $d(\mathbf{y}, w) \setminus Y$ that match $j \in \mathcal{J}$. We have $l(X_j) = l((X_j \cap Y) \cup (X_j \setminus Y)) = l(X_j \cap Y) + l(X_j \setminus Y) = l(X_j \cap Y) + l(Z_j)$. Hence $l(Z_j) = l(X_j) - l(X_j \cap Y) = \sum_h y_{jh} - \sum_h \gamma_{jh} = \sum_h y^*_{jh}$. Thus $\sum_h y^*_{jh} \le \lfloor w \rfloor$ and (4.96) is satisfied by the truncated solution.

Finally, the constraints (4.103) and (4.104). First we observe that $|\mathcal{G}_1| \le n - 1$ and $|\mathcal{G}_2| \le n - 1$. Otherwise $|\mathcal{G}_1| > n - 1$ or $|\mathcal{G}_2| > n - 1$ and since by the saturation $|\mathcal{G}_1| + |\mathcal{G}_2| \le n$ we would have $|\mathcal{G}_1| = 0$ or $|\mathcal{G}_2| = 0$ which contradicts the assumption of non-empty groups. Second, by summing up (4.23) side by side for **s** over all jobs and doing the same for (4.24) we get

$$\sum_{h \in \mathcal{G}_2} \sum_j b_{jh} - |\mathcal{G}_2| w + (1-n)\Delta(\mathcal{G}_1) + (n-1)r \le 0$$

and

$$\sum_{h \in \mathcal{G}_1} \sum_j b_{jh} - |\mathcal{G}_1| w + (1-n)\Delta(\mathcal{G}_2) + (n-1)r \le 0,$$

respectively. Since $|\mathcal{G}_1| \le n-1$ and $|\mathcal{G}_2| \le n-1$, we get

$$\sum_{h \in \mathcal{G}_2} \sum_j b_{jh} - |\mathcal{G}_2| \lfloor w \rfloor + (1-n)\Delta(\mathcal{G}_1) + (n-1)\lfloor r \rfloor \le 0$$

and

$$\sum_{h \in \mathcal{G}_1} \sum_j b_{jh} - |\mathcal{G}_1| \lfloor w \rfloor + (1-n)\Delta(\mathcal{G}_2) + (n-1)\lfloor r \rfloor \le 0.$$

By the machine saturation we have $|\mathcal{G}_2| \lfloor w \rfloor = \sum_{h \in \mathcal{G}_2} \sum_j t^*_{jh}$ and $|\mathcal{G}_1| \lfloor w \rfloor = \sum_{h \in \mathcal{G}_1} \sum_j z^*_{jh}$ which proves that (4.103) and (4.104) are satisfied by the truncated solution.

Therefore the truncated solution $(y^* = (z^*, t^*), \lfloor r \rfloor, \lfloor w \rfloor)$ is feasible for $Q$, and by Theorem 4.11 there exists $\mathbf{x}^*$ such that $(y^* = (z^*, t^*), \mathbf{x}^*, \lfloor r \rfloor, \lfloor w \rfloor)$ is feasible for $S$. Moreover $\lfloor r^* \rfloor \le \lfloor r \rfloor$, and $\lfloor w \rfloor - \lfloor r \rfloor = \lceil w^* - r^* \rceil$ since $\mathbf{s} = (\mathbf{y}, \mathbf{x}, r, w)$ is feasible for $\ell p$. Thus the solution $(y^* = (z^*, t^*), \mathbf{x}^*, \lfloor r \rfloor, \lfloor w \rfloor)$ is feasible for $\ell p$ and $lp = \lfloor r \rfloor$. For a feasible solution to $Q$ with integral $\lfloor w \rfloor$ and $\lfloor r \rfloor$ all lower and upper bounds in the network $Q$ are integral thus we can find in polynomial time an integral $\mathbf{y}^*$. Finally for given integral and fixed $\lfloor r \rfloor$, $\lfloor w \rfloor$, and $\mathbf{y}^*$ the $S$ becomes a network-flow model with integral lower and upper bounds on the flows. Thus we can find in polynomial time an integral $\mathbf{x}^*$ such that the integer solution $(\mathbf{y}^*, \mathbf{x}^*, \lfloor r \rfloor, \lfloor w \rfloor)$ is feasible for $lp$ and $lp = \lfloor r \rfloor$. □

Figure 4.5 gives an integral solution to $\mathcal{ILP}$ for the instance in Fig. 4.4. The solution has part (b) of size $\lfloor r \rfloor = 1$ that consists of job $J_1$ on $\mathcal{G}_2$ and job $J_6$ on $\mathcal{G}_1$. This part (b) is shorter than the part (b) in $\mathbf{s}$ which is of size $r = \frac{3}{2}$, see Fig. 4.4, and thus $\mathbf{s}$ cannot be an optimal solution to $\ell p$.

## 4.9 The Proof of the Conjecture

We are now ready to prove Theorem 4.3 which proves the conjecture.

***Proof*** For contradiction suppose the optimal value for $\ell p$ is fractional, $lp = r = \lfloor r \rfloor + \epsilon$, where $\epsilon > 0$. By Theorem 4.10 there is a feasible integral solution to $\ell p$

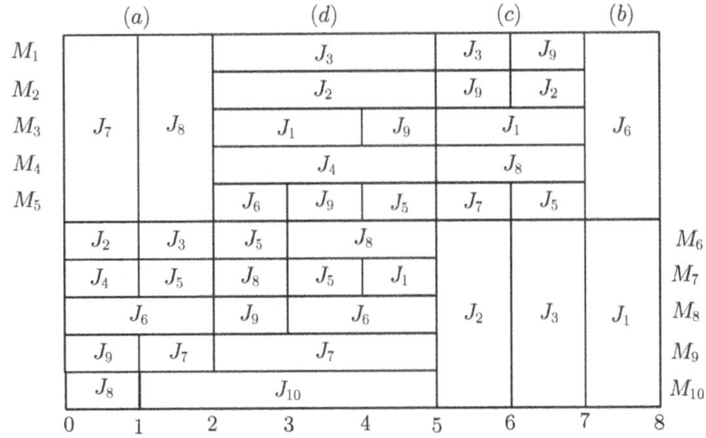


**Fig. 4.5** An integral solution ($\mathbf{y}^*$, $\mathbf{x}^*$, $\lfloor r \rfloor = 1$, $\lfloor w \rfloor = 3$) for $S$ in Fig. 4.4

with $lp = \lfloor r \rfloor$ for $\sum_{j \in B_1} \varepsilon_j = \epsilon$ or $\sum_{j \in B_2} \varepsilon_j = \epsilon$. By Theorem 4.12 there is a feasible integral solution to $\ell p$ with $lp = \lfloor r \rfloor$ for $\sum_{j \in B_1} \varepsilon_j > \epsilon$ and $\sum_{j \in B_2} \varepsilon_j > \epsilon$. Thus there is a feasible integral solution for $\ell p$ with $\lfloor r \rfloor < r$. Hence there is a feasible solution to $\ell p$ which is smaller than optimal $r$ which gives contradiction and proves the first part of the theorem. Thus optimal $\mathbf{s}$ has both $r$ and $w$ integer. The $\mathbf{s}$ is feasible for $S$ and thus it is feasible for $Q$ by Lemma 4.14. For a feasible solution to $Q$ with integral $w$ and $r$ all lower and upper bounds in the network $Q$ are integral thus we can find in polynomial time an integral $\mathbf{y}$. Finally for given integral and fixed $r$, $w$ and $\mathbf{y}$ the $S$ becomes a network with integral lower and upper bounds on the flows. Thus we can find in polynomial time an integral $\mathbf{x}$ such that the integer solution ($\mathbf{y}$, $\mathbf{x}$, $r$, $w$) is feasible for $lp$ and $lp = r$. □

The question remains whether there is a simpler, perhaps more direct (not using $LP$- relaxations), approach that would result in the polynomial-time algorithm for two groups, also another natural question remains whether there is a shorter proof of the conjecture. These two remain challenging questions worthy further investigation.

## 4.10 Complexity of Open Shop Scheduling with Preemptions Allowed at Any Points

The idea of using a linear program to find a schedule that minimizes makespan for open shop with multiprocessor operations has been introduced in Sect. 4.3 for two groups, $p = 2$. This idea has been extended in Ittig [14] to any fixed $p > 2$. The extension is presented in this section. We begin with $p = 3$. Then any schedule $S$

**Table 4.1** Possible intervals types in schedule $\mathcal{S}$; 0 and 1 in column $\mathcal{G}_\ell$, $\ell = 1, 2, 3$ denote individual and group operations on machines in $\mathcal{G}_\ell$, respectively

| Interval type | Types of operations on machines in | | Interval length | |
|---|---|---|---|---|
| | $\mathcal{G}_1$ | $\mathcal{G}_2$ | $\mathcal{G}_3$ | |
| 1 | 1 | 1 | 0 | $a_1$ |
| 2 | 0 | 0 | 1 | $a_2$ |
| 3 | 1 | 0 | 1 | $b_1$ |
| 4 | 0 | 1 | 0 | $b_2$ |
| 5 | 0 | 1 | 1 | $c_1$ |
| 6 | 1 | 0 | 0 | $c_2$ |
| 7 | 1 | 1 | 1 | $r$ |
| 8 | 0 | 0 | 0 | $w$ |

partitions the interval $[0, C_{\max}]$ into $2^p = 8$ disjoint interval types, some may be empty, listed in Table 4.1.

The interval of type (1) has group operations on both $\mathcal{G}_1$ and $\mathcal{G}_2$, thus 1 in the columns $\mathcal{G}_1$ and $\mathcal{G}_2$, and individual operations or idle time on $\mathcal{G}_3$, thus 0 in the column $\mathcal{G}_3$. The length of the interval of type (1) is denoted by $a_1$. Similarly the interval of type (2) has individual operations or idle time on both $\mathcal{G}_1$ and $\mathcal{G}_2$, thus 0 in the columns $\mathcal{G}_1$ and $\mathcal{G}_2$, and group operations on $\mathcal{G}_3$, thus 1 in the column $\mathcal{G}_3$. The length of the interval of type (2) is $a_2$. The other interval types should be clear from the table by now. Some of those interval types may be empty in $\mathcal{S}$, then their lengths equal 0. The interval types can be permuted in any possible way still giving the schedule with the same makespan as $\mathcal{S}$. In order to find the schedule that minimizes makespan we define variables as in Fig. 4.6, where the variables $x^i_{j\ell}$ and $y^i_{jh}$, for $J_j \in \mathcal{J}$, $\ell = 1, 2, 3$ and $M_h \in \mathcal{M}$, are introduced for pair $2i - 1$ and $2i$ of the interval types, $i = 1, 2, 3$. The two interval types in each pair complement one another; they partition the three groups into two disjoint sets. The variable $0 \le x^i_{j\ell}$ denotes the amount of job $J_j$ group operation $\hat{O}_{j\ell}$ processed on $\mathcal{G}_\ell$ in the intervals of types $(2i - 1)$ and $(2i)$, $i = 1, 2, 3$, and the variable $0 \le y^i_{jh}$ denotes the amount of job $J_j$ individual operation $O_{jh}$ processed on $M_h$ in the intervals of types $(2i - 1)$ and $(2i)$, $i = 1, 2, 3$. The remaining amount $0 \le a_{j\ell} - (x^1_{j\ell} + x^2_{j\ell} + x^3_{j\ell})$ of job $J_j$ group operation $\hat{O}_{j\ell}$ is left for the interval of type (7), and the remaining amount $0 \le b_{jh} - (y^1_{jh} + y^2_{jh} + y^3_{jh})$ of job $J_j$ individual operation $O_{jh}$ is left for the interval of type (8). The remaining non-negative variables $a_1, a_2, b_1, b_2, c_1, c_2, r$, and $w$ denote the lengths of the intervals (1) – (8), respectively.

The constraints for each interval need to ensure that each job is processed in the interval for *not* longer than the length of the interval, and each machine is occupied for *not* longer than the length of the interval. Thus the constraints ensure that a feasible schedule can be obtained for each interval using the algorithms for $O|\text{pmtn}|C_{\max}$ discussed earlier in Sect. 3.7.1. For the interval type (1) of length $a_1$ we thus have the following constraints:

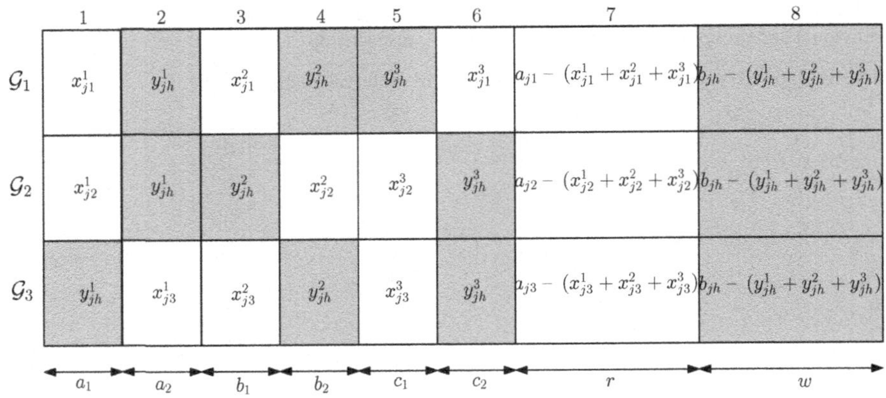


**Fig. 4.6** The variables and interval types used in the linear program to minimize makespan

$$x_{j1}^1 + x_{j2}^1 + \sum_{h \in \mathcal{G}_3} y_{jh}^1 \leq a_1 \qquad j \in \mathcal{J}$$

for the jobs, and the following:

$$\sum_j x_{j1}^1 = a_1$$

$$\sum_j x_{j2}^1 = a_1$$

$$\sum_j y_{jh}^1 \leq a_1 \qquad h \in \mathcal{G}_3$$

for the machines. The constraints for the interval types (2)–(6) can be readily obtained in a similar fashion. The reader is encouraged to write them down, see Problem 4.2. For the interval type (7) we have

$$(a_{j1} + a_{j2} + a_{j3}) - \sum_{i=1}^{3} \sum_{\ell=1}^{3} x_{j\ell}^i \leq r \qquad j \in \mathcal{J}$$

for the jobs, and

$$\sum_j a_{j1} - \sum_{i=1}^{3} \sum_j x_{j\ell}^i = r \qquad \ell = 1, 2, 3$$

for the groups. Finally, for the interval type (8) we have

$$\sum_h b_{jh} - \sum_{i=1}^{3} \sum_h y_{jh}^i \leq w \qquad j \in \mathcal{J}$$

for the jobs, and

$$\sum_j b_{jh} - \sum_{i=1}^{3} \sum_j y_{jh}^i \leq w \qquad h \in \mathcal{M}$$

for the machines. The makespan equals $a_1+a_2+b_1+b_2+c_1+c_2+r+w$. However we have the following equalities:

$$\Delta(\mathcal{G}_1) = a_1 + b_1 + c_2 + r,$$

$$\Delta(\mathcal{G}_2) = a_1 + b_2 + c_1 + r,$$

$$\Delta(\mathcal{G}_3) = a_2 + b_1 + c_1 + r,$$

which can be used to reduce the number of variables in the linear program. By eliminating the variables $a_2$, $b_2$, and $c_2$ we obtain the following objective for the linear program:

$$\min(w - 2r - a_1 - b_1 - c_1).$$

Following the idea of interval types, linear programs can be obtained in polynomial time for any fixed number $p$ of groups. All entries in the constraint matrix of those linear programs are 0, +1, or −1, thus the linear programs can be solved by a strongly polynomial algorithm. This proves that the makespan minimization for open shop scheduling problem with multiprocessor operations is polynomial for any fixed number of groups $p$. Observe that the number of interval types equals $2^p$, and thus it is exponential when $p$ is part of the input. Therefore problem complexity remains an open question for the case when $p$ is part of the problem input. A polynomial-time algorithm, if any exists, that would produce optimal schedules needs to somehow limit the number of possible interval types so that the number can be bounded by a polynomial of the input size. The question whether such a bound exists remains open.

**Problem 4.1** Is the problem of makespan minimization for preemptive scheduling of open shop with multiprocessor operations polynomial?

## 4.11 Integer Preemptions: Approximations

The solutions minimizing makespan for the open shop scheduling problem with multiprocessor operations and preemptions allowed at any points can be rounded in polynomial time to obtain optimal solutions with preemptions allowed at integer points only for $p = 2$. We presented this approach in Sects. 4.3–4.9 where we also proved that the minimum makespan for the latter problem equals $\lceil C_{\max} \rceil$, where $C_{\max}$ is the minimum makespan of the former. Though it may be tempting to think that the approach based on rounding in polynomial results in optimal solutions for other values of $p \geq 3$, this is unfortunately not the case. We showed in Sect. 4.2 that such rounding in polynomial time is impossible unless $NP = P$. However, the optimal solutions to the linear program for the problem with preemptions at any points can be rounded to provide *approximate* solutions to the problem with integral preemptions only for any fixed $p$. Ittig [14] has shown a polynomial-time rounding algorithm $A$ that gives solutions within a constant absolute error for any fixed number of groups $p$.

**Theorem 4.13** *Let $C$ be the makespan of the optimal solution with preemptions at any points, and let $C^A$ be the makespan of the solution with preemptions at integer points only obtained by the rounding algorithm $A$. We have*

$$C^A - C \leq 2p \cdot (2^{p-1} - 1) + 3.$$

Despite this constant absolute error obtained for the rounding algorithm we have the following implication of Theorem 4.1.

**Theorem 4.14** *If $P \neq NP$, then no polynomial-time algorithm for University timetabling for $p \geq 3$ exists with the worst case ratio less than $\frac{4}{3}$.*

***Proof*** Consider the set $\mathcal{I}$ of instances of University timetabling defined in the proof of Theorem 4.1. The problem $\Pi$ defined by $\mathcal{I}$ and the question whether $I \in \mathcal{I}$ has a schedule with makespan not exceeding 3 or not is $NP$-complete which follows immediately from the proof of Theorem 4.1. Suppose for contradiction that there is a polynomial-time algorithm $B$ such that $C^B_{\max}/C^*_{\max} < 4/3$ for any instance of University timetabling. Thus, in particular, $C^B_{\max}/C^*_{\max} < 4/3$ for any instance of $\Pi$. The algorithm $B$ can be used to solve $\Pi$ as follows. If $C^A_{\max} \leq 3$ for instance $I$, then the answer for $I$ is affirmative. Otherwise, if $C^B_{\max} > 3$ for $I$, then, since all processing times in $I$ are integer, we have $C^B_{\max} \geq 4$ and integer. Thus, since $C^*_{\max} > 3C^B_{\max}/4$, we get $C^*_{\max} > 3$ and the answer for $I$ is negative. Since $C^B_{\max}$ can be computed in polynomial time for each $I \in \mathcal{I}$, we have $\Pi$ in $P$. This implies $P = NP$ since $\Pi$ is $NP$-complete and gives contradiction. □

These results indicate that the rounding algorithm $A$ may give the ratios $\frac{C^A}{C^*}$, where $C^*$ is the makespan of optimal schedule with preemptions at integer points only, close to 1 for the instances with large instance degree $\Delta \leq C^*$ and fixed $p$. However the worst case ratios are not smaller than $\frac{4}{3}$ for the instances with short schedules, thus small instance degree $\Delta$, and arbitrary $p$. The inapproximability in Theorem 4.14 holds for open shops with 0-1 operations and no preemptions.

At the beginning of this chapter, we showed that the worst case ratio equals 2 for a simple decomposition algorithm. Asration and de Werra [1] give a polynomial-time algorithm with the worst case ratio $\frac{7}{6}$; however, their algorithm requires additional assumptions about the $\Delta$'s. Both approximations are for preemptive schedules with preemptions at integer points.

## 4.12 Other Models of Multiprocessor Operations

Brucker and Krämer [5] and Brucker [4] consider a different model of open shop with multiprocessor operations. Their model assumes the same subset of machines $\mathcal{M}_h$ for each operation $O_{i,h}$ regardless of the job $J_i$. The sets $\mathcal{M}_h$, $h = 1, \ldots, m$ may not be disjoint in which case they are called incompatible; disjoint sets are called compatible. They consider open shops with fixed $m$, which is called the number of stages. The stages form a compatibility graph with vertices corresponding to the stages and edges between the stages that are compatible. For unit-time operations they show that the open shop scheduling is polynomial for a number of objective functions including makespan, total weighted completion time, and weighted number of tardy jobs, see Brucker [4] for a complete list of results. The makespan minimization for three stages, $m = 3$, and arbitrary processing times reduces to either $O2||C_{\max}$ or to $O3||C_{\max}$ depending of the compatibility graph, see Brucker and Krämer [5].

## Problems

**4.1** Show that the open shop scheduling with multiprocessor operations is NP-hard in the strong sense for $p = 3$.

**4.2** Write down complete linear program for $p = 3$, and for $p = 4$.

## References

1. A.S. Asratian, D. de Werra, A generalized class-teacher model for some timtabling problems. Eur. J. Oper. Res. **143**, 531–542 (2002)
2. A.S. Asratian, R.R. Kamalian, Interval colorings of edges of a multigraph (in Russian). Appl Math (also arXiv:1401.8079v1) **5**, 25–34 (1987)
3. J.A. Bondy, U.S.R. Murty, *Graph Theory with Applications* (North-Holland Publishing, Amsterdam, 1976)
4. P. Brucker, *Scheduling Algorithms* (Springer, Berlin, 1995)

5. P. Brucker, A. Krämer, Schop scheduling with multiprocessor tasks on dedicated processors. Ann. Oper. Res. **57**, 13–27 (1995)
6. E. Cole, K. Ost, S. Schirra, Edge-coloring Bipartite Multigraphs in $O(E \log D)$. Combinatorica **21**, 5–12 (2001)
7. D. de Werra, An introduction to timetabling. Eur. J. Oper. Res. **19**, 151–162 (1985)
8. D. de Werra, A.S. Asratian, S. Durand, Complexity of some special types of timetabling problems. J. Sched. **5**, 171–183 (2002)
9. D. de Werra, T. Kis, W. Kubiak, Preemptive open shop scheduling with multiprocessors: polynomial cases and applications. J. Sched. **11**, 75–83 (2008)
10. S. Even, A. Itai, A. Shamir, On the complexity of timetable and multicomodity flow problems. SIAM J. Comput. **5**, 691–703 (1976)
11. H.N. Gabow, O. Kariv, Algorithms for edge coloring bipartite graphs and multigraphs. SIAM J. Comput. **11**, 117–129 (1982)
12. T. Gonzalez, S. Sahni, Open shop scheduling to minimize finish time. J. ACM **23**, 665–679 (1976)
13. C. Gotlieb, The construction of class-teacher time-tables, in *Proc. IFIP Congress 62, Munich* (North-Holland, Amsterdam, 1963)
14. O. Ittig, Scheduling with multiprocessors: a rounding network algorithm with a constant error. Master's thesis, Memorial University of Newfoundland and EPFL, 2005
15. T. Kis, D. de Werra, W. Kubiak, A projective algorithm for preemptive open shop scheduling with two multiprocessor groups. Oper. Res. Lett. **38**, 129–132 (2010)
16. W. Kubiak, On a conjecture for the university timetabling problem. Discrete Appl. Math. **299**, 26–49 (2021)
17. E.L. Lawler, *Combinatorial Optimization. Networks and Matroids* (Dover Publications, New York, 2001)
18. A. Schrijver, *Theory of Linear and Integer Programming* (Wiley, London, 1999)
19. É. Tardos, A strongly polynomial algorithm to solve combinatorial linear programs. Oper. Res. **34**, 250–256 (1986)

# Chapter 5
# Concurrent Open Shops

## 5.1 Introduction

The *concurrent* open shop scheduling permits processing more than one operation of the same job at a time, which is the main difference from traditional open shop scheduling. We use the abbreviation *cncnt* of the word concurrent to denote concurrent open shop scheduling in the extended Graham et al. notation [14]. Thus the problem $O|\text{cncnt}|\sum C_i$ denotes the problem of minimization total completion time for concurrent open shops. Ahmadi et al. [1], Leung et al. [19], Wagneur and Sriskandarajah [26], and Cheng et al. [5] provide a broad list of real-life applications of the concurrent open shop scheduling. We list some of them below:

- *Product design*. A product design team whose members independently design modules for different products. A product design is only completed once all its modules have been designed, Ahmadi et al. [1];
- *Audit*. A team of accountants auditing various parts of different companies. The audited company receives a final report once all accountants completed their audit, Ahmadi et al. [1];
- *Assembly*. Assembly of a final product often needs to wait until all parts for the assembly are available, Ahmadi et al. [1] and Framinan et al. [11];
- *Lenses production*. Each type of plastic lenses is produced on a dedicated production line. The manufacturer produces lenses based on confirmed customer orders. Each order consists of different quantities of various lens types (order parts). After completion on different production lines the components of a customer order are packaged and shipped to the customer as a complete order, Ahmadi et al. [1];
- *A car repair shop*, Leung et al. [19];
- *A paper converting facility*. The facility produces paper products of different types and sizes from large rolls of paper. Any product is produced on a dedicated machine. Orders are received from customers; each order specifies quantities

 
W. Kubiak, *A Book of Open Shop Scheduling*, International Series in Operations Research & Management Science 325,
https://doi.org/10.1007/978-3-030-91025-9_5

of different paper products. The customer receives its entire order in a single shipment in order to reduce transportation and order handling costs, Leung et al. [19];

- *Airplane maintenance*, Wagneur and Sriskandarajah [26].

This broad range of applications has been almost matched by a long list of the terms used for concurrent open shops starting with the Coordinated Scheduling of Customer Orders Problem or the scheduling of customer orders used by Ahmadi and Bagchi [2], see also Ahmadi et al. [1], open shops with job overlaps used by Wagneur and Sriskandarajah [26], concurrent open shop used by Roemer [24], or the order scheduling used by Framinan and Perez-Gonzalez [10]. Roemer [24] points out that the literature on the problem of scheduling customer orders and on concurrent open shops evolved in parallel and in isolation from one another which may have lead to some redundancy of research efforts, see Roemer [24] for a detailed account of those concurrent research efforts. This book uses the term concurrent open shops which seems to best reflect the concept which is not limited to customer orders and manufacturing only.

In all those real-life situations described above the customer preference for the objective function seems to prevail. A customer prefers to receive a *complete* order fast, thus the minimization of flow time or equivalently the minimization of total completion time, may be preferred over other objectives. At the same time it may not be of much importance to the customer when exactly particular components of its order are completed as long as the complete order is received fast. Similarly, a client is interested in the arrival time of the final audit report rather than in the completion times of partial reports for various department, like in the audit example mentioned above. Ahmadi et al. [1] provide a comprehensive discussion of the rationale behind the choice of objective functions for the concurrent open shop scheduling. We focus on the following three objective functions: total completion time, number of tardy jobs, and total tardiness in this chapter. Those functions seem to be well aligned with the real-life applications of the concurrent open shop scheduling. In Chap. 6 we introduce a special class of concurrent open shops where some operations of a job are *required* to be processed simultaneously at any time.

## 5.2 Complexity of Concurrent Open Shop Scheduling

We begin with concurrent open shop scheduling with 0-1 operations. The solution to $O|\text{cncnt}, p_{ij} = 0, 1|C_{\max}$ is trivial. The optimal $C_{\max}$ equals the maximum machine workload $L$ which occurs on machines with maximum number of operations. For any other regular objective function, optimal schedules exist among schedules with makespan on each machine being equal to the machine's workload. In other words among the schedules with no idle time on any machine. When the number of machines is part of the problem input the problem is NP-hard in the strong sense for total completion time and thus for total tardiness, and for the number of tardy jobs.

Somewhat surprisingly the proof of NP-hardness for $O|cncnt, p_{ij} = 0, 1|\sum C_i$ is much easier than for $O|p_{ij} = 0, 1|\sum C_i$, see Sect. 3.6.2. We have the following theorem.

**Theorem 5.1** *The problems* $O|cncnt, p_{ij} = 0, 1|\sum C_i, \sum U_i, \sum T_i$ *are NP-hard in the strong sense.*

***Proof*** The reduction is from the MAXIMUM INDEPENDENT SET problem, see Garey and Johnson [12]. Let a simple graph $G = (V, E)$ and a positive integer $k$ be an instance of the MAXIMUM INDEPENDENT SET problem. We set the set of jobs $\mathcal{J} = V$ and the set of machines $\mathcal{M} = E$ in the corresponding instance of the $O|cncnt, p_{ij} = 0, 1|\sum C_i$ problem. The job $v \in V$ has a unit-time operation on each machine $e \in E$ such that $v \in e$, and it is missing on all other machines. The number of unit-time operations of job $v$ equals the degree of vertex, deg($v$), in $G$. We are looking for a concurrent open shop schedule with the total completion time $\sum C_i$ being at most $2|V| - k$ (the number of tardy jobs being at most $|V| - k$ for the objective $\sum U_i$, and the total tardiness being at most $|V| - k$ for the objective $\sum T_i$).

Without loss of generality we can limit ourselves to schedules with $C_{\max} = 2$. The key observation is that in any such schedule all jobs that complete at time 1 form an independent set in $G$, and all remaining jobs finish at time 2. Thus having larger independent set results in smaller total completion time (fewer tardy jobs, and smaller total tardiness assuming due date $d = 1$ for each job).

Suppose that schedule $S$ has total completion time not exceeding $2|V| - k$. Without loss of generality we assume that each unit-time operation starts either at 0 or at 1 in $S$. Let $J$ be the set of all jobs with *all* their unit-time operations starting at 0. We have $C_v = 1$ for each job $v \in J$. The jobs in $V \setminus J$ complete at 2 each, thus $C_v = 2$ for each job $v \in V \setminus J$. Therefore $\sum C_i = |J| + 2(|V| - |J|) = 2|V| - |J| \leq 2|V| - k$ for $S$, which implies $k \leq |J|$. The set $J$ is independent, since otherwise there would be a machine $e = \{v, u\} \in E$ with $v, u \in J$, thus either $v$ or $u$ would need to complete at 2 in $S$ which leads to contradiction since both $v$ and $u$ complete at 1 in $S$. Thus $J$ is independent set in $G$ of size at least $k$.

Now suppose $U$ is an independent set of size at least $k$ in $G$. Start each job $v \in U$ on each machines $e = \{v, u\} \in E$ for some $u \in V$ at 0. There is no conflict since the request to start two jobs $u$ and $v$ from $U$ at 0 on the same machine $e$ implies that $e = \{v, u\} \in E$ which is a contradiction since $U$ is an independent set in $G$. Thus $C_v = 1$ for each $v \in U$. For each job $u \in J \setminus U$, start its unit-time operation at 1 on each machine $e = \{u, v\}$ for some $v \in U$ or either at 0 or at 1 on each machine $e = \{u, v\}$ for some $u \in J \setminus U$. Since there are exactly two unit-time operations of two different jobs on each machine $e$, a feasible concurrent schedule can easily be obtained, see Figs. 5.1 and 5.2 for an example. Observe that $C_u \leq 2$ for each $u \in J \setminus U$. Thus $\sum C_i \leq |U| + 2(|V| - |U|) = 2|V| - |U| \leq 2|V| - k$ since $|U| \geq k$. Hence we obtained a required schedule for an independent set of size at least $k$ in $G$. The proof for the other two objective functions $\sum U_i$ and $\sum T_i$ is similar. □

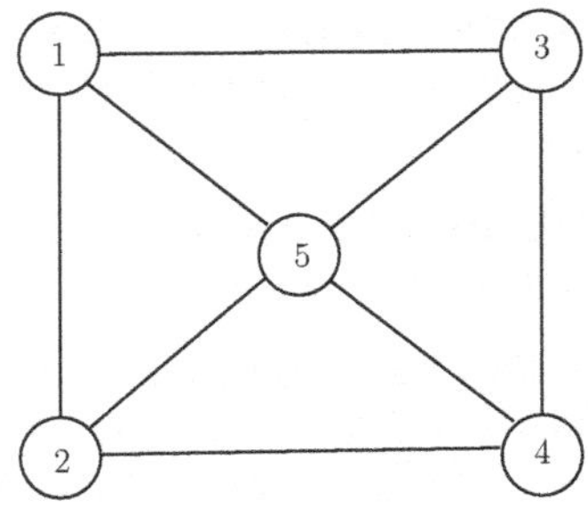


**Fig. 5.1** Graph $G$

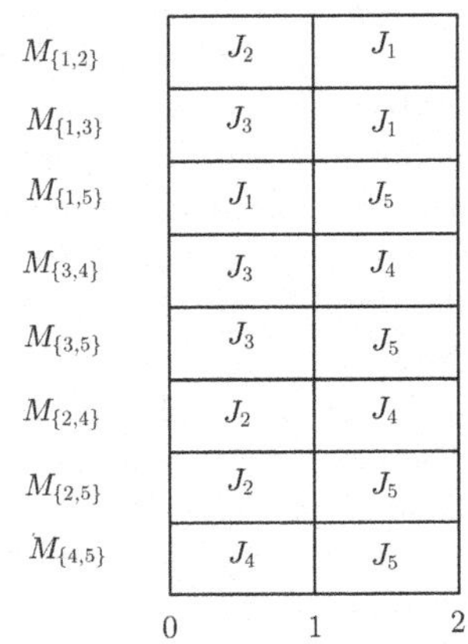


**Fig. 5.2** A schedule for graph $G$ and independent set $\{2, 3\}$

Roemer [24] settles the complexity status for the minimization of total completion time on two machines by proving the following theorem.

**Theorem 5.2** *The problem $O2|cncnt|\sum C_i$ is NP-hard in the strong sense.*

## 5.3 Permutation Schedules and Minimization of Maximum Lateness

A schedule of concurrent open shop is a *permutation schedule* if all operations are scheduled without preemption, idle time, and in the same order on each machine. Wagneur and Sriskandarajah [26], see also Mastrolilli et al. [20], prove that optimal schedules for regular objective functions can be found among permutation schedules.

**Theorem 5.3** *For each feasible schedule $S$ of a concurrent open shop there is a permutation schedule $S_\sigma$ that completes each job not later than in $S$.*

***Proof*** Let $S$ be a feasible schedule for an instance $I$ of $O|\text{cncnt}|\sum C_i$. Let $C_i$ be completion time of job $J_i$ in $S$. For $S$, consider $m$ instances $I_1, \ldots, I_m$, one for each machine in $\mathcal{M}$, of the single machine maximum lateness minimization problem $1||L_{\max}$. The instance $I_h$ for machine $M_h$ is made up of $n$ jobs $J_1, \ldots, J_n$ with processing times $p_{1,h}, \ldots, p_{n,h}$ and due dates $d_1 = C_1, \ldots, d_n = C_n$ respectively. The schedule $S$ on machine $M_h$, denoted by $S_h$, completes job $i$ at $C_{i,h} \leq C_i = d_i$ (we assume $C_{i,h} = 0$ if job $J_i$ is missing on $M_h$). Thus $S_h$ is a feasible schedule for $I_h$ with zero maximum lateness. Without loss of generality we assume that the schedule is non-preemptive. Therefore the maximum lateness of an optimal solution for $I_h$ is non-positive as well. The optimal solution $S_h^*$ to $1||L_{\max}$ orders the jobs in the Earliest Due Date Order, $\sigma$, see Jackson [17]. The same argument works for each instance $I_h$, $h = 1, \ldots, m$. Thus the same EDD sequence for each $I_h$ defines a permutation schedule $S_\sigma$ for $I$. Since the maximum lateness for each $S_h^*$ is non-positive, no job completes later in $S_\sigma$ than it does in $S$. □

Leung et al. [19] take this result even further to prove that the Earliest Due Date permutation is optimal for the minimization of maximum lateness.

**Theorem 5.4** *The Earliest Due Date permutation is optimal for* $O|cncnt|L_{\max}$.

***Proof*** Consider an optimal schedule $S$ for an instance $I$ of $O|cncnt|L_{\max}$. By Theorem 5.3 we may assume that the schedule is a permutation schedule $S_\sigma$ for some permutation of jobs $\sigma$. Suppose there exists position $k$, $k = 1, \ldots, n-1$, in $\sigma$ such that $i = \sigma(k)$ and $j = \sigma(k+1)$ and $d_i > d_j$ so that $\sigma$ is not $EDD$. We call the position $k$ a *violator*. Let $k$ be the violator in $S_\sigma$ with the largest value of $i$. Without loss of generality assume that $S_\sigma$ maximizes $i$, and among all permutation schedules with maximum $i$ has the maximum $k$. The exchange of $i$ and $j$ on machine $\mathcal{M}_h$ results in permutation $\sigma'$ such that

$$\max\{C_{i,h} - d_i, C_{j,h} - d_j\} > \max\{C'_{i,h} - d_i, C'_{j,h} - d_j\}$$

for $h = 1, \ldots, m$, where $C_{i,h}$ and $C'_{i,h}$ are completion times of job $J_i$ on machine $\mathcal{M}_h$ in $S_\sigma$ and $S_{\sigma'}$ respectively. Hence

$$\max\{\max_h\{C_{i,h} - d_i\}, \max_h\{C_{j,h} - d_j\}\} >$$
$$\max\{\max_h\{C'_{i,h} - d_i\}, \max_h\{C'_{j,h} - d_j\}\}$$

and the exchange reduces the maximum lateness for the pair $i$ and $j$, and leaves the maximum lateness for the remaining jobs unchanged. Therefore we get an optimal permutation schedule $S_{\sigma'}$ for $I$ with either smaller $i$ or the same $i$ but larger $k$ which contradicts the choice of $S_\sigma$ and proves the theorem. □

The EDD solution to $O|\text{cncnt}|L_{\max}$ is illustrated for an instance with $n = 7$ jobs and $m = 5$ machines given in Table 5.1. The EDD permutation is $J_2, J_5, J_1, J_4, J_7, J_6, J_3$ for the instance. The optimal schedule is given in Fig. 5.3 where $L_{\max} = L_3 = C_3 - d_3 = 18 - 11 = 7$.

**Table 5.1** An instance of $O|\text{cncnt}|L_{\max}$ with $m = 5$ machines

| Job ($i$) | $M_1$ | $M_2$ | $M_3$ | $M_4$ | $M_5$ | $d_i$ |
|---|---|---|---|---|---|---|
| 1 | 3 | 2 | 0 | 0 | 1 | 7 |
| 2 | 1 | 0 | 4 | 6 | 0 | 5 |
| 3 | 2 | 0 | 3 | 3 | 1 | 11 |
| 4 | 0 | 4 | 4 | 0 | 3 | 8 |
| 5 | 0 | 0 | 5 | 1 | 2 | 6 |
| 6 | 4 | 3 | 0 | 0 | 3 | 10 |
| 7 | 3 | 0 | 2 | 0 | 1 | 9 |

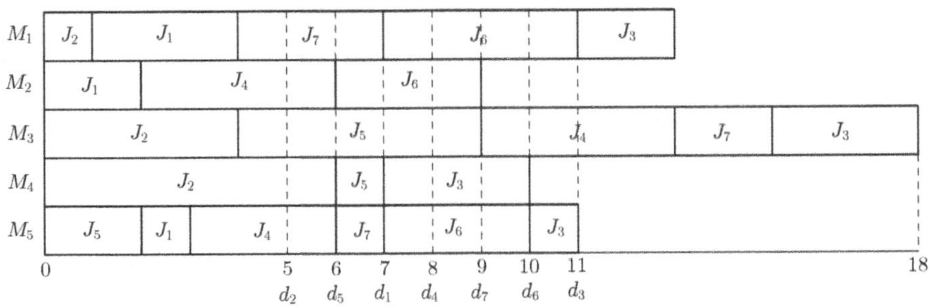


**Fig. 5.3** Optimal schedule with the EDD permutation $J_2, J_5, J_1, J_4, J_7, J_6, J_3$ for the instance in Table 5.1

## 5.4 2-Approximation Algorithm for $O|\text{cncnt}|\sum w_i C_i$

### 5.4.1 *The Algorithm*

In this section we present a 2-approximation algorithm for the problem $O|\text{cncnt}|\sum C_i$ developed by Mastrolilli et al. [20]. By Theorem 5.3 the search for optimal or approximate schedules can be limited to permutation schedules. All such schedules have the same makespan $C_{\max}$ which equals maximum machine workload $L$. Thus one knows that the job in position $n$ completes at $C_{\max}$ but one does not know which job that is, i.e., does not know $\sigma(n)$. The approximation algorithm considers all machines with the heaviest workload $C_{\max}$ and selects one of them, say $M_h$ (possible ties are broken arbitrarily). The machine then is used to select a job $\sigma(n)$ for the position $n$. The selected job has the minimum job weight to operation-processing-time on $M_h$ ratio $\frac{w_i}{p_{i,h}}$ (again possible ties are broken arbitrarily). The selection $\sigma(n)$ affects both the machines workloads, which are reduced by deleting job $\sigma(n)$ from each machine, and the weight $w_i$ of each remaining job $i$, which is reduced by $\frac{w_{\sigma(n)}}{p_{\sigma(n),h}} p_{i,h}$, prior to the next iteration. The next iteration, in order to find the job $\sigma(n-1)$ in position $n-1$, begins with these updated machine workloads and job weights and proceeds exactly in the same way as for the job in position $n$. The algorithm stops after $n$ iterations once the permutation $\sigma(1), \ldots, \sigma(n)$ has been found. Formally, the algorithm works with two main lists:

The list of machine workloads in iteration $k$, $k = n, \ldots, 1$,

$$L_1(k), \ldots, L_m(k),$$

and the list of job weights in iteration $k$

$$w_i(k) \qquad i \in J(k),$$

where $J(k)$ is the set of $k$ jobs left to schedule in iteration $k$. The algorithm starts with the following machine-workload list:

$$L_1(n) = L_1, \ldots, L_m(n) = L_m,$$

the following job weight list:

$$w_1(n) = w_1, \ldots, w_n(n) = w_n, \tag{5.1}$$

and the set of jobs

$$J(n) = N = \mathcal{J}.$$

In iteration $k$, $k = n, \ldots, 1$, the algorithm computes: The index of a machine with the *heaviest* workload in that iteration

$$\mu(k) = \arg\max_h\{L_h(k)\};$$

the index of a job with the *minimum* job weight to operation-processing-time ratio on that machine

$$\sigma(k) = \arg\min_i \left\{ \frac{w_i(k)}{p_{i,\mu(k)}} \right\};$$

and the *minimum* job weight to operation-processing-time ratio on that machine

$$\theta(k) = \frac{w_{\sigma(k)}(k)}{p_{\sigma(k),\mu(k)}}. \tag{5.2}$$

The job $\sigma(k)$ is then scheduled in position $k$ on each machine. To complete the iteration the algorithm reduces machine workloads by deleting job $\sigma(k)$ from each machine and sets

$$L_h(k-1) = L_h(k) - p_{\sigma(k),h} \qquad h = 1, \ldots, m,$$

and updates the job weights by setting

$$w_i(k-1) = w_i(k) - \theta(k) p_{i,\mu(k)} \qquad i \in J(k). \tag{5.3}$$

Finally

$$J(k-1) = J(k) \setminus \{\sigma(k)\},$$

and the next iteration starts if $k > 1$. Otherwise, the algorithm stops. The algorithm runs in $O(n(n+m))$ time.

### 5.4.2 *The Proof*

Following Mastrolilli et al. [20], we now show that the algorithm is a 2-approximation algorithm for $O|\text{cncnt}|\sum_i w_i C_i$. We begin with the following observation about the job weights produced by the algorithm:

**Observation 5.5** *By (5.1) and (5.3) we have*

$$w_j(k-1) = w_j - \sum_{l=k}^{n} p_{j,\mu(l)} \theta(l)$$

*and by (5.2) and (5.3)*

$$w_{\sigma(k)}(k-1) = 0.$$

The observation implies the following observation about the weight $w_{\sigma(k)}$ of the job in position $k$.

**Observation 5.6** *For any job* $\sigma(k)$

$$\begin{aligned}
&\sum_{h=1}^{m} p_{\sigma(k),h} \sum_{S \subseteq N : \sigma(k) \in S} y_{S,h} \\
&= p_{\sigma(k),\mu(k)} y_{J(k),\mu(k)} + \cdots + p_{\sigma(n),\mu(n)} y_{J(n),\mu(n)} \\
&= p_{\sigma(k),\mu(k)} \theta(k) + \cdots + p_{\sigma(n),\mu(n)} \theta(n) \\
&= w_{\sigma(k)} - w_{\sigma(k)}(k-1) \\
&= w_{\sigma(k)},
\end{aligned}$$

*where*

$$y_{S,h} = \begin{cases} \theta(k) & \text{if } h = \mu(k) \text{ and } S = J(k) \text{ for some } k = 1, \ldots, n, \\ 0 & \text{otherwise.} \end{cases} \tag{5.4}$$

Finally, we observe the following for the job completion times.

**Observation 5.7**

$$C_{\sigma(k)} = \sum_{i \in J(k)} p_{i,\mu(k)} = \sum_{i=1}^{k} p_{\sigma(i),\mu(k)} \qquad k = 1, \ldots, n$$

*and*

$$C_{\sigma(1)} \leq \cdots \leq C_{\sigma(n)}.$$

We are now ready to prove the main theorem of this section.

**Theorem 5.8** *The algorithm is a 2-approximation algorithm for $O|cncnt|\sum w_i C_i$.*

***Proof*** The following LP- relaxation of $O|\text{cncnt}|\sum w_i C_i$

$$LP : \min \sum_{i=1}^{n} w_i C_i$$

$$\text{s. t.} \sum_{i \in S} p_{i,h} C_i \geq f_h(S) \qquad \text{for all } h = 1, \ldots, m \text{ and } S \subseteq N,$$

where

$$f_h(S) = \frac{1}{2} \sum_{i \in S} p_{i,h}^2 + \frac{1}{2} \Big( \sum_{i \in S} p_{i,h} \Big)^2 \tag{5.5}$$

was given in Chen and Hall [4], see also Mastrolilli et al. [20]. Its dual was given in Mastrolilli et al. [20]

$$D : \max \sum_{h=1}^{m} \sum_{S \subseteq N} f_h(S) y_{S,h}$$

$$\text{s. t} \sum_{h=1}^{m} p_{i,h} \sum_{S \subseteq N : i \in S} y_{S,h} = w_i \qquad \text{for all } i = 1, \ldots, n,$$

$$y_{S,h} \geq 0 \qquad \text{for all } h = 1, \ldots, m \text{ and } S \subseteq N.$$

For the function $f_h(S)$ in both linear programs, Schulz [25] proves the following inequality:

$$\Big( \sum_{i \in S} p_{i,h} \Big)^2 \leq \Big( 2 - \frac{2}{n-1} \Big) f_h(S) \quad \text{for any } h = 1, \ldots, m \text{ and } S \subseteq N. \tag{5.6}$$

We are now ready to complete the proof. We have

$$\begin{aligned}
\sum_{k=1}^{n} w_{\sigma(k)} C_{\sigma(k)} &=^{(1)} \sum_{k=1}^{n} \Big( \sum_{h=1}^{m} p_{\sigma(k),h} \sum_{S \subseteq N : \sigma(k) \in S} y_{S,h} \Big) C_{\sigma(k)} \\
&=^{(2)} \sum_{k=1}^{n} \Big( \sum_{h=1}^{m} p_{\sigma(k),h} \big( y_{J(k),h} + \cdots + y_{J(n),h} \big) \Big) C_{\sigma(k)} \\
&= \sum_{k=1}^{n} \Big( p_{\sigma(k),\mu(k)} y_{J(k),\mu(k)} + \cdots + p_{\sigma(k),\mu(n)} y_{J(n),\mu(n)} \Big) C_{\sigma(k)} \\
&= \sum_{k=1}^{n} y_{J(k),\mu(k)} \big( p_{\sigma(1),\mu(k)} C_{\sigma(1)} + \ldots p_{\sigma(k),\mu(k)} C_{\sigma(k)} \big) \\
&\leq^{(3)} \sum_{k=1}^{n} y_{J(k),\mu(k)} \Big( C_{\sigma(k)} \big( p_{\sigma(1),\mu(k)} + \ldots p_{\sigma(k),\mu(k)} \big) \Big) \\
&=^{(4)} \sum_{k=1}^{n} y_{J(k),\mu(k)} \big( p_{\sigma(1),\mu(k)} + \ldots p_{\sigma(k),\mu(k)} \big)^2 \\
&\leq^{(5)} \Big( 2 - \frac{2}{n+1} \Big) \sum_{k=1}^{n} y_{J(k),\mu(k)} f_{\mu(k)}(J(k)) \\
&\leq^{(6)} \Big( 2 - \frac{2}{n+1} \Big) \sum_{k=1}^{n} w_i C_i^{LP} \\
&\leq^{(7)} \Big( 2 - \frac{2}{n+1} \Big) \sum_{k=1}^{n} w_i C_i^{*},
\end{aligned}$$

where the equality (1) follows from Observation 5.6, the equality (2) follows from (5.4), the inequalities (3) and (4) follow from Observation 5.7, the inequality (5) follows by (5.6), the inequality (6) holds since the solution $y_{S,h}$ is feasible for the dual $D$, and $C_i^{LP}$ is an optimal solution to the primal $LP$, finally (7) holds since $C_i^*$ is an optimal solution to $O|\text{cncnt}|\sum w_i C_i$. □

Mastrolilli et al. [20] prove that the performance guarantee of the algorithm cannot be better than $2 - \frac{2}{n+1}$.

### 5.4.3 An Example

We illustrate the algorithm run on an instance of $O|\text{cncnt}|\sum w_i C_i$ with $n = 7$ jobs and $m = 5$ machines specified in Table 5.2.

In iteration $k = 7$ we have the following list of machine workloads in Table 5.3 thus $\mu(7) = 3$, and the list of job weights is shown in Table 5.4 thus $\sigma(7) = 5$, and $\theta(7) = 0.4$. The iteration $k = 6$ starts with the following machine workloads in Table 5.5. Thus $\mu(6) = 3$, recall that the ties can be broken arbitrarily. The weights for the remaining jobs in $J(6)$ are given in Table 5.6. Thus $\sigma(6) = 2$ and $\theta(6) = \frac{3.4}{4} = 0.85$.

The iteration $k = 5$ starts with the machine workloads in Table 5.7. Thus $\mu(5) = 1$.

The weights for the remaining jobs in $J(5)$ are given in Table 5.8. Thus $\sigma(5) = 1$ and $\theta(5) = \frac{3}{3} = 1$.

The iteration $k = 4$ starts with the machine workloads in Table 5.9. Thus $\mu(4) = 3$, again ties can be broken arbitrarily.

The weights for the remaining jobs in $J(4)$ are given in Table 5.10. Thus $\sigma(4) = 4$ and $\theta(4) = \frac{3}{4} = 0.75$.

The iteration $k = 3$ starts with the machine workloads in Table 5.11. Thus $\mu(3) = 1$.

The weights for the remaining jobs in $J(3)$ are given in Table 5.12. Thus $\sigma(3) = 6$ and $\theta(3) = \frac{3}{4} = 0.75$.

Finally, the iteration $k = 2$ starts with the following machine workloads in Table 5.13. Thus $\mu(2) = 3$.

The weights for the remaining jobs in $J(2)$ are given in Table 5.14. Thus $\sigma(2) = 3$ and $\theta(2) = \frac{2.5}{3}$.

**Table 5.2** An instance of $O|\text{cncnt}|\sum w_i C_i$ with $m = 5$ machines

| Job ($i$) | $M_1$ | $M_2$ | $M_3$ | $M_4$ | $M_5$ | $w_i$ |
|---|---|---|---|---|---|---|
| 1 | 3 | 2 | 0 | 0 | 1 | 3 |
| 2 | 1 | 0 | 4 | 6 | 0 | 5 |
| 3 | 2 | 0 | 3 | 3 | 1 | 10 |
| 4 | 0 | 4 | 4 | 0 | 3 | 8 |
| 5 | 0 | 0 | 5 | 1 | 2 | 2 |
| 6 | 4 | 3 | 0 | 0 | 3 | 7 |
| 7 | 3 | 0 | 2 | 0 | 1 | 11 |

**Table 5.3** Machine workloads in iteration $k = 7$: $\mu(7) = 3$

| | $M_1$ | $M_2$ | $M_3$ | $M_4$ | $M_5$ |
|---|---|---|---|---|---|
| Load | 13 | 9 | 18 | 10 | 11 |

**Table 5.4** Job weights and operation processing times on machine $\mu(7) = 3$ in iteration $k = 7$

| | $J_1$ | $J_2$ | $J_3$ | $J_4$ | $J_5$ | $J_6$ | $J_7$ |
|---|---|---|---|---|---|---|---|
| Weight | 3 | 5 | 10 | 8 | 2 | 7 | 11 |
| $M_3$ | 0 | 4 | 3 | 4 | 5 | 0 | 2 |

**Table 5.5** Machine workloads in iteration $k = 6$: $\mu(6) = 3$

| | $M_1$ | $M_2$ | $M_3$ | $M_4$ | $M_5$ |
|---|---|---|---|---|---|
| Load | 13 | 9 | 13 | 9 | 9 |

**Table 5.6** Job weights and operation processing times on machine $\mu(6) = 3$ in iteration $k = 6$

| | $J_1$ | $J_2$ | $J_3$ | $J_4$ | $J_6$ | $J_7$ |
|---|---|---|---|---|---|---|
| Weight | 3 | 3.4 | 8.8 | 6.4 | 7 | 10.2 |
| $M_3$ | 0 | 4 | 3 | 4 | 0 | 2 |

**Table 5.7** Machine workloads in iteration $k = 5$: $\mu(5) = 1$

| | $M_1$ | $M_2$ | $M_3$ | $M_4$ | $M_5$ |
|---|---|---|---|---|---|
| Load | 12 | 9 | 9 | 3 | 9 |

**Table 5.8** Job weights and operation processing times on machine $\mu(5) = 1$ in iteration $k = 5$

| | $J_1$ | $J_3$ | $J_4$ | $J_6$ | $J_7$ |
|---|---|---|---|---|---|
| Weight | 3 | 6.25 | 3 | 7 | 8.5 |
| $M_1$ | 3 | 2 | 0 | 4 | 3 |

**Table 5.9** Machine workloads in iteration $k = 4$: $\mu(4) = 3$

| | $M_1$ | $M_2$ | $M_3$ | $M_4$ | $M_5$ |
|---|---|---|---|---|---|
| Load | 9 | 7 | 9 | 3 | 8 |

**Table 5.10** Job weights and operation processing times on machine $\mu(4) = 3$ in iteration $k = 4$

| | $J_3$ | $J_4$ | $J_6$ | $J_7$ |
|---|---|---|---|---|
| Weight | 4.25 | 3 | 3 | 5.5 |
| $M_3$ | 3 | 4 | 0 | 2 |

**Table 5.11** Machine workloads in iteration $k = 3$: $\mu(3) = 1$

| | $M_1$ | $M_2$ | $M_3$ | $M_4$ | $M_5$ |
|---|---|---|---|---|---|
| Load | 9 | 3 | 5 | 3 | 5 |

**Table 5.12** Job weights and operation processing times on machine $\mu(3) = 1$ in iteration $k = 3$

| | $J_3$ | $J_6$ | $J_7$ |
|---|---|---|---|
| Weight | 4 | 3 | 4 |
| $M_1$ | 2 | 4 | 3 |

**Table 5.13** Machine workloads in iteration $k = 2$: $\mu(2) = 3$

| | $M_1$ | $M_2$ | $M_3$ | $M_4$ | $M_5$ |
|---|---|---|---|---|---|
| Load | 5 | 0 | 5 | 3 | 2 |

**Table 5.14** Job weights and operation processing times on machine $\mu(2) = 3$ in iteration $k = 2$

| | $J_3$ | $J_7$ |
|---|---|---|
| Weight | 2.5 | 1.75 |
| $M_3$ | 3 | 2 |

Therefore the algorithm produces the following permutation: $J_7, J_3, J_6, J_4, J_1, J_2, J_5$ of jobs on each machine. The schedule for this permutation is shown in Fig. 5.4; the total weighted completion time for the schedule $\sum w_i C_i$ equals 355.

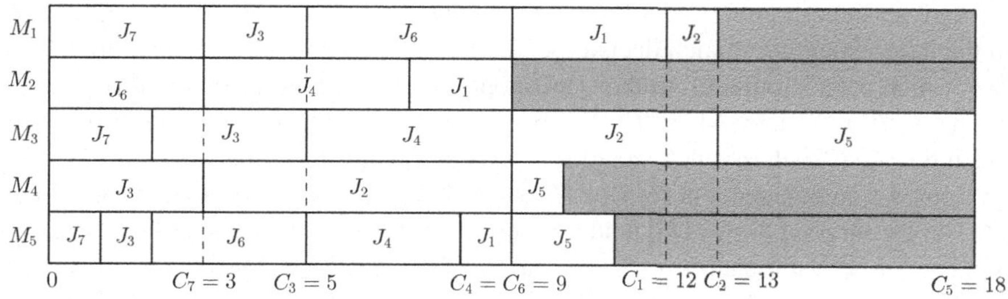


**Fig. 5.4** 2-approximation solution for the instance in Table 5.2

## 5.5 Hardness of Approximation

This section concentrates on the hardness of approximation for the problems of minimization of total completion time, $O|\text{cncnt}|\sum C_i$, total tardiness, $O|\text{cncnt}|\sum T_i$, and number of tardy jobs, $O|\text{cncnt}|\sum U_i$, for concurrent open shops. We begin by introducing a key inapproximability result for independent sets of uniform hypergraphs.

A pair $H = (N, E)$ where $N$ is a finite set of vertices and $E$ is a family of subsets (hyperedges) of $N$ each of size exactly $r$ is called an $r$-uniform hypergraph. An *independent set* of an $r$-uniform hypergraph $H = (N, E)$ is a subset $I$ of vertices that does not completely include any of the hyperedges in $E$, i.e., $e \setminus I \neq \emptyset$ for each hyperedge $e \in E$. Dinur et al. [7] prove the following.

**Theorem 5.9** *For any $\gamma \in (0, 1)$, $\delta \in (0, 1/2)$, and integer $r \geq 3$ the following problem is $NP$-hard. Given an $r$-uniform hypergraph $H = (N, E)$ decide whether*

*(i) There exists an independent set of $H$ of size at least $\left(1 - \frac{1}{r-1} - \delta\right)|N|$, or*
*(ii) Each independent set of $H$ has size strictly less than $\gamma|N|$.*

### 5.5.1 $O|cncnt|\sum C_i$

Mastrolilli et al. [20] make the decision problem in Theorem 5.9 a point of departure to prove the following limit on the approximation guarantee for polynomial-time algorithms for $O|\text{cncnt}|\sum C_i$.

**Theorem 5.10** *The problem $O|cncnt|\sum C_i$ is NP-hard to approximate within a factor $\frac{6}{5} - \epsilon$ for any $\epsilon > 0$, unless $P = NP$.*

***Proof*** By Theorem 5.9, for any $\gamma \in (0, 1)$, $\delta \in (0, \frac{1}{2})$, and integer $r \geq 3$, there is a class of $r$-uniform hypergraphs where the size of a maximum independent set is either greater or equal $(1 - \frac{1}{r-1} - \delta)|N|$ or less than $\gamma|N|$, and it is NP-hard to decide for a given hypergraph $H = (N, E)$ in that class whether it falls in the former or in the latter category. If such decision could be made in polynomial time, then $P = NP$ which seems unlikely. In the corresponding instance $f(H)$ of $O|\text{cncnt}|\sum C_i$, we have $\mathcal{J} = N$ and $\mathcal{M} = E$. A job $v \in N$ has a unit-time operation on each machine $e \in E$ such that $v \in e$, and it is missing on any other machine.

Consider an independent set $I$ of $H$. On each machine $e \in E$ schedule the jobs in $e \cap I$ in the interval $[0, |e \cap I|]$, and the jobs in $e \cap (N \setminus I)$ in the interval $[|e \cap I|, r]$. Since $e \setminus I \neq \emptyset$ for each $e \in E$, we have $|e \cap I| \leq r - 1$. Moreover, since each machine's $e$ workload is exactly $r$ for $r$-uniform hypergraphs, each job in $N \setminus I$ can be completed by $r$. Therefore, we can readily obtain a feasible schedule where each job in $I$ completes by $r - 1$, and each job in $N \setminus I$ completes by $r$. Therefore, $(i)$ in Theorem 5.9 can be used to calculate an upper bound $U$ on the value $OPT(f(H))$ of minimum total completion time for the instance $f(H)$ with $H$ in the category $(i)$ as follows. If $(i)$ holds, then

$$\begin{aligned} OPT(f(H)) &\leq (r-1)|I| + r(|N| - |I|) \\ &= r|N| - |I| \\ &\leq r|N| - \left(1 - \frac{1}{r-1} - \delta\right)|N| \\ &= \left(r - 1 + \frac{1}{r-1} + \delta\right)|N| \\ &= U. \end{aligned}$$

On the other hand, in an optimal schedule for $f(H)$ the set of all jobs that complete by $r - 1$ is an independent set $I_O$ of $H$, and all jobs in $N \setminus I_O$ complete at $r$. Therefore, $(ii)$ in Theorem 5.9 can be used to calculate a lower bound $L$ on the value $OPT(f(H))$ of minimum total completion time for $f(H)$ with $H$ in the category $(ii)$ as follows. If $(ii)$ holds, then

$$\begin{aligned} OPT(f(H)) &= r(|N| - |I_O|) + |I_O| \\ &= r|N| - (r-1)|I_O| \\ &> r|N| - (r-1)\gamma|N| \\ &= \Big(r - (r-1)\gamma\Big)|N| \\ &= L. \end{aligned}$$

Suppose $A$ is a $\left(\frac{6}{5} - \epsilon\right)$-approximation algorithm for $O|\text{cncnt}|\sum C_i$, for some $0 < \epsilon < 1$, which runs in polynomial time. Consider the class $C$ of $r$-hypergraphs with $r = 3, \gamma = \frac{\epsilon}{4}, \delta = \frac{\epsilon}{2}$. We have $U = \frac{5+\epsilon}{2}$ and $L = 3 - \frac{\epsilon}{2}$. Run $A$ on $f(H)$ where $H \in C$, if $A(f(H)) \leq L$, then $OPT(f(H)) \leq A(f(H)) \leq L$ and the condition $(ii)$ does not hold for $H$. Thus, the size of maximum independent set is greater or equal $\left(1-\frac{1}{r-1}-\delta\right)|N|$. If $\frac{A(f(H))}{\frac{6}{5}-\epsilon} > U$, then $OPT(f(H)) \geq \frac{A(f(H))}{\frac{6}{5}-\epsilon} > U$ and $(i)$ does not hold for $H$. Thus, the size of maximum independent set is less than $\gamma|N|$. Finally there is no instance $f(H)$ such that

$$L < A(f(H)) \leq \left(\frac{6}{5}-\epsilon\right)U \tag{5.7}$$

since

$$\frac{L}{U} = \frac{6-\epsilon}{5+\epsilon} > \frac{6}{5} - \epsilon \tag{5.8}$$

for $H \in C$. Therefore $A$ could distinguish between the two categories of hypergraphs in the class $C$ in polynomial time which leads by Theorem 5.9 to contradiction if $P \neq NP$, and proves the theorem for $O|\text{cncnt}|\sum C_i$. □

The factor $\frac{6}{5} - \epsilon$ in Theorem 5.10 can be strengthened to $2 - \epsilon$ under the assumption that the Unique Games Conjecture holds, see Khot [18], and Bansal and Khot [3] for the conjecture. The following inapproximability result of Bansal and Khot [3] is key for the proof of the factor $2 - \epsilon$.

**Theorem 5.11** *Assuming the Unique Games Conjecture holds, for any* $\delta \in (0, 1)$, $\gamma \in (0, 1)$, *and integer* $r \geq 2$ *the following problem is* $NP$*-hard. Given an* $r$*-uniform hypergraph* $H = (N, E)$ *decide whether*

*(iii) There exist disjoint subsets* $N_1, \ldots, N_r \subseteq N$, *satisfying* $|N_i| \geq \frac{1-\delta}{r}|N|$ *and such that* $|e \cap N_i| \leq 1$ *for* $e \in E$ *and* $i = 1, \ldots, r$, *or*

*(iv) Each independent set of* $H$ *has size at most* $\gamma|N|$.

This result is then used by Bansal and Khot [3] to prove the $2 - \epsilon$ factor.

**Theorem 5.12** *Assuming the Unique Games Conjecture,* $O|cncnt|\sum C_i$ *is hard to approximate within a factor* $2 - \epsilon$ *for any* $\epsilon > 0$, *unless* $P = NP$.

***Proof*** We use the same transformation $f(H)$ as in the proof of Theorem 5.10. If $(iii)$ in Theorem 5.11 holds for $H$, then schedule job $v \in N_i$ in time slot $[i-1, i]$ of each machine $e$ such that $v \in e \cap N_i$. The schedule thus obtained is feasible since the sets $N_1, \ldots, N_r$ are disjoint, and $|e \cap N_i| \leq 1$ for each $e \in E$ and $i = 1, \ldots, r$. Therefore at least $\frac{1-\delta}{r}|N|$ jobs complete at $i$ for $i = 1, \ldots, r$ in the schedule. Complete the schedule by scheduling the remaining jobs from $N \setminus (N_1 \cup \cdots \cup N_r)$ in the available unit-time slots in the interval $[0, r]$ on each machine $e$. Each of those jobs completes by $r$. Hence we get the following upper bound on the total completion time $OPT(f(H))$ for each hypergraph $H$ in the category $(iii)$

$$\begin{aligned} OPT(f(H)) &\leq \left(\frac{1-\delta}{r}(1+\cdots+r)+\delta r\right)|N| \\ &= \left(\frac{1-\delta}{2}(r+1)+\delta r\right)|N| \end{aligned}$$

$$= \left(\frac{1+\delta}{2}r + \frac{1-\delta}{2}\right)|N|$$
$$\leq \left(\frac{1+\delta}{2}(r+1)\right)|N|$$
$$= U.$$

If $(iv)$ in Theorem 5.11 holds for $H$, the lower bound is calculated in a similar way as for $(ii)$ in the proof of Theorem 5.10

$$OPT(f(H)) \geq \Big(r - (r-1)\gamma\Big)|N|$$
$$= L.$$

Suppose there is a $(2-\epsilon)$-approximation polynomial-time algorithm $A$ for $O|\text{cncnt}|\sum C_i$. Without loss of generality $0 < \epsilon < 1$. Consider the class $C$ of $r$-hypergraphs with $\delta \leq \frac{1}{r+1}$, $\gamma \leq \frac{1}{r-1}$, and $r \geq \frac{6}{\epsilon} - 1$. Run $A$ on an instance $f(H)$ with $H \in C$. If $A(f(H)) < L$, then $OPT(f(H)) \leq A(f(H)) < L$ and the condition $(iv)$ does not hold for $H$. Thus, $H$ falls in the category $(iii)$. If $\frac{A(f(H))}{2-\epsilon} > U$, then $OPT(f(H)) \geq \frac{A(f(H))}{2-\epsilon} > U$ and $(iii)$ does not hold for $H$. Thus, the size of maximum independent set is less than $\gamma|N|$ and $H$ falls in the category $(iv)$. Finally there is no $f(H)$ such that

$$L \leq A(f(H)) \leq (2-\epsilon)U \tag{5.9}$$

since

$$\frac{L}{U} = 2\frac{(r-(r-1)\gamma)}{(r+1)(1+\delta)} > 2-\epsilon \tag{5.10}$$

for the class $C$. Therefore $A$ could distinguish between the two categories of hypergraphs from $C$ in polynomial time. This, assuming the Unique Games Conjecture holds, leads by Theorem 5.11 to contradiction if $P \neq NP$ and proves the theorem.□

### 5.5.2 $O|cncnt|\sum T_i$, and $O|cncnt|\sum U_i$

The problems $O|\text{cncnt}|\sum T_i$ and $O|\text{cncnt}|\sum U_i$ are harder to approximate than $O|\text{cncnt}|\sum C_i$. Polynomial-time algorithms cannot guarantee approximations within a factor $(1-c)\ln m$ for any constant $c > 0$ for those two due date based problems.

To prove this we first recall a hard to approximate SET COVER problem which becomes a point of departure in the hardness proof for $O|\text{cncnt}|\sum U_i$ and $O|\text{cncnt}|\sum T_i$. An instance of the SET COVER problem is made up of a collection

$V_1, \ldots, V_\ell$ of subsets of a set $V = \{v_1, \ldots, v_n\}$. The SET COVER problem is the problem of selecting as few as possible subsets from the collection $V_1, \ldots, V_\ell$ such that every $v \in V$ is included into at least one of the selected subsets, Garey and Johnson [12]. Feige [9] shows that the problem cannot be approximated within a factor $(1 - c) \ln n$ for any $c > 0$ in polynomial time, unless $NP$ includes slightly superpolynomial problems. Dinur and Steurer [8] strengthen this result by proving that the problem cannot be approximated within a factor $(1-c) \ln n$ for any $c > 0$ in polynomial time, unless $P = NP$. The following result of Dinur and Steurer [8] is key to showing that both $O|\text{cncnt}| \sum U_i$ and $O|\text{cncnt}| \sum T_i$ are hard to approximate within a factor $(1 - c) \ln m$ for any $c > 0$, unless $P = NP$.

**Theorem 5.13** *Set cover is $NP$-hard to approximate within a factor $(1-c) \ln n$ for any $c > 0$, unless $P = NP$.*

We have the following inapproximability result for $O|\text{cncnt}| \sum U_i$ and $O|\text{cncnt}| \sum T_i$.

**Theorem 5.14** *$O|cncnt| \sum U_i$ and $O|cncnt| \sum T_i$ are hard to approximate within a factor $(1 - c) \ln m$ for any $c > 0$, unless $P = NP$.*

***Proof*** Let collection $V_1, \ldots, V_\ell$ of subsets of $V$ be an instance $I$ of the SET COVER problem. For each $v \in V$ define the set $S_v = \{i : v \in V_i, i = 1, \ldots \ell\}$ and $S = \{1, \ldots, \ell\}$. We assume that all sets $S_v$ have the same cardinality $r = \max_v\{|S_v|\}$. Otherwise, we can add a collection of $r - |S_v| < r$ copies of the subset $\{v\}$ to the collection $V_1, \ldots, V_\ell$, thus creating a regular instance $I_r$. The sizes of minimum set cover are the same in both $I$ and $I_r$. Each $S_v$ corresponds to machine $S_v$ in the concurrent open shop. There are $m = n = |V|$ machines in the concurrent open shop instance $I_O$. Each $i \in S$ corresponds to a job with a unit-time operation on each machine $S_v$ such that $i \in S_v$ and missing operations on any machine $S_v$ such that $i \notin S_v$. There are $\ell$ jobs in $I_O$. Set $d_i = r - 1$ for each job in $I_O$.

Let the family $\mathcal{V}$ be a minimum set cover. Then $S_v \cap \mathcal{V} \neq \emptyset$ for each $v \in V$ so that each machine has a job from $\mathcal{V}$. Consider a schedule where all the jobs from $\mathcal{V}$ are scheduled at the end of the schedule on each machine. Thus each job in $\{1, \ldots, \ell\} \setminus \mathcal{V}$ completes by $r - 1$ in the schedule, and each job in $\mathcal{V}$ completes at $r$. Thus exactly $\ell - |\mathcal{V}|$ jobs completes by their due dates in the schedule and the number of tardy jobs equals the size of the subset cover $|\mathcal{V}|$, i.e., $\sum U_i = |\mathcal{V}|$.

On the other hand, for a schedule with the number of tardy jobs equal to $\sum U_i$ this number equals the number of jobs that complete at $r$ in the schedule. Let $S'$ be the set of those jobs. Thus for each machine $S_v$ there exists job $i \in S'$, or in other words for each $v$ there is $V_i$ such that $v \in V_i$. Thus $S'$ is a subset cover, and $\sum U_i = |S'|$.

Suppose there is $(1 - c) \ln m$-approximation polynomial-time algorithm $A$ for $O|\text{cncnt}| \sum U_i$ for some $c > 0$, then we have

$$\frac{\sum U_i^A(I_O)}{\sum U_i^{OPT}(I_O)} \leq (1 - c) \ln m, \tag{5.11}$$

where $\sum U_i^A(I_O)$ and $\sum U_i^{OPT}(I_O)$ are the numbers of tardy jobs in a schedule produced by $A$ and in an optimal schedule, respectively. We have $\sum U_i^A(I_O) = |S^A(I_r)|$ and $\sum U_i^{OPT}(I_O) = |S^{OPT}(I_r)|$ for some set covers $S^A(I_r)$ and $S^{OPT}(I_r)$ for $I_r$. However, $|S^{OPT}(I_r)|$ is the cardinality of minimum set cover $C^*(I_r)$ for $I_r$ and thus the cardinality of minimum set cover $C^*(I)$ for $I$. Finally, there is set cover $S(I)$ for $I$ such that $|S^A(I_r)| \geq |S(I)|$. Therefore

$$\frac{|S(I)|}{|C^*(I)|} \leq \frac{\sum U_i^A(I_O)}{|C^*(I)|} \leq (1-c)\ln m.$$

Since the number of machines $m$ in the concurrent open shop instance $I_O$ equals $n = |V|$ we have

$$\frac{|S(I)|}{|C^*(I)|} \leq (1-c)\ln n,$$

which proves that $A$ is a $(1-c)\ln n$-approximation algorithm for the set cover problem which runs in polynomial time. This however contradicts Theorem 5.13. □

A similar inapproximability result was obtained by Ng et al. [22] under a stronger than the $P \neq NP$ assumption. The result was based on the inapproximability result for set cover obtained by Feige [9]. Garg et al. [13] present further complexity results for $O|\text{cncnt}|\sum w_i C_i$ and its special cases.

## 5.6 Fixed Number of Machines and Special Cases

Cheng et al. [5] give a PTAS for the problem $Om|\text{cncnt}|\sum w_i C_i$ where the number of machines $m$ is not part of the input. Ahmadi et al. [1] report a $\frac{\sqrt{5}+3}{\sqrt{5}+1}$-approximation for the two-machine problem $O2|\text{cncnt}|\sum w_i Ci$, see also Roemer [23]. Ahmadi et al. [1] propose heuristics and report computational experiments with the heuristics for $O|\text{cncnt}|\sum w_i C_i$. Cheng and Wang [6] give a pseudopolynomial-time algorithm for the problem $Om|\text{cncnt}|\sum w_i U_i$ where the number of machines $m$ is not part of the input. This implies that $Om|\text{cncnt}, p_{ij} = 0, 1|\sum U_i$ is polynomial. Leung et al. [19] give a polynomial-time algorithm $Om|\text{cncnt}|\sum U_i$ for job-ordered open shops, see Sect. 7.4 for definition of job-ordered open shops. We observe that $O1|\text{cncnt}|\sum U_i$ is the same as the single machine problem $1||\sum U_i$ which is solved by Hodgson–Moore algorithm, see Moore [21], in $O(n \log n)$ time. Besides, we observe that operation processing times follow the same order on each machine in job-ordered open shops. These two observations suffice to prove that $Om|\text{cncnt}|\sum U_i$ for job-ordered open shops, see Problem 5.5. Leung et al. [19] give an exact algorithm based on constraint propagation and bounding approach for $O|\text{cncnt}|\sum U_i$ and report on computational experiments with the algorithm. Framinan and Perez-Gonzalez [10] propose heuristics for $O|\text{cncnt}|\sum T_i$ and report on computational experiments with the heuristics.

## 5.7 Conflict Graphs and the Classification of Open Shops

The open shop scheduling problems can be defined and classified by using the concept of operation conflict graphs introduced in Chap. 2. The operation conflict graph determines which two operations cannot be done in parallel in any feasible schedule. That is, if $(o, o')$ is an edge in the graph, then operations $o$ and $o'$ can never be processed in parallel in a feasible schedule. The classification of open shops can then be done according the characteristic of the set of edges of the conflict graph. To be more precise, let $\mathcal{O}$ be the set of all operations $O_{i,h}$, where $J_i \in \mathcal{J}$ and $M_h \in \mathcal{M}$. An operation conflict graph is a simple graph $C = (\mathcal{O}, \mathcal{E})$ with the set of vertices $\mathcal{O}$ and the set of edges in $\mathcal{E}$ linking operations in $\mathcal{O}$. We suggest the following classification depending on $\mathcal{E}$. The classification can be easily extended to other classes of $\mathcal{E}$.

- Let $\mathcal{O}_h = \{O_{1,h}, \ldots, O_{n,h}\}$, and let $\mathcal{K}_h$ be a clique of size $n$ on $\mathcal{O}_h$, $h = 1, \ldots, m$. The union $\mathcal{E} = \mathcal{K}_1 \cup \cdots \cup \mathcal{K}_m$ of $m$ disjoint cliques is a conflict graph of concurrent open shop.
- Let $\mathcal{J}_i = \{O_{i,1}, \ldots, O_{i,m}\}$, and let $\mathcal{G}_i$ be a clique of size $m$ on $\mathcal{J}_i$, $i = 1, \ldots, n$. The union $\mathcal{E} = \mathcal{K}_1 \cup \cdots \cup \mathcal{K}_m \cup \mathcal{G}_1 \cup \cdots \cup \mathcal{G}_n$ of $n + m$ cliques is a conflict graph of open shop.
- Let $\mathcal{R}$ be any non-empty set of edges $(O_{i,h}, O_{j,\ell})$ such that $i \neq j$ and $h \neq \ell$. The union $\mathcal{E} = \mathcal{K}_1 \cup \cdots \cup \mathcal{K}_m \cup \mathcal{G}_1 \cup \cdots \cup \mathcal{G}_n \cup (\mathcal{O}, \mathcal{R})$ is a conflict graph of open shop with additional resources in Chap. 2.
- Let $\mathcal{P}$ be any non-empty set of edges $(O_{i,h}, O_{j,\ell})$ such that $h \neq \ell$. The union $\mathcal{E} = \mathcal{K}_1 \cup \cdots \cup \mathcal{K}_m \cup (\mathcal{O}, \mathcal{P})$ is a conflict graph of partially concurrent open shop. Partially concurrent open shops are studied by Ilani et al. [16] and Grinshpoun et al. [15].

## Problems

**5.1** Prove Theorem 5.1 for $\sum_i U_i$ and $\sum_i T_i$.

**5.2** Show that the performance guarantee of the 2-approximation algorithm in Sect. 5.4 cannot be better than $2 - \frac{2}{n+1}$.

**5.3** Find an optimal schedule for an instance in Table 5.2.

**5.4** Proof Theorem 5.14 for $O|cncnt|\sum T_i$.

**5.5** Show that the problem $Om|\text{cncnt}|\sum U_i$ is polynomial for job-ordered open shops.

## References


1. R.H. Ahmadi, U. Bagchi, T.A. Roemer, Coordinated scheduling of customer orders for quick response. Naval Res. Logist. **52**, 483–512 (2005)
2. R.H. Ahmadi, U. Bagchi, *Scheduling of Multi-Job Customer Orders in Multimachine Environments* (ORSA/TIMS, Philadelphia, 1990)
3. N. Bansal, S. Khot, Inapproximability of hypergraph vertex cover and applications to scheduling problems, in *Automata, Languages and Programming. ICALP 2010*, ed. by S. Abramsky, C. Gavoille, C. Kirchner, F. Meyer auf der Heide, P.G. Spirakis. Lecture Notes in Computer Science, vol. 6198 (Springer, Berlin, 2010), pp. 250–261
4. Z.L. Chen, N.G. Hall, Supply chain scheduling: assembly systems. Working paper, Department of Systems Engineering, University of Pennsylvania, 2001
5. T.C.E. Cheng, Q. Nong, C.T. Ng, Polynomial-time approximation scheme for concurrent open shop scheduling with a fixed number of machines to minimize the total weighted completion time. Naval Res. Logist. **58**, 763–770 (2011)
6. T.C.E. Cheng, G. Wang, Customer order scheduling on multiple facilities. Working paper no. 11/98-9, Faculty of Business and Information Systems, The Hong Kong Polytechnic University, 1999
7. I. Dinur, V. Guruswami, S. Khot, O. Regev, A new multilayered PCP and the hardness of hypergraph vertex cover. SIAM J. Comput. **34**, 1129–1146 (2005)
8. I. Dinur, D. Steurer, Analytical approach to parallel repetition, in *Proceedings of the 2014 ACM symposium on Theory of Computing. STOC'14* (ACM, Berlin, 2014), pp. 624–633
9. U. Feige, A threshold of $\ln n$ for aproximating set cover. J. ACM **45**, 634–652 (1998)
10. J.M. Framinan, P. Perez-Gonzalez, Order scheduling with tardiness objective: improved approximate solutions. Eur. J. Oper. Res. **266**, 840–850 (2018)
11. J.M. Framinan, P. Perez-Gonzalez, V. Fernandez-Viagas, Deterministic assembly scheduling problems: a review and classification of concurrent-type scheduling models and solution procedures. Eur. J. Oper. Res. **273**, 401–417 (2019)
12. M.R. Garey, D.S. Johnson, *Computers and Intractability: A Guide to the Theory of NP-Completeness* (W. H. Freeman, San Francisco, 1979)
13. N. Garg, A. Kumar, V. Pandit, Order scheduling models: hardness and algorithms, in *Lecture Notes in Computer Science 4855* (Springer, Berlin, 2007), pp. 96–107
14. R.L. Graham, E.L. Lawler, J.K. Lenstra, A.H.G. Rinnooy Kan, Optimization and approximation in deterministic sequencing and scheduling: a survey. Ann. Discr. Math. **5**, 287–326 (1979)
15. T. Grinshpoun, H. Ilani, E. Shufan, The representation of partially-concurrent open shop problems. Ann. Oper. Res. **252**, 455–469 (2017)
16. H. Ilani, E. Shufan, T. Grinshpoun, Partially concurrent open shop scheduling with integral preemtions. Ann. Oper. Res. **259**, 157–171 (2017)
17. J.R. Jackson, Scheduling a production line to minimize maximum tardiness. Management science research project, research report 43, UCLA, 1955
18. S. Khot, On the power of unique 2-prover 1-round games, in *Proceedings pf the 34th Annual ACM Symposium on Theory of Computing* (2002), pp. 767–775
19. J.Y.-T. Leung, H. Li, M. Pinedo, Scheduling orders for multiple product types with due date related objectives. Eur. J. Oper. Res. **168**, 370–389 (2006)
20. M. Mastrolilli, M. Queyranne, A.S. Schulz, O. Svensson, N.A. Uhan, Minimizing the sum of weighted completion times in a concurrent open shop. Oper. Res. Lett. **38**, 390–395 (2010)

21. J.M. Moore, An *n* job, one machine sequencing algorithm for minimizing the number of late jobs. Manag. Sci. **15**, 102–109 (1968)
22. C.T. Ng, T.C.E. Cheng, J.J. Yuan, Concurrent open shop scheduling to minimize the weighted number of tardy jobs. J. Sched. **6**, 405–412 (2003)
23. T.A. Roemer, A note on establishing heuristic bounds by instance construction. Technical report, Sloan School at MIT, Cambridge, MA, 2004
24. T.A. Roemer, A note on the complexity of the concurrent open shop scheduling problem. J. Sched. **9**, 389–396 (2006)
25. A.S. Schulz, Scheduling to minimize total weighted completion time: performance guarantees of lp-based heuristics and lower bounds, in *Integer Programming and Combinatorial Optimization, IPCO 1996*, ed. by W.H. Cunningham, S.T. McCormick, M. Queyranne. Lecture Notes in Computer Science (Springer, Berlin, 1996), pp. 301–315
26. E. Wagneur, C. Sriskandarajah, Open shops with jobs overlap. Eur. J. Oper. Res. **71**, 366–378 (1993)

# Chapter 6
# Open Shop Scheduling with Simultaneity Constraints

## 6.1 Open Shop with Simultaneity Constraints

The open shop with simultaneity constraints has been introduced to enrich the features of class–teacher timetabling model in order to better reflect requirements imposed by real-life scheduling problems; see Even et al. [12], de Werra [4], de Werra and Erschler [5], and de Werra et al. [6]. In addition to a bipartite multigraph $G = (\mathcal{J}, \mathcal{M}, E)$ there is a set $C$ of pairs $\{e, f\}$ of non-adjacent edges in an instance of the problem with simultaneity constraints. The problem is to find an edge coloring $c$ of $G$ using as few colors as possible and such that the edges $e$ and $f$ in each pair $\{e, f\} \in C$ are colored with the same color, i.e., $c(e) = c(f)$ for each $\{e, f\} \in C$. This last condition is called simultaneity constraints; see de Werra [4], de Werra and Erschler [5], and de Werra et al. [6]. The constraints are quite powerful since they make the open shop with simultaneity constraints to include the edge-coloring problem for any multigraph as a subproblem. We prove this now.

For a multigraph $(G = (V, E), \mathrm{mp})$ with the set of vertices $V = \{1, \ldots, n\}$, define a set of jobs $\mathcal{J} = \{J_1, \ldots, J_n\}$, a set of machines $\mathcal{M} = \{M_1, \ldots, M_n\}$, and a bipartite multigraph $(\mathcal{J}, \mathcal{M}, F)$ where for each $e = (i, j) \in E$ the edges $(J_i, M_j)$ and $(J_j, M_i)$ belong to $F$. Moreover, $\mathrm{mp}(e) = p_{i,j} = \mathrm{mp}((J_i, M_j)) = \mathrm{mp}((J_j, M_i))$. Finally, the pair $\{(J_i, M_j), (J_j, M_i)\} \in C$, i.e., the edges in the pair are required to be colored with the same color. This constraint applies to all $p_{i,j}$ copies of the edge $(i, j)$. Thus we obtain an instance of an open shop with simultaneity constraints. We need to show that there is an edge coloring of $G$ with $k$ colors if and only if there is an edge coloring of $(\mathcal{J}, \mathcal{M}, F)$ with $k$ colors that meets simultaneity constraints.

Suppose that there is $k$-edge coloring of a multigraph $G$. The coloring is a collection of matchings $E_1, \ldots, E_k$ in $G$ that covers each edge of $G$ exactly once. For each $\ell = 1, \ldots, k$, define $F_\ell = \{(J_i, M_j), (J_j, M_i) : (i, j) \in E_\ell\}$. Each $F_\ell$ is a matching in $(\mathcal{J}, \mathcal{M}, F)$. The collection of matchings $F_1, \ldots, F_k$ covers each edge



W. Kubiak, *A Book of Open Shop Scheduling*, International Series in Operations Research & Management Science 325,
https://doi.org/10.1007/978-3-030-91025-9_6

of $(\mathcal{J}, \mathcal{M}, F)$ exactly once. Thus we obtain a $k$-edge coloring of $(\mathcal{J}, \mathcal{M}, F)$. The coloring meets the simultaneity constraints since each pair $\{(J_i, M_j), (J_j, M_i)\} \in C$ is colored with the same color as the edge $(i, j) \in E$.

Now, suppose that there is $k$-edge coloring of $(\mathcal{J}, \mathcal{M}, F)$ that meets simultaneity constraints in $C$. The coloring is a collection of matchings $F_1, \ldots, F_k$ in $(\mathcal{J}, \mathcal{M}, F)$ that covers each edge of $(\mathcal{J}, \mathcal{M}, F)$ exactly once. Because the simultaneity constraints are met, each pair $\{(J_i, M_j), (J_j, M_i)\} \in C$ belongs to exactly one $F_\ell$ for some $\ell = 1, \ldots, k$. Let $E_\ell$ be obtained by replacing each pair $\{(J_i, M_j), (J_j, M_i)\} \in C$ in $F_\ell$ by $(i, j)$. Each $E_\ell$ is a matching in $G$, and the collection $E_1, \ldots, E_k$ in $G$ covers each edge of $G$ exactly once. Thus there is $k$-edge coloring of $G$.

Figure 6.1 illustrates this reduction for multigraph $G$ given in Fig. 6.1a. The corresponding bipartite multigraph $(\mathcal{J}, \mathcal{M}, F)$ is given in Fig. 6.1b where the simultaneity constraints include pairs $\{(J_1, M_4), (J_4, M_1)\}$ and $\{(J_2, M_3), (J_3, M_2)\}$ among others. We leave it to the reader as an exercise to write down a complete set $C$. The schedule with makespan $C_{\max} = 6$ for $(\mathcal{J}, \mathcal{M}, F)$ that corresponds to the 6-edge coloring of $G$ is given in Fig. 6.1c. Observe that due to the simultaneity constraints job $J_4$ is scheduled on $M_1$ in the interval $[0, 1]$, and job $J_1$ is scheduled on $M_4$ in the interval $[0, 1]$. Similarly, job $J_3$ is scheduled on $M_2$ in the interval $[0, 1]$, and job $J_2$ is scheduled on $M_3$ in the interval $[0, 1]$.

By the Holyer [19] complexity result for edge coloring of graphs, the open shop scheduling problem with simultaneity constraints is NP-hard in the strong sense. de Werra [4] and de Werra and Erschler [5] give some polynomial cases of the open shop scheduling problem. Coffman et al. [3], Gandhi et al. [13], and Khuller et al. [21] apply the problem to file transfer scheduling and data migration; see Chap. 11 for details. In this chapter, which is based on Dereniowski et al. [7], we concentrate on a class of graphs recently introduced in the context of scheduling in wireless networks with primary interference. The graphs are introduced in Sect. 6.2 and characterized in Sect. 6.6.1.

## 6.2 Wireless Networking with Primary Interference

We begin with introducing a wireless networking problem and reviewing the recent results in this field. The problem provides the main source of motivation for the scheduling problems we study in this chapter.

Consider a graph $G = (V, E)$, in which the vertices in set $V$ represent agents (i.e., transmitters and receivers) in a communication network, and $E \subseteq \{(i, j) \colon i, j \in V, i \neq j\}$ is a set of wireless connections representing pairs of agents between which data flow can occur. At each vertex $v \in V$ of the network, information packets are received over time and these packets must be transmitted to their destinations, which correspond in our model to the neighbors of that vertex $v$ (such a model is called *single hop*). We assume that time is slotted and that packets are of equal size, each packet requiring one time slot of service across a

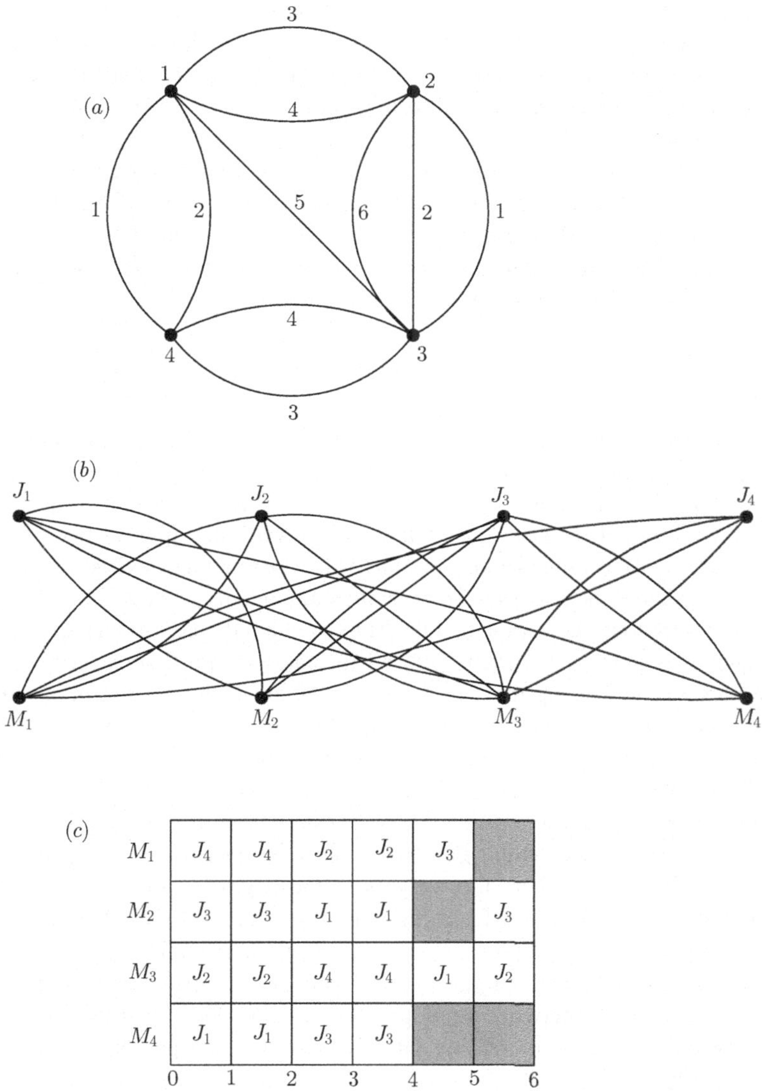


**Fig. 6.1** (**a**) Multigraph $G$ with 6-edge coloring; (**b**) its corresponding open shop with simultaneity constraints; (**c**) the schedule with $C_{\max} = 6$ that meets simultaneity constraints

link. A stochastic queue is associated with each edge in the network, representing the packets waiting to be transmitted on this link. We assume that the stochastic arrivals to edge $(i, j)$ have long-term rates $\lambda_{ij}$ and are independent of each other. We denote by $\lambda$ the vector of the arrival rates $\lambda_{ij}$ for every edge $(i, j)$.

An important issue in operating a wireless communication network is that two connections might interfere with each other. We focus on the simplest interference model that can be found in the literature, which is known as the *primary interference model*. This model states that two connections interfere with each other if and only if the corresponding edges share a vertex in $G$. Thus, at every time slot, the set of connections that are activated should form a matching.

A *scheduling algorithm* selects a set of edges to activate at each time slot and transmits packets on those edges. The goal is to find a scheduling algorithm that, informally speaking, keeps the sizes of the queues from growing unboundedly when this algorithm is adopted. Clearly, if the values of the $\lambda_{ij}$'s are chosen sufficiently large, no algorithm can attain this. Therefore, usually, one defines the *stability region* $\Lambda^*$ of a graph $G$ (with respect to rates $\lambda$) as

$$\Lambda^* = \left\{\lambda \in \mathbb{R}^{E(G)} : \lambda < \mathbf{u} \text{ for some } \mathbf{u} \in \mathbf{conv}(\mathcal{M}_G)\right\},$$

where $\mathcal{M}_G$ is the set of all matchings in $G$ and $\mathbf{conv}(\mathcal{M}_G)$ is the convex hull of the characteristic (0–1) vectors of the elements of $\mathcal{M}_G$ (and the "$<$" sign is componentwise). It should be clear that, for any $\lambda$ that is not an element of the closure of $\Lambda^*$, there is no algorithm that prevents the queues from growing unboundedly. On the other hand, one may ask whether there exists a scheduling algorithm that keeps the queues from growing unboundedly when $\lambda \in \Lambda^*$. Formally, this is equivalent to the condition that the Markov chain that represents the evolution of the queues is positive recurrent (i.e., all its states are positive recurrent) for all arrival rates $\lambda \in \Lambda^*$, using the scheduling algorithm. If this condition is satisfied, then we say that the scheduling algorithm *achieves* 100% *throughput on* $G$. For more details regarding the queue evolution process under this model, see Brzezinski et al. [2], Dimakis and Walrand [8], and Joo et al. [20].

A good first choice for a scheduling algorithm turns out to be the *Maximum Weight Matching (MWM) algorithm* that selects, in each time slot, a maximum weight matching in $G$, where the weights of the edges are given by the current queue lengths. It was shown in Tassiulas and Ephremides [31] that MWM achieves 100% throughput on any graph $G$. However, it is not a tractable algorithm in many situations because to find an optimal solution it needs centralized computing and full knowledge of both the network topology and all queue lengths at every time slot. Hence, there has been an increasing interest in simple and potentially distributed algorithms. One example of such an algorithm is known as the *Greedy Maximal Scheduling (GMS) algorithm*; see Hoepman [18] and Lin and Shroff [26]. This algorithm effectively selects the set of links served in a greedy fashion according to the queue lengths at these links (i.e., GMS greedily selects a maximal weight matching). Note that GMS can be implemented in a distributed setting; see Khun et al. [23] and Leconte et al. [24]. A drawback of using this algorithm is that, in general, it does not achieve 100% throughput for every graph $G$. However, Dimakis and Walrand [8] gave a sufficient condition on network graphs (in the primary interference model) for which the GMS algorithm does achieve 100% throughput. We say that a graph $G$ is *OLoP* (*OLoP* stands for *Overall Local Pooling*) if, for every subgraph $G'$ of $G$, there exists a function $w\colon E(G') \to [0, 1]$ (that depends on $G'$) such that every inclusion-wise maximal matching $M$ in $G'$ satisfies $\sum_{e \in M} w(e) = 1$. The sufficient condition from Dimakis and Walrand [8] is the following.

**Theorem 6.1** *GMS achieves 100% throughput on every OLoP graph G.*

In Birand et al. [1], a complete structural characterization of all OLoP graphs was given. We repeat this characterization in Sect. 6.6.1 for completeness.

## 6.3 Scheduling Links to Satisfy Link Demand and Related Problems

We consider problems related to the following graph-theoretical question:

MATCH
Input: An undirected, simple graph $G = (V, E)$ and a function $r\colon E \to [0, 1] \cap \mathbb{Q}$.
Question: Does there exist a sequence of matchings $M_1, M_2, \ldots$ in $G$ such that

$$\liminf_{n \to \infty} \frac{|\{i\colon e \in M_i, i = 1, 2, \ldots, n\}|}{n} \geq r(e) \quad \text{for all } e \in E? \tag{6.1}$$

In other words, we are asking whether, for a given graph $G$ with edge weights $r(e)$, there exists a (possibly infinite) sequence of matchings so that each edge $e$ is hit by at least a fraction $r(e)$ of the matchings.

The problem MATCH falls into the general framework of a problem earlier studied by Grötschel et al. in [16] and by Hajek and Sasaki in [17]. The problem studied by these authors, which is also known as "scheduling links to satisfy link demand" (see Sect. 6.2), is the following. For a graph $G$, let $\mathcal{M}_G$ be the collection of all its matchings. Given a graph $G$ and a function $r\colon E(G) \to \mathbb{Q}_+$, find a function $w\colon \mathcal{M}_G \to \mathbb{Q}$ that satisfies $\sum_{M \in \mathcal{M}_G : e \in M} w(M) \geq r(e)$ for all $e \in E$ and has minimum total weight $\sum_{M \in \mathcal{M}_G} w(M)$. A polynomial-time algorithm for this problem was given in Grötschel et al. [16]. Hajek and Sasaki [17] improve this result by giving a $O(|E| \cdot |V|^5)$-time algorithm. (The minimum total weight itself can be determined in $O(|V|^5)$-time; Hajek and Sasaki [17].)

It is not hard to see that the problem MATCH is equivalent to the question whether the minimum total weight of the problem above is at most 1. The solution to this latter problem, however, does not say anything about the number of matchings required in MATCH. This motivates our interest in studying the following three related problems in this chapter:

**Problem 1 (K-MATCH)** Given an undirected, simple graph $G = (V, E)$, a function $r\colon E \to [0, 1] \cap \mathbb{Q}$, and a positive integer $k$, do there exist $k$ matchings $M_1, \ldots, M_k$ in $G$ such that

$$|\{i\colon e \in M_i\}| \geq kr(e) \quad \text{for all } e \in E? \tag{6.2}$$

**Problem 2 (FIND-K-MATCH)** Given an undirected, simple graph $G = (V, E)$, a function $r\colon E \to [0, 1] \cap \mathbb{Q}$, and a positive integer $k$ for which K-MATCH is affirmative, find $k$ matchings $M_1, \ldots, M_k$ in $G$ that witness K-MATCH.

**Problem 3 (MIN-MATCH)** Given an undirected, simple graph $G = (V, E)$ and a function $r\colon E \to [0, 1] \cap \mathbb{Q}$ for which K-MATCH is affirmative, find the smallest $k$ such that the answer to K-MATCH is affirmative.

The remainder of the chapter is organized as follows: Sect. 6.4 explains the relationship between the problem MATCH and fractional edge coloring, and the relationship of the problems K-MATCH and FIND-K-MATCH with edge coloring.

The computational complexity of MATCH and K-MATCH is briefly analyzed in Sect. 6.5. We will show there that the problem K-MATCH is $\mathcal{NP}$-complete for general graphs. A natural subclass of graphs to consider is the *OLoP* class of graphs defined in Dimakis and Walrand [8]. An appealing feature of OLoP graphs is that they result in 100% throughput for some simple distributed algorithms like the greedy maximal scheduling algorithm (Dimakis and Walrand [8]) in wireless networks (alluded to in Sect. 6.2). Essentially, the greedy maximal scheduling algorithm solves the FIND-K-MATCH for some $k$ for an OLoP graph, provided that MATCH is affirmative for the graph. However, the value of $k$, referred to as the schedule length (makespan), given by the algorithm does not have to be the smallest possible. To address the problem of schedule makespan minimization for the OLoP class, we chose this class of graphs as an input to the problems MATCH, K-MATCH, and FIND-K-MATCH and show that this restriction renders all three polynomial. Birand et al. [1] provide a structural characterization of these graphs by showing that their blocks are either $K_2$ or $K_4$ or are obtained from certain graphs on 7, 5, and 3 vertices by iterative non-adjacent cloning of some vertices of degree 2. We leave the details of the characterization of OLoP graphs to Sect. 6.6.1. We then focus on a much broader class of graphs, and we refer to them as *Generalized OLoP*(b), or *GOLoP(b)* graphs. The blocks of *GOLoP(b)* are obtained from *any* connected graph on at most $b$ vertices by iterative non-adjacent cloning of some vertices of degree 2. We show that restricting the input to the problems MATCH, K-MATCH, and FIND-K-MATCH to *GOLoP(b)* graphs with a fixed, that is not being part of the input, $b$ renders all three problems polynomial. Finally, realizing that the problems MATCH, K-MATCH, and FIND-K-MATCH are only stepping stones in schedule makespan minimization, we focus on upper bounds on schedule length for OLoP and *GOLoP(b)* graphs. We show that the least common denominator $d$ for the rate function $r$ is an upper bound for the former and a constant multiple of $d$, where the constant depends on $b$ but not on $r$, is an upper bound for the latter. We conjecture that the constant is actually equal to 2.

In particular, the main contributions of this chapter are as follows:

1. A linear-time algorithm to solve K-MATCH and MATCH for GOLoP graphs (Sects. 6.6.2 and 6.6.3).
2. A $O(|V(G)|^2)$ running time algorithm for FIND-K-MATCH on GOLoP graphs. The algorithm uses $O(|V(G)|)$ distinct matchings (Sect. 6.6.4).

3. We show that, for OLoP graphs, the least common denominator $d$ for the rate function $r$ is an upper bound for the length of the shortest sequence of matchings that forms a solution to K-MATCH, given that the problem K-MATCH has an affirmative answer for some $k$ (Sect. 6.7.1).
4. We give an upper bound on the integer $k$ for which K-MATCH has an affirmative answer in the class of GOLoP graphs. A conjecture of Seymour [30] implies that this bound does not exceed $2d$ in general. We prove that this is true for all graphs with $|V(G)| \leq 10$ (Sect. 6.7.2).
5. Using the bound from Sect. 6.7.2, we construct a pseudopolynomial-time algorithm for the shortest schedule for GOLoP graphs (Sect. 6.7.3).

Finally, Sect. 6.8 concludes the chapter and presents some open questions.

## 6.4 Relationship with Edge Coloring

For a given rate function $r$ and an integer $k$, we will write $\lceil kr \rceil$ for the function $e \mapsto \lceil kr(e) \rceil$. We have the following equivalence; see Dereniowski et al. [7] for proof.

*Claim 1* Let $G$ be a graph, let $r$ be a rate function for $G$, and let $k$ be an integer. The answer to K-MATCH is YES if and only if $\chi'(G, \lceil kr \rceil) \leq k$.

We also have the following equivalence:

*Claim 2* Let $G$ be a graph, let $r$ be a rate function for $G$, and let $k$ be an integer. The following three statements are equivalent:

1. $\chi'_f(G, r) \leq 1$.
2. The answer to MATCH is YES.
3. The answer to K-MATCH is YES for some $k$.

## 6.5 Complexity of MATCH and K-MATCH for General Graphs

In this section we will consider the problems MATCH and K-MATCH from a complexity point of view. For the former problem, we show that it is polynomial-time solvable. Though this was already shown in Hajek and Sasaki [17], we present here a new algorithm that improves the complexity of the one given in Hajek and Sasaki [17], where a $O(|V(G)|^5)$-time algorithm has been given.

*Claim 3* The problem MATCH can be solved in $O(|V(G)|^4)$ time.

***Proof*** Let $\mathcal{P}$ be the matching polyhedron corresponding to $G$ (as defined by Edmonds [9]), i.e., $\mathcal{P} = \mathbf{conv}\{\mathbf{e}_M : M \in \mathcal{M}_G\}$, where $\mathbf{e}_M$ is the $|E(G)|$-dimensional 0–1 characteristic vector of $M$. We claim that $\chi'_f(G, r) \leq 1$ if and only if $r \in \mathcal{P}$. This is sufficient due to Claim 2. To prove the claim, suppose that $r \in \mathcal{P}$. Then, $r = \sum_{M \in \mathcal{M}_G} p(M)\mathbf{e}_M$ for some $p\colon \mathcal{M}_G \to [0, 1]$ with $\sum_{M \in \mathcal{M}_G} p(M) = 1$. It follows that $\{p(M)\}$ is a feasible solution to (1.9) (see Sect. 1.2), ensuring that $\chi'_f(G, r) \leq 1$. Conversely, suppose that $\chi'_f(G, r) \leq 1$. Let $\{p(M)\}$ be a solution to (1.9) (see Sect. 1.2), such that $\sum_{M \in \mathcal{M}_G} p(M) = 1$. Such a solution can always be obtained by sufficiently increasing $p(\emptyset)$ if needed. By (1.9), we have $\sum_{M \in \mathcal{M}_G(e)} p(M) = r(e)$ for all $e \in E$. Thus, $r = \sum_{M \in \mathcal{M}_G} p(M)\mathbf{e}_M$, which shows that $r \in \mathcal{P}$. This proves the claim. Finally, the test whether $r \in \mathcal{P}$ can be done in polynomial time using the Padberg-Rao [27] separation algorithm. Moreover, Letchford, Reinelt, and Theis [25] proved that the test can be done in $O(|V(G)|^4)$ time. □

Although the complexity of MATCH is $O(|V(G)|^4)$ for general graphs, we will see in Sect. 6.6.3 that MATCH can be solved in linear time if $G$ is a GOLoP graph (see Sect. 6.6.1 for the definition of a GOLoP graph). For the K-MATCH problem, we show the following using Claim 1.

*Claim 4* K-MATCH is $\mathcal{NP}$-complete in the strong sense.

***Proof*** We will use a transformation from the $k$-edge-coloring problem, which is known to be $\mathcal{NP}$-complete (see Holyer [19]). Consider the $k$-edge-coloring problem on a graph $G = (V, E)$. We define a rate function $r$ on $E$ such that $r(e) = \frac{1}{k}$ for each edge $e \in E$. Thus we obtain an instance of the K-MATCH problem. Now using Claim 1, we immediately conclude that both problems are equivalent and hence K-MATCH is $\mathcal{NP}$-complete in the strong sense. □

## 6.6 GOLoP Graphs

We have seen that the problems MATCH, K-MATCH, and FIND-K-MATCH are all computationally intractable when the input is allowed to be any graph. The current section deals with these problems when the input is restricted to a smaller class of graphs, namely the OLoP and GOLoP graphs. The remaining subsections consist of showing that the problems MATCH, K-MATCH, and FIND-K-MATCH can indeed be solved in polynomial time when the input is restricted to GOLoP graphs.

### *6.6.1 Characterization of OLoP Graphs; GOLoP Graphs*

For a formal definition of the OLoP and GOLoP graphs, we need the following notation. Let $G = (V, E)$ be a connected graph. We call $x \in V$ a *cut-vertex of* $G$ if $G - x$ is not connected. We call a maximal connected induced subgraph $B$ of $G$ such that $B$ has no cut-vertex a *block of* $G$. Let $B_1, B_2, \ldots, B_q$ be the blocks of $G$. We call the collection $\{B_1, B_2, \ldots, B_q\}$ the *block decomposition of* $G$. It is known that the block decomposition is unique and that $E(B_1), E(B_2), \ldots, E(B_q)$ (where $E(B_i)$ denotes the set of edges in $B_i$, $i = 1, \ldots, q$) form a partition of $E$ (see, for instance, West [32]). Furthermore, the vertex sets $V(B_i)$ and $V(B_j)$ of every two blocks $B_i$ and $B_j$, $i, j = 1, \ldots, q$, $i \neq j$, intersect in at most one vertex and this vertex is a cut-vertex of $G$. Block decompositions give a tree-like decomposition of a graph in the following sense. Construct the *block-cutpoint graph of* $G$ by keeping the cut-vertices of $G$ and replacing each block $B_i$ of $G$ by a vertex $b_i$. Make each cut-vertex $x$ adjacent to $b_i$ if and only if $x \in V(B_i)$. It is known that the block-cutpoint graph of $G$ forms a tree (e.g., West [32]). With this tree-like structure in mind, we say that a block $B_i$ is a *leaf block* if it contains at most one cut-vertex of $G$. Clearly, if $q \geq 2$, then $\{B_i\}_{i=1}^q$ contains at least two leaf blocks.

As some blocks, the complete and complete bipartite graphs will be used. Formally, $K_i$, $i \geq 1$, denotes the complete graph on $i$ vertices in which every pair of vertices is adjacent, and $K_{i_1,i_2}$ is a complete bipartite graph with two vertex partitions $V_1$ and $V_2$ of size $i_1$ and $i_2$, respectively, such that $u$ and $v$ are adjacent for each $u \in V_1$ and $v \in V_2$ and no two vertices in $V_1$ or in $V_2$ are adjacent in $K_{i_1,i_2}$.

We start with a characterization of OLoP graphs (see Birand et al. [1]). We will do this in terms of the block decomposition introduced above. It turns out that the block decomposition of an OLoP graph is relatively simple in the sense that there are only two types of blocks. The types are defined by the following two families of graphs.

$\mathcal{B}_1$: Construct $\mathcal{B}_1$ as follows. Let $H$ be a graph with $V(H) = \{c_1, c_2, \ldots, c_k\}$, with $k \in \{5, 7\}$, such that

1. $c_1 - c_2 - \cdots - c_k - c_1$ is a cycle;
2. if $k = 5$, then the other adjacencies are arbitrary; if $k = 7$, then all other pairs are non-adjacent, except possibly $\{c_1, c_4\}$, $\{c_1, c_5\}$, and $\{c_4, c_7\}$.

Then, $H \in \mathcal{B}_1$.

Now iteratively perform the following operation. Let $H' \in \mathcal{B}_1$, and let $v \in V(H')$ with $\deg(v) = 2$. Construct $H''$ from $H'$ by adding a vertex $v'$ such that $N(v') = N(v)$. $v'$ is called a *non-adjacent clone* of $v$. Then, $H'' \in \mathcal{B}_1$. We say that a graph is *of the* $\mathcal{B}_1$ *type* if it is isomorphic to a graph in $\mathcal{B}_1$.

$\mathcal{B}_2$: Let $\mathcal{B}_2 = \{K_2, K_3, K_4\} \cup \left\{K_{2,t}, K_{2,t}^+ : t \geq 2\right\}$, where $K_{2,t}^+$ is constructed from $K_{2,t}$ by adding an edge between the two vertices on the side of the bipartition that has cardinality 2. We say that a graph is *of the* $\mathcal{B}_2$ *type* if it is isomorphic to a graph in $\mathcal{B}_2$.

In simple words, graphs of the $\mathcal{B}_1$ type are constructed as follows. Starting with a cycle of length five or seven. Then we may add some additional edges between vertices of the cycle, subject to some constraints. Finally, we may iteratively take a vertex $v$ of degree 2 and add a non-adjacent clone $v'$ of $v$. The following result characterizes OLoP graphs.

*Claim 5 (Birand et al. [1])* Let $G = (V, E)$ be a graph, and let $\{B_1, \ldots, B_q\}$ be its block decomposition. $G$ is an OLoP graph if and only if at most one block of $G$ is of the $\mathcal{B}_1$ type and all other blocks are of the $\mathcal{B}_2$ type.

It follows from Claim 5 that OLoP graphs can be constructed by starting with a block that is either of the $\mathcal{B}_1$ or of the $\mathcal{B}_2$ type and then iteratively adding a block of the $\mathcal{B}_2$ type by "glueing" it on an arbitrary vertex.

This motivates the following definition of a generalized OLoP graph. Let $b \geq 1$ be an integer. A graph $G$ is called GOLoP($b$) if every block of $G$ can be obtained from a connected graph on at most $b$ vertices by iteratively non-adjacent cloning a vertex of degree two. We say that a multigraph $H = (G, \text{mp})$ is GOLoP($b$) if the graph $G$ is GOLoP($b$). It is not hard to see that every OLoP graph is also a GOLoP(7) graph.

To deal with GOLoP graphs, we will frequently use the following notation. Let $G$ be a GOLoP graph. Let $C_1, \ldots, C_p$ be maximal sets of vertices of degree two (in $G$) such that $|C_i| \geq 2$ and all vertices in set $C_i$ have the same two neighbors $u_i$ and $v_i$. We refer to these sets as non-adjacent clones in $G$. Choose $p$ maximal as well. Consider the auxiliary graph $G'$ constructed from $G - \bigcup_{i=1}^{p} C_i$ by adding new vertices $a_1, \ldots, a_p$ such that, for $i \in [p] = \{1, \ldots, p\}$, $a_i$ is adjacent to precisely $u_i$ and $v_i$, and by adding new edges $u_i v_i$ for all $i \in [p]$. Let $W = \{a_1, \ldots, a_p\}$. We call the pair $(G', W)$ the *collapsed graph* associated with $G$. For $i \in [p]$, let $F_i = H[C_i \cup \{u_i, v_i\}]$; see Fig. 6.2.

It was shown in Birand et al. [1] that OLoP graphs have $O(|V(G)|)$ edges. The proof of this result generalizes easily to the setting of GOLoP($b$). We include the generalization for completeness:

*Claim 6* Let $b$ be a fixed integer, and let $G$ be a GOLoP($b$) graph. Then, $|E(G)| = O(|V(G)|)$.

Since finding the block decomposition of a graph $G$ can be done in $O(|V(G)| + |E(G)|)$ time (see, e.g., Gross and Yellen [15]), Claim 6 has the following corollary:

*Claim 7* Let $b$ be a fixed integer, and let $G$ be a GOLoP($b$) graph. Finding the block decomposition of $G$ can be done in $O(|V(G)|)$ time.

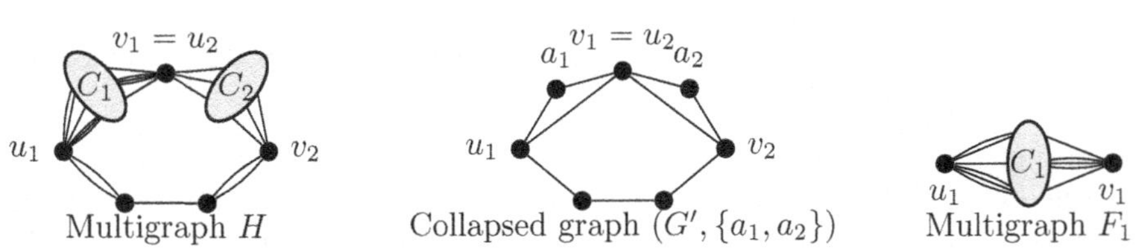


**Fig. 6.2** The multigraph $H$ and the sets $C_i$ (left), the corresponding collapsed graph $(G', W)$ (middle), and one of the multigraphs $F_1$, $F_2$ (right). In this figure, we added multiple edges to represent the values of the function mp

### 6.6.2 K-MATCH *for GOLoP Graphs*

In this section, it will be notationally more convenient to think of an edge coloring of a multigraph $H = (G, \mathrm{mp})$ as a schedule. We need to introduce a bit more notation. A *schedule* (of length $k$) is a function $S\colon \{1, \dots, k\} \to \mathcal{M}_H$. For $e \in E(H)$, let $T_S(e) = \sum_{t=1}^{k} \mathbb{1}(e \in S(t))$ (where $\mathbb{1}$ is the indicator function). Informally speaking, $T_S(e)$ is the total amount of time schedule $S$ spends on edge $e$. Likewise, for vertices $u, v \in V(H)$, let $T_S(u, v) = \sum_{t=1}^{k} \mathbb{1}(S(t)$ covers both $u$ and $v)$ and, for a matching $M \in \mathcal{M}_H$, let $T_S(M) = \sum_{t=1}^{k} \mathbb{1}(S(t) = M)$. A schedule $S$ is said to be *feasible for* $H$ if $T_S(e) = \mathrm{mp}(e)$ for all $e \in E(H)$. We state the following two observations without a proof:

*Claim 8* Let $H$ be a multigraph, and let $k$ be an integer. Then, $\chi'(H) \le k$ if and only if there exists a feasible schedule of length $k$ for $H$.

*Claim 9* Let $H = (G, \mathrm{mp})$ be a multigraph, and let $x$ be a cut-vertex of $H$. Let $K^1, \dots, K^p$ be the connected components of $H - x$. Then,

$$\chi'(H) = \max\left[\deg(x),\ \max_{i=1,\dots,p}\left\{\chi'(H[V(K^i) \cup \{x\}])\right\}\right].$$

This latter observation allows us to concentrate on the blocks of GOLoP multigraphs. To deal with the sets of clones in GOLoP multigraphs, we start with a lemma for bipartite multigraphs in which one side of the bipartition has exactly two vertices (two-machine open shop).

**Lemma 6.1** *Let $F$ be a bipartite multigraph on vertex sets $X, Y$ with $X = \{u, v\}$. Then, for $\tau \in \mathbb{Z}_+$, there exists a feasible schedule $S$ for $F$ such that $T_S(u, v) = \tau$ if and only if*

$$\tau \le \deg(u) + \deg(v) - \Delta(F). \qquad (6.3)$$

*Moreover, for all $\tau$ satisfying* (6.3), *there exists a feasible schedule of length* $\deg(u) + \deg(v) - \tau$ *such that* $T_S(u, v) = \tau$.

***Proof*** See Dereniowski et al. [7] for a complete proof. □

Having dealt with the sets of clones in GOLoP multigraphs, we can now prove the following.

**Lemma 6.2** *Let $b$ be a fixed integer, and let $H$ be a GOLoP($b$) multigraph. Then, $\chi'(H)$ can be determined in $O(|V(H)|)$ time.*

***Proof*** Let $H = (G, \mathrm{mp}_H)$. Since, by Claim 7, the block decomposition of a GOLoP($b$) graph can be found in $O(|V(G)|) = O(|V(H)|)$ time, it follows from Claim 9 that it suffices to prove the lemma for the blocks of $H$. So we may assume that $H$ is 2-connected.

Let $(G', W)$ be the collapsed graph associated with $G$, and let $p, C_1, \ldots, C_p$, $F_1, \ldots, F_p$ as in the definition of the collapsed graph. By the maximality of the sets $C_1, \ldots, C_p$, every set of clones $C_i$ has a unique pair of common neighbors $\{u_i, v_i\}$. Since $u_i$ and $v_i$ are the vertices of the graph on at most $b$ vertices from which $H$ was constructed by iteratively cloning the vertices of degree two, there are at most $\binom{b}{2}$ choices of $u_i$ and $v_i$ and, hence, $p \leq \binom{b}{2}$. Let $T_i^* = \deg_{F_i}(u_i) + \deg_{F_i}(v_i) - \Delta(F_i)$.

For conciseness, we will write $\mathcal{M}'$ for $\mathcal{M}_{G'}$ and $\mathcal{M}'(e)$ for $\mathcal{M}_{G'}(e)$. We will construct an integer linear programming problem whose objective value is $\chi'(H)$ and whose variables correspond to the matchings in $G'$. The idea is that the edges $u_i v_i$ in matchings in $G'$ will play the role of pairs of edges $\{u_i c, v_i c'\}$ in matchings in $H$ with distinct $c, c' \in C_i$. Consider the following integer linear programming problem:

$$z^* = \min_{\mathbf{w} \in \mathbb{Z}_+^{\mathcal{M}'}} \sum_{M \in \mathcal{M}'} w(M) \tag{6.4}$$

$$\text{s.t.} \sum_{M \in \mathcal{M}'(e)} w(M) = \mathrm{mp}_H(e) \qquad \text{for all } e \in E(G') \setminus \bigcup_{i=1}^{p} \{u_i a_i, v_i a_i, u_i v_i\} \tag{6.4a}$$

$$\sum_{M \in \mathcal{M}'(z a_i) \cup \mathcal{M}'(u_i v_i)} w(M) = \sum_{c \in C_i} \mathrm{mp}_H(zc) \qquad \text{for all } z \in \{u_i, v_i\}, i \in [p] \tag{6.4b}$$

$$\sum_{M \in \mathcal{M}'(u_i v_i)} w(M) \leq T_i^* \qquad \text{for all } i \in [p]. \tag{6.4c}$$

Constructing the problem needs calculating the values of $T_i^*$, which can clearly be done in $O(|C_i|)$ time. Since $p \leq \binom{b}{2}$, the problem is an integer linear programming problem with $O(1)$ variables and constraints. Using Eisenbrand's algorithm for integer linear programming in fixed dimension (Eisenbrand [11]), this problem can be solved in $O(1)$ time. Thus, the overall complexity of computing $z^*$ is $O(|V(H)|)$.

We claim that $z^* = \chi'(H)$. First, to prove that $z^* \geq \chi'(H)$, we claim that any solution to (6.4) can be turned into a feasible schedule for $H$ of length $z^*$ with the help of Lemma 6.1. To see this, consider an optimal solution $\{w(M)\}_{M \in \mathcal{M}'}$ of (6.4). We can, by the constraints of (6.4), construct a function $S' \colon \{1, \ldots, z^*\} \to \mathcal{M}'$ such that $T_{S'}(e) = \mathrm{mp}_H(e)$ for all $e \in E(G') \setminus \bigcup_{i=1}^{p} \{u_i a_i, v_i a_i, u_i v_i\}$, $T_{S'}(z a_i) + T_{S'}(u_i v_i) = \sum_{c \in C_i} \mathrm{mp}_H(zc)$ for all $z \in \{u_i, v_i\}, i \in [p]$, and $T_{S'}(u_i v_i) \leq T_i^*$ for all $i \in [p]$. $S'$ is not a schedule because $S'$ is defined on the matchings of $G'$ and not on the matchings of $H$.

We will turn $S'$ into a schedule for $H$ as follows. Let $S^{(0)} = S'$. We will iteratively construct a sequence $S^{(1)}, \ldots, S^{(p)}$ of functions from $\{1, \ldots, z^*\}$ to $\mathcal{M}' \cup \mathcal{M}_H$, the last of which will be a schedule for $H$. For $i \in [p]$, do the following. Let $\tau_i = \sum_{M \in \mathcal{M}'(u_i v_i)} w(M)$. By (6.4c), $\tau_i \leq T_i^*$. It follows from Lemma 6.1 that there exists a feasible schedule $S_i \colon \{1, \ldots, \deg_{F_i}(u_i) + \deg_{F_i}(v_i) - \tau_i\} \to \mathcal{M}_{F_i}$ such that

$T_{S_i}(u_i, v_i) = \tau_i$. Let $Z_i = \{t \mid S^{(i-1)}(t) \in \mathcal{M}'(u_ia_i) \cup \mathcal{M}'(v_ia_i) \cup \mathcal{M}'(u_iv_i)\}$. It follows from (6.4b) that

$$|Z_i| = \sum_{c \in C_i} \mathrm{mp}_H(u_ic) + \sum_{c \in C_i} \mathrm{mp}_H(v_ic) - \tau_i = \deg_{F_i}(u_i) + \deg_{F_i}(v_i) - \tau_i.$$

Let $\phi_i : Z_i \to \{1, \deg_{F_i}(u_i) + \deg_{F_i}(v_i) - \tau_i\}$ be a bijection such that, for all $t \in Z_i$, $S_i(\phi_i(t))$ covers both $u_i$ and $v_i$ if and only if $u_iv_i \in S^{(i-1)}(t)$. Now, for all $t \in \{1, \ldots, z^*\}$, set

$$S^{(i)}(t) = \begin{cases} \left(S^{(i-1)}(t) \setminus \{u_ia_i, v_ia_i, u_iv_i\}\right) \cup S_i(\phi_i(t)) & \text{for } t \in Z_i, \\ S^{(i-1)}(t) & \text{otherwise.} \end{cases}$$

Observe that $T_{S^{(i)}}(u_ic) = \mathrm{mp}_{F_i}(u_ic)$ and $T_{S^{(i)}}(v_ic) = \mathrm{mp}_{F_i}(v_ic)$ for all $c \in C_i$. Moreover, $T_{S^{(i)}}(e) = T_{S^{(i-1)}}(e)$ for all $e \in E(H) \cup E(G') \setminus (E(F_i) \cup \{u_ia_i, v_ia_i, u_iv_i\})$.

After having done this for all $i \in [p]$, let $S = S^{(p)}$. Then, $S(t) \in \mathcal{M}_H$ for all $t \in \{1, \ldots, z^*\}$ and $T_S(e) = \mathrm{mp}_H(e)$ for all $e \in E(H)$. Thus, $S$ is a feasible schedule of length $z^*$ for $H$, implying by Claim 8 that $\chi'(H) \leq z^*$.

To prove that $\chi'(H) \geq z^*$, consider a schedule $S\colon [\chi'(H)] \to \mathcal{M}_H$ of length $\chi'(H)$. Such a schedule exists because of Claim 8. For all $t \in [\chi'(H)]$, let $I_u(t) \subseteq [p]$ (resp., $I_v(t)$) be the set of all indices $i$ such that $u_ic \in S(t)$ (resp., $v_ic \in S(t)$) for some $c \in C_i$. Define

$$S'(t) = \Big(S(t) \setminus \bigcup_{i=1}^{p} E(F_i)\Big) \cup \Big(\bigcup_{i \in I_u(t) \setminus I_v(t)} \{u_ia_i\}\Big) \cup \Big(\bigcup_{i \in I_v(t) \setminus I_u(t)} \{v_ia_i\}\Big) \cup \Big(\bigcup_{i \in I_u(t) \cap I_v(t)} \{u_iv_i\}\Big)$$

and, for $M \in \mathcal{M}'$, let $w(M) = T_{S'}(M)$. We claim that $\mathbf{w}$ is a solution of (6.4). It is straightforward to see that $S'(t) \in \mathcal{M}'$ for all $t$. Next, observe that, for all $e \in E(G') \setminus \cup_{i=1}^{p} \{u_ia_i, v_ia_i, u_iv_i\}$,

$$\sum_{M \in \mathcal{M}'(e)} w(M) = T_{S'}(e) = \left|\{t : e \in S'(t)\}\right| = |\{t : e \in S(t)\}| = T_S(e) = \mathrm{mp}_H(e).$$

Moreover, we have, for each $za_i$ with $z \in \{u_i, v_i\}$ and $i \in [p]$,

$$\sum_{\substack{M\in\mathcal{M}'(za_i)\\ \cup\mathcal{M}'(u_iv_i)}} w(M) = T_{S'}(za_i) + T_{S'}(u_iv_i) = |\{t : zc \in S(t), c \in C_i\}| =$$

$$\sum_{c\in C_i} T_S(zc) = \sum_{c\in C_i} \mathrm{mp}_H(zc).$$

Next, observe that the schedule $S$ implies a schedule $S_i$ for $F_i$. It follows from Lemma 6.1 that $T_{S_i}(u_i, v_i) \leq T_i^*$. Therefore,

$$\sum_{M\in\mathcal{M}'(u_i,v_i)} w(M) = T_{S'}(u_i, v_i) = |\{t : i \in I_u(t) \cap I_v(t)\}| = T_{S_i}(u_i, v_i) \leq T_i^*.$$

This proves that $w(M)$ is a solution to (6.4), thereby proving that $z^* \leq \chi'(H)$. This proves Lemma 6.2. □

This resolves our second problem.

**Theorem 6.2** *Let $b$ and $k$ be fixed integers. Let $G$ be a GOLoP($b$) graph, and let $r$ be a rate function for $G$. Then* K-MATCH *can be solved in $O(|V(G)|)$ time.*

***Proof*** It follows from Claim 1 that the answer to K-MATCH is YES if and only if $\chi'(G, \lceil kr \rceil) \leq k$. The theorem follows from Lemma 6.2. □

### *6.6.3* MATCH *for GOLoP Graphs*

In this section, we focus on MATCH for GOLoP graphs. We will use an algorithm that is very similar to the algorithm used for the K-MATCH problem in the previous section. We will do this by proving continuous versions of the results from the previous section.

By duplicating edges, Claim 9 generalizes easily to the setting of the fractional chromatic index of weighted graphs. To be precise, we have the following.

*Claim 10* Let $F$ be a graph, let $r$ be a rate function for $E(F)$, and let $K^1, \ldots, K^p$ be the connected components of $F - x$. Then, letting $r_i = r|_{E(F|K_i)}$ for $i \in [p]$, it holds that

$$\chi'_f(F, r) = \max\left[\sum_{e\in\delta(x)} r(e),\ \max_{i=1,\ldots,p}\left\{\chi'_f(F[V(K^i) \cup \{x\}], r_i)\right\}\right].$$

As in the previous section, it will be notationally more convenient to think of a fractional edge coloring of a weighted graph as a schedule. Other than in the previous section, our schedule will now be a function that is defined on a continuous-time range. Let us make some definitions. For $T \geq 0$, a *schedule* (of length $T$) is a piecewise constant function $S: [0, T] \to \mathcal{M}_G$. For $e \in E(G)$, let $T_S(e) = \int_0^T \mathbb{1}(e \in S(t))\mathrm{d}t$ (where $\mathbb{1}$ is the indicator function). Informally speaking, $T_S(e)$ is the total amount of time schedule $S$ spends on edge $e$. Likewise, for vertices $u, v \in V(G)$, let $T_S(u, v) = \int_0^T \mathbb{1}(S(t)$ cover both $u$ and $v\mathrm{d}t$ and, for a matching $M \in \mathcal{M}_G$, let $T_S(M) = \int_0^T \mathbb{1}(S(t) = M)\mathrm{d}t$. A schedule $S$ is said to be *r-feasible (for G)* if $T_S(e) = r(e)$ for all $e \in E(G)$.

We state the following obvious result without a proof:

*Claim 11* $\chi'_f(G, r) \le t$ if and only if there exists an $r$-feasible schedule of length $t$ for $G$.

We have the following fractional version of König's edge-coloring theorem, which easily follows from König's original edge-coloring theorem by standard compactness arguments.

*Claim 12* Let $G$ be a bipartite graph, and let $r$ be a rate function for $G$. Then $\chi'_f(G, r) = \max_{u \in V(G)}\{r(u)\}$.

We start with the following continuous version of Lemma 6.1.

**Lemma 6.3** *Let $F$ be a bipartite graph on vertex sets $X, Y$ with $X = \{u, v\}$, and let $r$ be a rate function for $H$. Then there exists an $r$-feasible schedule $S$ for $F$ such that $T_S(u, v) = \tau$ if and only if $0 \le \tau \le r(u) + r(v) - \chi'_f(F, r)$. Moreover, if $0 \le \tau \le r(u) + r(v) - \chi'_f(F, r)$, then there exists an $r$-feasible schedule of length $r(u) + r(v) - \tau$ such that $T_S(u, v) = \tau$.*

The previous claim allows us to deal with the blocks of GOLoP graphs.

**Lemma 6.4** *Let $G$ be a graph in GOLoP(b), and let $r$ be a rate function for $G$. Then, $\chi'_f(G, r)$ can be determined in $O(|V(G)|)$ time.*

It now follows immediately from Claim 2 and Lemma 6.4 that MATCH can be solved in linear time for GOLoP graphs.

**Theorem 6.3** *Let $b \ge 1$. Let $G$ be a GOLoP(b) graph, and let $r$ be a rate function for $G$. The problem* MATCH *can be solved in $O(|V(G)|)$ time.*

### *6.6.4* FIND-K-MATCH *for GOLoP Graphs*

We have described, in Sect. 6.6.2, a linear-time algorithm to verify whether the chromatic index of a GOLoP multigraph $(G, \lceil kr \rceil)$ is at most $k$. In this section, we show that this algorithm can be turned into a quadratic-time algorithm to find a schedule of length $k$ for $G$ with respect to $r$, if such a schedule exists. In order to make the computation efficient, we store each schedule as a set of pairs $(M, \ell)$, where $M$ is a matching in $G$ and $\ell$ is the number of occurrences of $M$ in the schedule.

We start by obtaining a pseudo-schedule $\mathcal{Z}$ being a collection of schedules for all blocks of $G$ and then show how to assemble the schedule for $G$ using $\mathcal{Z}$. We begin with the details of $\mathcal{Z}$. Let $G$ be a graph, let $B^1, \ldots, B^q$ be the blocks of $G$, and let $C_1^j, \ldots, C_{p_j}^j$ be the sets of non-adjacent clones in block $B^j$. For $j \in [q]$ and $i \in [p_j]$, let $(B^{j\prime}, C_i^j)$ be the collapsed graph associated with $B^j$, and let $u_i^j, v_i^j$ be the common neighbors of the vertices in $C_i^j$. Finally, let $F_i^j = G[C_i^j \cup \{u_i^j, v_i^j\}]$. Define a *pseudo-schedule* for $G$ as $\mathcal{Z} = \big((Z^1, Z_1^1, \ldots, Z_{p_1}^1), \ldots, (Z^q, Z_1^q, \ldots, Z_{p_q}^q)\big)$, where $Z^j$ is a schedule for $B^{j\prime}$ and $Z_i^j$ is a schedule for $F_i^j$. Note that each $Z^j$ and $Z_i^j$ is a list of pairs $(M, \ell)$, with $M$ being a matching in the relevant graph and $\ell$ being the number of occurrences of $M$ in the corresponding schedule. Define

$$c(\mathcal{Z}, x) = \sum_{j\in[q]} \sum_{\substack{(M,\ell)\in Z^j\\ \text{and } M \text{ covers } x}} \ell, \quad \text{for any cut-vertex } x \text{ of } G,$$

$$|Z^j| = \sum_{(M,\ell)\in Z^j} \ell, \quad \text{for } j \in [q],$$

$$|\mathcal{Z}| = \max\left\{\max_{j\in[q]} |Z^j|, \quad \max\{c(\mathcal{Z}, x) \colon x \text{ is a cut-vertex of } G\}\right\}.$$

Here, $|Z^j|$ denotes the length of schedule $Z^j$, $c(\mathcal{Z}, x)$ denotes the number of matchings in $\mathcal{Z}$ that cover $x$, and $|\mathcal{Z}|$ denotes the length of pseudo-schedule $\mathcal{Z}$.

First we give an algorithm that computes $\mathcal{Z}$ of length $k$ in linear time.

---

**Algorithm 1:** Find-pseudo-schedule

---

Input: A GOLoP graph $G$ and a rate function $r$ for $E(G)$.
Output: Either a pseudo-schedule $\mathcal{Z}$ of length at most $k$ for $G$ with respect to $r$ or the message that no such pseudo-schedule exists.

1. Find the block decomposition $B^1, \ldots, B^q$ of $G$.
2. If any cut-vertex $x \in V(G)$ satisfies $\sum_{e\in\delta(x)} \lceil r(e)k \rceil > k$, then terminate because no schedule of length at most $k$ exists for $G$ with respect to $r$.
3. For each $j \in [q]$:
   a. Construct $H^j = (B^j, \lceil kr_j \rceil)$, where $r_j = r|_{E(B_j)}$.
   b. Identify the sets $C_1^j, \ldots, C_{p_j}^j$ and the vertices $u_i^j$ and $v_i^j$, $i \in [p_j]$.
   c. Solve the integer linear program (6.4) corresponding to $B^{j\prime}$ to construct $Z^j$. If the optimal value of the IP is greater than $k$, terminate because no schedule of length at most $k$ exists for $G$ with respect to $r$.
   d. For $i \in [p_j]$, construct a schedule $Z_i^j$ of length $\deg_{F_i^j}(u_i^j) + \deg_{F_i^j}(v_i^j) - \tau_i^j$ for $F_i^j$ such that $T_{Z_i^j}(u_i^j, v_i^j) = \tau_i^j$.

---

Before we prove the correctness of this algorithm, we need a small technical lemma that allows us to easily construct the sets $C_1, \ldots, C_p$.

**Lemma 6.5** *Let $G$ be a graph, let $c \in V(G)$ be such that $\deg_G(c) = 2$, and let $u, v$ be the neighbors of $c$. Let $G'$ be obtained from $G$ by non-adjacent cloning of $c$. If $G'$ has no cut-vertex, then either $\min\{\deg_{G'}(u), \deg_{G'}(v)\} \geq 3$, or $G'$ is a 4-cycle.*

***Proof*** Let $c_1 \neq c$ be a clone of $c$. If $V(G') = \{u, v, c, c_1\}$, then $G'$ is a 4-cycle and the claim holds. So we may assume that there exists $v' \in V(G') \setminus \{u, v, c, c_1\}$. Because $G'$ is 2-connected, it follows that there exist paths $P_1, P_2$ from $v'$ to $c_1$ in $G'$ such that $V(P_1) \cap V(P_2) = \{v', c_1\}$. Since $N_{G'}(c_1) = \{u, v\}$, it follows that one of $P_1, P_2$ contains $u$ and not $v$, and the other contains $v$ and not $u$. This implies that $\deg_{G'}(u) \geq 3$ and $\deg_{G'}(v) \geq 3$. □

This allows us to prove the correctness and the complexity of algorithm FIND-PSEUDO-SCHEDULE.

**Lemma 6.6** *Let $b \geq 1$, let $G$ be a GOLoP($b$) graph and let $r$ be a rate function for $G$. Then, algorithm* FIND-PSEUDO-SCHEDULE *either finds a pseudo-schedule of length at most $k$ for $G$ with respect to $r$, or determines that no such pseudo-schedule exists. Its running time is $O(|V(G)|)$.*

***Proof*** See Dereniowski et al. [7] for the proof. □

Now we prove that, given a pseudo-schedule $\mathcal{Z}$ of length at most $k$, we can assemble a sequence of at most $k$ matchings in $G$. Furthermore, this sequence requires at most $O(|V(G)|)$ distinct matchings in $G$, which permits a succinct encoding of schedule of $G$ that specifies a list of pairs $(M, \ell)$, with $M$ being a matching in the $G$ and $\ell$ being the number of occurrences of $M$ in the schedule.

We begin by giving an informal description of the method. Consider a block-cutpoint tree $T$ of $G$ rooted at any vertex. We start by ordering the blocks of $G$, which is equivalent to ordering the vertices of $T$, so that for each $i \geq 1$ the first $i$ blocks of the order induce a connected subgraph of $G$. This can be achieved by, for example, taking any depth-first-search ordering of the vertices in $T$. Recall that $\mathcal{Z}$ gives a feasible schedule, or equivalently a sequence of matchings, for each block of $G$. Each matching of $G$ is then assembled from the block matchings as follows. We start with an empty matching $M$ of $G$ and then process the blocks according to their previously fixed order to construct $M$ iteratively. If the cut-vertex connecting the current block with its "parent" block is an end vertex of an edge already in $M$, then we try to find a matching in the current block not saturating the cut-vertex. If such a matching exists, then its edges are added to $M$. If there is no edge in $M$ that saturates the cut-vertex, then we take a matching, if any, in the current block that saturates the cut-vertex and we add its edges to $M$. The matchings in the blocks are obtained from the corresponding schedules for the blocks. When all blocks are processed, then we add the matching $M$ thus assembled to the output sequence of matchings and simultaneously update the block schedules corresponding to the block matchings used by $M$. We then proceed to the next iteration to find the next matching of $G$. In Theorem 6.4 we prove that this method gives the desired sequence of $k$ matchings for $G$.

We are now ready to give a more formal pseudo-code of an algorithm that constructs the desired sequence of matchings from the pseudo-schedule $\mathcal{Z}$.

---

**Algorithm 2:** Recover-schedule

---

Input: A GOLoP graph $G$, the block decomposition of $G$, and a pseudo-schedule $\mathcal{Z}$ of length $k \leq |\mathcal{Z}|$.
Output: A schedule $S$ for $G$ given as a set $\{(M, \ell) \colon M \in \mathcal{M}_G, \ell \in \mathbb{Z}_+\}$.

1. Order the blocks $B^1, \ldots, B^q$ so that $G\big[\bigcup_{l=1}^{j-1} V(B^l)\big]$ is connected for all $j \in [q]$. Set $S := \emptyset$.
2. While there exists $j \in [q]$ such that $Z^j \neq \emptyset$, do the following:

   a. Set $M := \emptyset$.
   b. For each $j \in [q]$ with $Z^j \neq \emptyset$, do the following:

      - If $j = 1$, choose $M^1$ such that $(M^1, \ell) \in Z^1$ and set $\ell' := \ell$.
      - If $j > 1$, then let $x^j$ be the unique cut-vertex of $G\big[\bigcup_{l=1}^{j-1} V(B^l)\big]$ that lies in $V(B^j)$. If $M$ saturates $x^j$, then choose any $M^j$ such that $(M^j, \ell) \in Z^j$ does not saturate $x^j$, if any. If $M$ does not saturate $x^j$, then choose $M^j$ that saturates $x^j$, if any. If no $M^j$ saturating $x^j$ exists, then take any $M^j$. If no such $M^j$ exists, go to step (2b) and consider the next value of $j$; otherwise, set $\ell' := \min\{\ell', \ell\}$.
      - For all $i \in [p_j]$ such that $M^j \cap \{u_i^j v_i^j, u_i^j a_i^j, v_i^j a_i^j\} \neq \emptyset$, do the following:

        – If $u_i^j v_i^j \in M^j$, then choose $(M_i^j, \ell) \in Z_i^j$ such that $M_i^j$ covers both $u_i^j$ and $v_i^j$ and set $\ell' := \min\{\ell', \ell\}$.
        – If $u_i^j a_i \in M^j$, then choose $(M_i^j, \ell) \in Z_i^j$ such that $M_i^j$ covers $u_i^j$ but not $v_i^j$ and set $\ell' := \min\{\ell', \ell\}$.
        – If $v_i^j a_i \in M^j$, then choose $(M_i^j, \ell) \in Z_i^j$ such that $M_i^j$ covers $v_i^j$ but not $u_i^j$ and set $\ell' := \min\{\ell', \ell\}$.
        – Set $M^j := \Big(M^j \setminus \{u_i^j v_i^j, u_i^j a_i^j, v_i^j a_i^j\}\Big) \cup M_i^j$.

      - Let $M := M \cup M^j$.

   c. For each $M^j$ chosen in step (2b), replace $(M^j, \ell)$ with $(M^j, \ell - \ell')$ in $Z^j$, and if $\ell - \ell' = 0$, then delete $(M^j, \ell - \ell')$ from $Z^j$. Similarly, for each $M_i^j$ chosen in step (2b), replace $(M_i^j, \ell)$ with $(M_i^j, \ell - \ell')$ in $Z_i^j$, and delete $(M_i^j, \ell' - \ell)$ from $Z_i^j$ if $\ell - \ell' = 0$. Let $S := S \cup \{(M, \ell')\}$.

---

Notice that since the number of distinct matchings in $Z^j$ is constant and the number of matchings $Z_i^j$ is $O(|V(F_i^j)|)$, it follows that each iteration of the main loop can be done in $O(|V(G)|)$ time. Thus, by Lemma 6.6 and by the fact that each iteration "eliminates" at least one matching in a block of $G$, the overall complexity of algorithm Recover-Schedule is $O(|V(G)|^2)$.

**Theorem 6.4** *Let $b \geq 1$, and let $G$ be a GOLoP($b$) with a rate function $r$. There exists a $O(|V(G)|^2)$-time algorithm that finds a schedule $S$ for $G$. Moreover, the number of pairwise different matchings in $S$ is $O(|V(G)|)$.*

***Proof*** The schedule $S$ is a result of the execution of FIND-PSEUDO-SCHEDULE for $G$ and $r$ (which produces a pseudo-schedule $\mathcal{Z}$) and the execution of RECOVER-SCHEDULE for $\mathcal{Z}$, $G$, and the block decomposition.

Now, observe that it follows from the ordering of the blocks $B^1, \ldots, B^q$ that for each $j > 1$, there exists a unique cut-vertex $x^j$ of $\bigcup_{l=1}^{j-1} V(B^l)$ that lies in $V(B^j)$. To prove the correctness of this algorithm, it suffices to show that after each iteration we have $|\mathcal{Z}| \leq k - t$, where $t \in [k]$ is the sum of the multiplicities of the matchings added to $S$ prior to and in this iteration. We prove this by induction on the number of iterations. The statement is clearly true at the beginning of the first iteration. Now let $\bar{\mathcal{Z}}$ be the pseudo-schedule, and let $S$ be the schedule at the beginning of an iteration. Furthermore, let $t = \sum_{(M,\ell)\in S} \ell$, and let $\mathcal{Z}$ be the pseudo-schedule at the end of the iteration. We denote by $\ell'$ the multiplicity of the matching chosen in the iteration.

First, suppose for a contradiction that $|Z^j| \geq k - t + 1$ for some $j \in [q]$. Since by induction $|\bar{Z}^j| \leq k - t + \ell'$, it follows that $|Z^j| = |\bar{Z}^j|$. Hence no matching $M^j$ was chosen in this iteration. Thus, $j > 1$. This means that for some $j' < j$, $M^{j'}$ saturates $x^j$, and every matching in $\bar{Z}^j$ saturates $x^j$, because $\bar{Z}^j \neq \emptyset$. But this implies that $c(\bar{\mathcal{Z}}, x^j) \geq \ell' + |\bar{Z}^j| > k - t + \ell'$, contrary to the inductive hypothesis. Thus, $|Z^j| \leq k - t$ for all $j \in [q]$.

Second, suppose for a contradiction that $c(\mathcal{Z}, x) \geq k - t + 1$ for some cut-vertex $x$ of $G$. By induction, we have $c(\bar{\mathcal{Z}}, x) \leq k - t + \ell'$. Therefore, $c(\mathcal{Z}, x) = c(\bar{\mathcal{Z}}, x)$. This implies that none of the matchings $M^j$ chosen in the iteration saturates $x$. Therefore, all matchings $M$ that saturate $x$ are already included in $S$ and hence $c(\mathcal{Z}, x) = 0$. This, however, contradicts $c(\bar{\mathcal{Z}}, x) \geq k - t + 1 > 0$. Thus, $c(\bar{\mathcal{Z}}, x) \leq k - t$ for every cut-vertex $x$ of $G$. This proves that $|\mathcal{Z}| \leq k - t$. This completes the proof of Lemma 6.4. □

## 6.7 An Upper Bound on the Schedule Length

In this section we are interested in finding, for a given graph $G$ and rates $r$ for which K-MATCH is affirmative for some $k$, an upper bound for the length of a shortest schedule. To get a handle on this bound, we will focus on the following property. We say that a graph $G$ has the *lcd-property (with constant $C$)* if for every rate function $r$, it holds that if K-MATCH has an affirmative answer for some $k$, then K-MATCH with $k = Cd$ also has an affirmative answer, where $d$ is the least common denominator for the rate function $r$. Notice that, in this definition, the value of $C$ depends only on the graph $G$ and not on the rate function $r$. Notice also that a graph having the lcd-property with constant $C$ does not necessarily have the lcd-property with constant $C + 1$. We do, however, have the following property.

*Property 6.1* Let $G$ be a graph having the lcd-property with constant $C$. Then $G$ also has the lcd-property with constant $tC$ for any positive integer $t$.

We first show that the lcd-property is equivalent to a property that relates the fractional chromatic index and the chromatic index of multigraphs associated with $G$.

*Claim 13* Let $G$ be a graph, and let $C \geq 1$ be an integer. The following two statements are equivalent:

1. $G$ has the lcd-property with constant $C$;
2. $\lceil \chi'_f(G, \mathrm{mp}) \rceil = \lceil \frac{1}{C}\chi'(G, C \cdot \mathrm{mp}) \rceil$ for every function mp.

***Proof*** (1) $\Longrightarrow$ (2): Let $(G, \mathrm{mp})$ be given. Since $\chi'(H) \geq \chi'_f(H)$ for every multigraph $H$, we have

$$\left\lceil \frac{1}{C}\chi'(G, C \cdot \mathrm{mp}) \right\rceil \geq \left\lceil \frac{1}{C}\chi'_f(G, C \cdot \mathrm{mp}) \right\rceil = \left\lceil \chi'_f(G, \mathrm{mp}) \right\rceil.$$

To prove the inequality in the other direction, let $p = \lceil \chi'_f(G, \mathrm{mp}) \rceil$. Set $r = \mathrm{mp}/p$ and $d = p/\gcd(p, \mathrm{mp})$. Then, $d$ is the least common denominator of $r$, and we may write $p = td$ for some integer $t \geq 1$. We have

$$\chi'_f(G, r) = \chi'_f\left(G, \frac{\mathrm{mp}}{p}\right) = \frac{\chi'_f(G, \mathrm{mp})}{p} \leq 1.$$

Thus, by Claim 2, it follows that K-MATCH has an affirmative answer for some $k$. By statement (1), K-MATCH with $k = Cd$ has an affirmative answer. This implies that

$$\frac{1}{C}\chi'(G, C \cdot \mathrm{mp}) \leq \frac{t}{C}\chi'\left(G, \frac{C \cdot \mathrm{mp}}{t}\right) = \frac{t}{C}\chi'(G, Cdr) \leq \frac{tCd}{C} = p.$$

Since $p$ is an integer, it follows that in fact $\lceil \frac{1}{C}\chi'(G, C \cdot \mathrm{mp}) \rceil \leq p$. This proves that (2) holds.

(2) $\Longrightarrow$ (1): Let $r$ be such that K-MATCH has an affirmative answer for some $k$. It follows from Claim 2 that $\chi'_f(G, r) \leq 1$. Let $d$ be the least common denominator for $r$. It follows that $\chi'_f(G, d \cdot r) \leq d$. Therefore, by (2),

$$\left\lceil \frac{1}{C}\chi'(G, Cd \cdot r) \right\rceil = \left\lceil \chi'_f(G, d \cdot r) \right\rceil \leq d,$$

which implies that $\frac{1}{C}\chi'(G, Cd \cdot r) \leq d$, as required. □

In Sect. 6.7.1, we will prove that every OLoP graph has the lcd-property with constant $C = 1$. However, the following example shows that not every graph has the lcd-property with constant 1.

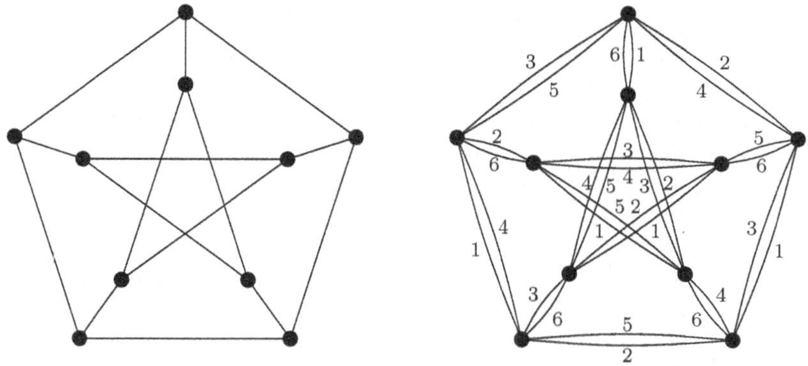

(a) (b)

**Fig. 6.3** (**a**) The Petersen graph $\widetilde{P}$; (**b**) a 6-edge coloring of $\widetilde{P}_2$

*Example 6.1* Let $\widetilde{P}$ be the Petersen graph (see Fig. 6.3a), and let $r(e) = \frac{1}{3}$ for each $e \in E(\widetilde{P})$. The answer to K-MATCH is NO for each $k \leq 3$, because $(\widetilde{P}, \lceil \frac{k}{3} \rceil)$ is isomorphic to $\widetilde{P}$ and $\chi'(\widetilde{P}) = 4$. Note that for each $k = 4, 5, 6$, the multigraph $(\widetilde{P}, \lceil \frac{k}{3} \rceil)$ is isomorphic to $\widetilde{P}_2$, i.e., the graph obtained by replacing each edge in $\widetilde{P}$ by two parallel edges. It is enough to argue that $\chi'(\widetilde{P}_2) \geq 6$. This, however, follows from the fact that each matching in the Petersen graph consists of at most 5 edges, and consequently, $E(\widetilde{P}_2)$ cannot be partitioned into at most 5 matchings, each of size at most 5, because $|E(\widetilde{P}_2)| = 30$. Thus, the answer to K-MATCH is NO for each $k \leq 5$.

Figure 6.3b gives a 6-edge coloring of $\widetilde{P}_2$, which implies that the answer to K-MATCH with $k = 2d$ is YES, where $d = 3$ is the least common denominator of the rates. This shows that for the Petersen graph, $C \geq 2$.

With the Petersen graph in mind, the following question arises: is it true that every graph has the lcd-property for some, graph but not rate function dependent, constant $C$? And if so, what is the smallest value of $C$? In Sect. 6.7.2, we will prove that the former question always has an affirmative answer, and that, for fixed $b$, every GOLoP($b$) graph satisfies the lcd-property with some constant that depends only on $b$. Finally, with Claim 13 in mind, we point out that the following conjecture of Seymour implies that every graph has the lcd-property with constant $C \leq 2$.

*Conjecture 6.1 (Seymour [30])* For every multigraph $H$, it holds $\lceil \chi'_f(H) \rceil = \lceil \frac{1}{2}\chi'(H_2) \rceil$, where $H_2$ is the multigraph obtained from $H$ by replacing each edge with two parallel edges.

This conjecture follows easily from the work of Plantholt and Tipnis [29] for graphs on at most 10 vertices, which we now show.

**Theorem 6.5** *Let $H = (G, mp)$ be a multigraph such that $|V(G)| \leq 10$. Then,* $\left\lceil \frac{1}{2}\chi'(G, 2mp) \right\rceil = \left\lceil \chi'_f(G, mp) \right\rceil$.

***Proof*** We first claim that is suffices to prove that $\chi'(H) = \left\lceil \chi'_f(H) \right\rceil$ for any $H = (G, \text{mp})$ such that $|V(G)| \leq 10$ and $\text{mp}(e)$ is even for all $e \in E(G)$. Indeed, if $\chi'(G, 2\text{mp}) = \left\lceil \chi'_f(G, 2\text{mp}) \right\rceil$, then

$$\left\lceil \frac{1}{2}\chi'(G, 2\text{mp}) \right\rceil = \left\lceil \frac{1}{2} \left\lceil \chi'_f(G, 2\text{mp}) \right\rceil \right\rceil = \left\lceil \frac{1}{2}\chi'_f(G, 2\text{mp}) \right\rceil = \left\lceil \chi'_f(G, \text{mp}) \right\rceil,$$

as required. Here, we use the fact that $\lceil \lceil x \rceil /2 \rceil = \lceil x/2 \rceil$ for all $x \in \mathbb{R}$.

Thus, let $H = (G, \text{mp})$ be a multigraph such that $|V(G)| \leq 10$ and $\text{mp}(e)$ is even for all $e \in E(G)$. Suppose for a contradiction that $\chi'(H) \neq \left\lceil \chi'_f(H) \right\rceil$. Thus, by Theorem 1.2 $\chi'(H) \neq \max\{\Delta(H), \lceil t(H) \rceil\}$. Then, it follows from Theorem 2 of Plantholt and Tipnis [29] that there exist a multigraph $H' = (G', \text{mp}')$ and a vertex $v \in V(G')$ such that (i) $G'$ is isomorphic to the Petersen graph, (ii) $H'$ is regular, (iii) there exists a 5-cycle $C$ in $H'$ that has an odd number of edges, (iv) $H$ is a submultigraph of $H'$, and (v) $H' - v$ is a submultigraph of $H$. Conditions (iv) and (v) imply that $\text{mp}(e) = \text{mp}'(e)$ for all $e \in E(G) \setminus \delta(v)$. Now let $u \in V(G) \setminus N(v)$ ($u$ exists because $G'$ is isomorphic to the Petersen graph). Since $\text{mp}(e)$ is even for all $e \in \delta(u)$ and $H'$ is regular, it follows that $\deg_{H'}(u)$ is even. Next, consider the 5-cycle $C$. Clearly, since $\text{mp}(e) = \text{mp}'(e)$ is even for all $e \in E(G) \setminus \delta(v)$, $C$ contains an edge $vv'$ such that $\text{mp}'(vv')$ is odd. But because $\text{mp}'(e)$ is even for all $e \in \delta(v') \setminus \{vv'\}$, this implies that $\deg_{H'}(v')$ is odd, a contradiction. This proves the theorem. □

By this theorem, all graphs with at most 10 vertices have the lcd-property with constant $C$ either 1 or 2.

### 6.7.1 OLoP Graphs

In this section, we prove that every OLoP graph has the lcd-property with $C = 1$. Our approach is to prove (2) in Claim 13 with $C = 1$ for any OLoP graph $G$. That is we need to prove the following theorem.

**Theorem 6.6** *For every OLoP multigraph $H = (G, mp)$, it holds that $\chi'(G, mp) = \left\lceil \chi'_f(G, mp) \right\rceil$.*

Because of Claims 9 and 10, it suffices to consider blocks of OLoP graphs. Indeed, suppose that Theorem 6.6 holds for blocks, and let $x$ be a cut-vertex in an OLoP multigraph $H$. Let $K^1, \ldots, K^p$ be the connected components of $H - x$. Then,

$$\chi'(H) = \max\left[\deg(x), \max_{i=1,\dots,p}\left\{\chi'(H[V(K^i)\cup\{x\}])\right\}\right]$$
$$= \max\left[\deg(x), \max_{i=1,\dots,p}\left\lceil\chi'_f(H[V(K^i)\cup\{x\}])\right\rceil\right] = \left\lceil\chi'_f(H)\right\rceil.$$

Thus, we concentrate on the blocks of OLoP graphs. We show that for a multigraph $H = (G, \text{mp})$ such that $G$ is a block of an OLoP graph, it holds that $\chi'(H) = \left\lceil\chi'_f(H)\right\rceil$. The proof relies on the observation that all blocks of an OLoP graph, with the exception of few small blocks with $|V(H)| \leq 5$, are either bipartite or nearly bipartite or can be easily "reduced" to a nearly bipartite graphs. We begin by briefly reviewing the main results for nearly bipartite multigraphs that are of interest to us.

A multigraph $H = (G, \text{mp})$ is called *nearly bipartite* if there exists a vertex $v \in V(G)$ such that $G - v$ is bipartite.

Eggan and Plantholt [10] proved that $\chi'(H) = \max\{\Delta(H), \lceil t(H)\rceil\}$ for every nearly bipartite multigraph $H$. Thus, by Theorem 1.2 we readily obtain the following.

**Theorem 6.7 (Eggan and Plantholt [10])** *If $H$ is a nearly bipartite multigraph, then $\chi'(H) = \left\lceil\chi'_f(H)\right\rceil$.*

Moreover, the following result was shown in Plantholt [28] (see also Plantholt and Tipnis [29]).

**Theorem 6.8 (Plantholt [28])** *If $H$ is a multigraph with $|V(H)| \leq 8$, then $\chi'(H) = \left\lceil\chi'_f(H)\right\rceil$.*

In the following two results, Claims 14 and 15, we will use these two theorems to prove Theorem 6.6. We start with the easier case.

*Claim 14* Every multigraph $H$ of the $\mathcal{B}_2$ type satisfies $\chi'(H) = \left\lceil\chi_f(H)\right\rceil$.

***Proof*** Let $H = (G, \text{mp})$ be a multigraph of the $\mathcal{B}_2$ type. First, if $G$ is isomorphic to $K_2$, $K_3$, or $K_4$, then the claim holds by Theorem 6.8. Next, if $G$ is isomorphic to $K_{2,t}$ $(t \geq 2\}$, then $H$ is bipartite and the result follows by König's edge-coloring theorem [22]. Finally, let $G$ be isomorphic to $K^+_{2,t}$ $(t \geq 2\}$. Let $u, v$ be the two vertices on the side of cardinality 2. Then $H - u$ is isomorphic to $K_{1,t}$. Hence $H$ is nearly bipartite and the result follows from Theorem 6.7. □

This leaves blocks of the $\mathcal{B}_1$ type.

*Claim 15* Every multigraph $H$ of the $\mathcal{B}_1$ type satisfies $\chi'(H) = \left\lceil\chi'_f(H)\right\rceil$.

***Proof*** First, consider a multigraph $H$ of the $\mathcal{B}_1$ type with $|V(H)| \geq 6$. Let us start with the case where $H$ is constructed from a 7-cycle.

*Claim 16* If $H$ contains a cycle of length seven, then $H$ is nearly bipartite. □

***Proof*** Consider $H$, and let $c_1 - c_2 - \cdots - c_7 - c_1$ be the vertices of a cycle of length seven in $H$. From the definition of graphs of the $\mathcal{B}_1$ type, it follows that we may assume that all pairs of vertices $c_i, c_j$ with $|i - j| \geq 2$ are non-adjacent except possibly $\{c_1, c_4\}$, $\{c_1, c_5\}$, and $\{c_4, c_7\}$. If both $c_1$ and $c_4$ have clones, then $H$ is a multi-cycle of length seven, and thus the result holds (for instance, $H - c_2$ is bipartite). From the symmetry, we may assume now that no vertex is a clone of $c_1$. We claim that $H - c_1$ is bipartite. Let $C(c_i)$ be the set of clones of vertex $c_i$, for $i = 2, \ldots, 7$, and let $C[c_i] = C(c_i) \cup \{c_i\}$. Notice that some of the sets $C(c_i)$ are necessarily empty, since only vertices of degree 2 may admit clones. Then the bipartition $V_1, V_2$ of $H - c_1$ is obtained as follows: $V_1 = \{C[c_3], C[c_5], C[c_7]\}$ and $V_2 = \{C[c_2], C[c_4], C[c_6]\}$. □

By Claim 16 and Theorem 6.7, we may assume that $H$ is not nearly bipartite and is constructed from a 5-cycle, say $c_1 - c_2 - \cdots - c_5 - c_1$. If two vertices of the cycle, say $c_1$ and $c_3$, admit clones, then $H - c_2$ is bipartite and thus $H$ is nearly bipartite, a contradiction. Thus, since $|V(H)| \geq 6$, exactly one vertex of the cycle admits clones. We may assume without loss of generality that $c_1$ admits clones in $H$ (see Fig. 6.4a). Furthermore, the pairs $\{c_2, c_5\}$, $\{c_2, c_4\}$, and $\{c_3, c_5\}$ are adjacent, because otherwise $H$ would be nearly bipartite. Notice that since $c_1$ admits clones, the pairs $\{c_1, c_3\}$ and $\{c_1, c_4\}$ are non-adjacent. Let $a = \text{mp}(c_2c_5)$ and $b = \text{mp}(c_3c_4)$. We distinguish two cases:

(i) $a \geq b$: Consider the graph $H'$ obtained from $H$ by deleting all edges between $c_3$ and $c_4$ (see Fig. 6.4b).
(ii) $a < b$: Consider the graph $H'$ obtained from $H$ by deleting all edges between $c_2$ and $c_5$ and deleting $a$ edges between $c_3$ and $c_4$ (see Fig. 6.4c).

We obtain the following result.

*Claim 17* $H'$ is nearly bipartite and

$$\chi'(H') = \begin{cases} \chi'(H) & \text{if } a \geq b, \\ \chi'(H) - a & \text{if } a < b, \end{cases} \quad \text{and } \chi'_f(H') = \begin{cases} \chi'_f(H) & \text{if } a \geq b, \\ \chi'_f(H) - a & \text{if } a < b. \end{cases}$$

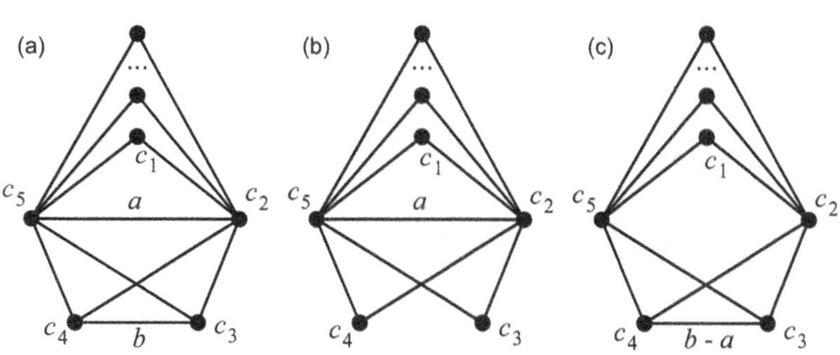


**Fig. 6.4** (**a**) The block $H$; (**b**) $H'$ when $a \geq b$; (**c**) $H'$ when $a < b$

***Proof*** Suppose that $a \geq b$. Clearly $\chi'(H') \leq \chi'(H)$. Consider now an optimal edge coloring of $H'$. Notice that the colors used to color the edges $(c_2c_5)^1, \ldots, (c_2c_5)^a$ are not used to color any other edges in $H'$. Thus we obtain an edge coloring of $H$ from the edge coloring of $H'$ by using the colors of the edges $(c_2c_5)^1, \ldots, (c_2c_5)^a$ to color the edges $(c_3c_4)^1, \ldots, (c_3c_4)^b$. This is possible since $a \geq b$. Thus we obtain a feasible edge coloring of $H$ using $\chi'(H')$ colors. Hence $\chi'(H') = \chi'(H)$. Clearly $H'$ is nearly bipartite since, for instance, $H' - c_2$ is bipartite.

Now suppose that $a < b$. Consider an optimal edge coloring of $H$. By re-coloring some of the edges $(c_3c_4)^1, \ldots, (c_3c_4)^b$, if necessary, we may assume that the edges $(c_2c_5)^1, \ldots, (c_2c_5)^a$ are colored with colors that are also used to color the first $a$ edges of $(c_3c_4)^1, \ldots, (c_3c_4)^b$. These colors are not used for any other edges in the graph. Now in $H'$, we may assume without loss of generality that exactly those $a$ edges of $(c_3c_4)^1, \ldots, (c_3c_4)^b$ have been deleted. Thus we obtain a feasible edge coloring of $H'$ with $\chi'(H) - a$ colors. We claim that this coloring is optimal. Indeed, if $\chi'(H') < \chi'(H) - a$, then we would obtain a feasible edge coloring of $H$ with strictly less than $\chi'(H)$ colors by coloring the edges between vertices $c_2$ and $c_5$ as well as the added edges between vertices $c_3$ and $c_4$ with $a$ new colors, a contradiction. Clearly $H'$ is nearly bipartite since, for instance, $H' - c_3$ is bipartite.

The proof for the fractional chromatic index of $H$ is similar and thus it is omitted. This proves Claim 17. □

Since both $a$ and $b$ in the definition of $H'$ are integers, Claim 17 implies that $\chi'(H) = \left\lceil \chi'_f(H) \right\rceil$, proving Claim 15 for $|V(H)| \geq 6$. Finally, Claim 15 holds for $|V(H)| = 5$ by Theorem 6.8. This proves the lemma. □

We just showed that for a multigraph $H = (G, \mathrm{mp})$ such that $G$ is a block of an OLoP graph, it holds that $\lceil \chi'_f(H) \rceil = \chi'(H)$. This allows us to finish up the proof of Theorem 6.6.

*Proof of Theorem 6.6* Let $H = (G, \mathrm{mp})$ be an OLoP multigraph. By Claim 9, $\chi'(G, \mathrm{mp})$ equals either the maximum cut-vertex degree or the maximum block chromatic index. The same holds for $\chi'_f(G, \mathrm{mp})$. The maximum cut-vertex degree is integral, and by Claims14 and 15, we have $\lceil \chi'_f(B) \rceil = \chi'(B)$ for each block of $H$. This proves that $\lceil \chi'_f(H) \rceil = \chi'(H)$ are required. □

### *6.7.2 GOLoP Graphs*

For any fixed integer $b \geq 1$, every GOLoP($b$) has the lcd-property with constant $C$, where $C$ only depends on $b$.

**Theorem 6.9** *There exists a function $C(b)$ such that every GOLoP($b$) graph has the lcd-property with constant at most $C(b)$.*

***Proof*** Please see Dereniowski et al. [7] for a complete proof. □

### 6.7.3 Minimizing the Schedule Length

After having found a linear (resp., a quadratic) algorithm for the problem K-MATCH (resp., FIND-K-MATCH), a natural next problem to take on is MIN-MATCH. That is, the problem of finding the smallest $k$ such that the answer to K-MATCH is affirmative. While trying to solve this problem, we run into two difficulties. The first difficulty lies in the fact that the smallest value of $k$ might be quite large. The second difficulty follows from the fact that K-MATCH having an affirmative answer does not necessarily imply that K-MATCH$[k']$ for $k' > k$ has an affirmative answer. Thus, a straightforward binary search does not work. It seems that the best we can do is a full search of all values of $k$ up to the upper bound that is given by Theorem 6.9. This leads to the following result.

*Claim 18* Let $b \geq 1$ be a constant integer. Let $G$ be a GOLoP($b$) graph, and let $r$ be a rate function for $G$. Then MIN-MATCH can be solved in $O(|V(G)|d)$ time, where $d$ is the least common denominator of the rates.

***Proof*** Let $C = C(b)$ as in Theorem 6.9. Notice that $C$ is a constant. For each block $B$ of $G$, let $K(B)$ be the set of values $k = 1, \ldots, Cd$ such that there exists a schedule of length $k$. These sets can be constructed in $O(|V(G)|d)$ total running time. Next, choose the smallest value $k$ such that $k \in K(B)$ for all blocks $B$, which also takes $O(|V(G)|d)$ time. □

Although the algorithm above is efficient in the theoretical sense, we do note that the constant $C$ given by Theorem 6.9 is quite large. However, Conjecture 6.1 suggests that actual constant is as small as 2. Moreover, in the case of OLoP graphs, Theorem 6.6 allows us to use $C = 1$.

## 6.8 Conclusions

While GOLoP($b$) graphs are important due to their natural connection to OLoP graphs, it would be interesting to derive the complexity results for more general classes of graphs; Golumbic [14]. This seems to be a promising direction for further research.

Finally, since the potential applications of the results presented in this chapter lie in the area of wireless networks where the centralized algorithms are difficult or impossible to implement, an interesting research direction is the investigation of the schedule makespan minimization considered here in a distributed setting.

# References

1. B. Birand, M. Chudnovsky, B. Ries, P. Seymour, G. Zussman, Y. Zwols, Analyzing the performance of greedy maximal scheduling via local pooling and graph theory, in *INFOCOM, 2010 Proceedings IEEE* (IEEE, 2010), pp. 1–9
2. A. Brzezinski, G. Zussman, E. Modiano, Enabling distributed throughput maximization in wireless mesh networks - a partitioning approach, in *Proc. ACM MOBICOM'06*, Sept. 2006
3. E.G. Coffman Jr., M.R. Garey, D.S. Johnson, A.S. LaPaugh, Scheduling file transfers. SIAM J. Comput. **14**, 744–780 (1985)
4. D. de Werra, Extensions of coloring models for scheduling purposes. Eur. J. Oper. Res. **92**, 474–492 (1996)
5. D. de Werra, J. Erschler, Open shop with some additional constraints. Graphs Combin. **12**, 81–93 (1996)
6. D. de Werra, N.V.R. Mahadev, U. Peled, Edge chromatic scheduling with simultaneity constraints. SIAM J. Disc. Math., 631–641 (1993)
7. D. Dereniowski, W. Kubiak, B. Ries, Y. Zwols, Optimal edge-coloring with edge rate constraints. Networks **63**, 165–182 (2013)
8. A. Dimakis, J. Walrand, Sufficient conditions for stability of longest queue first scheduling: second order properties using fluid limits. Adv. Appl. Probab. **38**(2), 505–521 (2006)
9. J. Edmonds, Maximum matching and a polyhedron with 0, 1-vertices. J. Res. Natl. Bur. Stand. **69**(1-2), 125–130 (1965)
10. L. Eggan, M. Plantholt, The chromatic index of nearly bipartite multigraphs. J. Combin. Theory B **40**(1), 71–80 (1986)
11. F. Eisenbrand, Fast integer programming in fixed dimension, in eds. by G. Di Battista, U. Zwick *Proceedings of the 11th Annual European Symposium on Algorithms*, volume 2832 of *LNCS* (Springer, 2003), pp. 196–207
12. S. Even, A. Itai, A. Shamir, On the complexity of timetable and multicommodity flow problems. SIAM J. Comput. **5**, 691–703 (1976)
13. R. Gandhi, M.M. Halldórsson, G. Kortsarz, H. Shachnai, Improved results for data migration and open shop scheduling. ACM Trans. Algorithms **2**, 116–129 (2006)
14. M.C. Golumbic, *Algorithmic Graph Theory and Perfect Graphs* (North-Holland Publishing Co., 2004)
15. J.L. Gross, J. Yellen, *Graph Theory and Its Applications* (CRC press, 2006)
16. M. Grötschel, L. Lovász, A. Schrijver, The ellipsoid method and its consequences in combinatorial optimization. Combinatorica **1**, 169–197 (1981)
17. B. Hajek, G. Sasaki, Link scheduling in polynomial time. IEEE Trans. Inf. Theory **34**, 910–917 (1988)
18. J.-H. Hoepman, Simple distributed weighted matchings. eprint cs.DC/0410047, Oct. 2004
19. I. Holyer, The NP-completeness of edge-colouring. SIAM J. Comput. **10**(4), 718–720 (1981)
20. C. Joo, X. Lin, N.B. Shroff, Performance limits of greedy maximal matching in multi-hop wireless networks, in *Proc. IEEE CDC'07*, Dec. 2007
21. S. Khuller, Y. Kim, Y.C. Wan, Algorithms for data migration with cloning, in *Proc. of the 22nd ACM Symposium on Principles od Database Systems*, pp. 27–36 (2003)
22. D. König, Über graphen und ihre anwendung auf determinantentheorie und mengenlehre. Mathematische Annalen **77**, 453–465 (1916)
23. F. Kuhn, T. Moscibroda, R. Wattenhofer, What cannot be computed locally! in *PODC*, pp. 300–309 (2004)
24. M. Leconte, J. Ni, R. Srikant, Improved bounds on the throughput efficiency of greedy maximal scheduling in wireless networks. IEEE/ACM Trans. Netw. **19**(3), 709–720 (2011)
25. A.N. Letchford, G. Reinelt, D.O. Theis, A faster exact separation algorithm for blossom inequalities. Integer Programm. Combin. Optim. 19–52 (2004)
26. X. Lin, N.B. Shroff, The impact of imperfect scheduling on cross-layer rate control in wireless networks. IEEE/ACM Trans. Netw. **14**(2), 302–315 (2006)

27. M.W. Padberg, M.R. Rao, Odd minimum cut-sets and b-matchings. Math. Oper. Res. **7**(1), 67–80 (1982)
28. M.J. Plantholt, An order-based upper bound on the chromatic index of a multigraph. J. Combin. Inf. Syst. Sci. **16**, 271–280 (1991)
29. M.J. Plantholt, S.K. Tipnis, The chromatic index of multigraphs of order at most 10. Discrete Mathematics **177**, 185–193 (1997)
30. P.D. Seymour, On multi-colourings of cubic graphs, and conjectures of Fulkerson and Tutte. Proc. Lond. Math. Soc.(3) **38**, 423–460. Citeseer, 1979
31. L. Tassiulas, A. Ephremides, Stability properties of constrained queueing systems and scheduling policies for maximum throughput in multihop radio networks. IEEE Trans. Automat. Contr. **37**(12), 1936–1948 (1992)
32. D.B. West, *Introduction to Graph Theory* (Prentice Hall Upper Saddle River, NJ, 2001)

# Chapter 7
# Proportionate and Ordered Open Shops

## 7.1 Introduction

The proportionate open shops have evolved along two parallel paths. The *machine-proportionate* open shops can be traced back to the paper by Dror [5] who refers to them as open shops with machine-dependent processing times. Naderi et al. [16] use the same terminology. The idea is that each job in the machine-proportionate open shop has the same length (total of its operation processing times), and the workload on each machine is shared equally by all jobs. The other path, the *job-proportionate* open shops was introduced by Koulamas and Kyparisis [10] who trace it back to Liu and Bulfin [15] where the concept of ordered open shops was introduced. The idea is that each machine in the job-proportionate has the same workload, and the same proportion of that workload on each machine belongs to job $J_i$. The main focus has been on the minimization of makespan which somewhat surprisingly turns out NP-hard. In fact the machine-proportionate and job-proportionate open shops are equivalent with respect to the minimization of makespan.

Dror [5], and Vakharia and Çatay [19] study total completion time for machine-proportionate open shops with two machines.

We use the notation introduced in Koulamas and Kyparisis [10] for the job-proportionate open shops, where $O3|\text{prpt}|C_{\max}$ denotes job-proportionate open shop makespan minimization on three-machines problem. The same notation will be used for machine-proportionate open shops. It should be clear from the context which model the notation refers to. The applications of proportionate open shops are given in Chap. 5. It is worth observing that open shops with unit-time operations discussed in Sect. 3.5 are both job-proportionate and machine-proportionate open shops.

 
W. Kubiak, *A Book of Open Shop Scheduling*, International Series in Operations Research & Management Science 325,
https://doi.org/10.1007/978-3-030-91025-9_7

## 7.2 Job-Proportionate Open Shops

A job-proportionate open shop is an open shop where $p_{i,h} = p_i > 0$ for each job $J_i$, $i = 1, \ldots, n$ and machine $M_h$, $h = 1, \ldots, m$. The $n \times m$ matrix of a job-proportionate open shop is shown below

$$\mathbb{P} = \begin{bmatrix} p_1 \cdots p_1 \\ p_2 \cdots p_2 \\ \cdot \;\cdots\; \cdot \\ p_i \;\cdots\; p_i \\ \cdot \;\cdots\; \cdot \\ p_n \cdots p_n \end{bmatrix}.$$

All machine loads of a job-proportionate open shop are equal, $L_1 = \cdots = L_m$, and equal to $L = \sum_{i=1}^{n} p_i$. The proportion $\frac{p_i}{L}$ of the total workload $L$ on each machine comes from job $J_i$. We assume that the jobs are ordered in non-decreasing order of processing times $p_1 \geq \cdots \geq p_n$, which is the same as non-decreasing order of the workload proportions. It is worth mentioning that the processing time matrix of job-proportional open shop admits a succinct encoding: One only needs the processing times $p_1, \ldots, p_n$ and the number of machines $m$ to specify the input.

### *7.2.1 Three-Machine Job-Proportionate Open Shop: Complexity and Approximation*

We study the three-machine job-proportionate open shop scheduling to minimize makespan, $O3|\text{prpt}|C_{\max}$, in this section. Koulamas and Kyparisis [10] present a $\frac{7}{6}$-approximation algorithm for this problem which is $NP$-hard in the ordinary sense. We now present their result however with a different proof. For the sake of presentation we begin by ignoring machine $M_2$ for the time being, which machine we chose to ignore is clearly irrelevant by definition of the job-proportionate open shop. We also assume without loss of generality that $2p_1 \leq L$. We will show no loss of generality resulting from this assumption later in the proof. Observe also that Koulamas and Kyparisis [10] show a polynomial-time algorithm for $2p_1 + p_2 \geq L$, see Problem 7.1.

The schedule on $M_1$ and $M_3$ that minimizes makespan is shown in Fig. 7.1. The schedule is feasible since each job $J_j$, $j = 2, \ldots, n$ is scheduled in the interval $[\sum_{i=1}^{j-1} p_i, \sum_{i=1}^{j} p_i]$ on $M_1$, and in the interval $[\sum_{i=2}^{j-1} p_i, \sum_{i=2}^{j} p_i]$ on $M_3$. These intervals are disjoint since $p_j \leq p_1$. Moreover, the longest job $J_1$ is scheduled in the interval $[0, p_1]$ on $M_1$, and in the interval $[\sum_{i=2}^{n} p_i, L]$ on $M_3$. The intervals are disjoint since $2p_1 \leq L$. We now include machine $M_2$ back and extend the schedule in Fig. 7.1 to include schedule on $M_2$. The extension does not always give

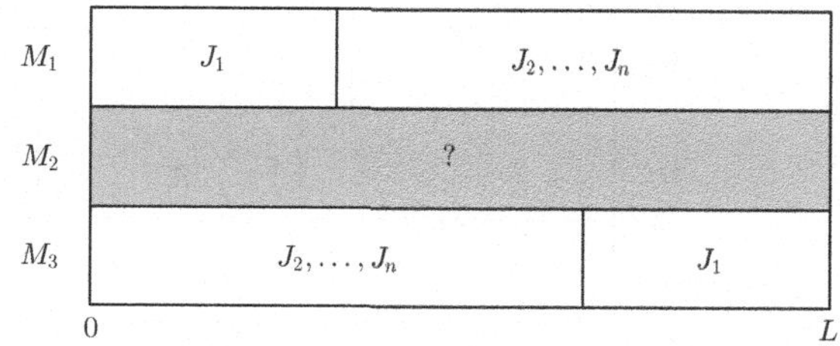


**Fig. 7.1** A feasible schedule on $M_1$ and $M_3$ with makespan $L$

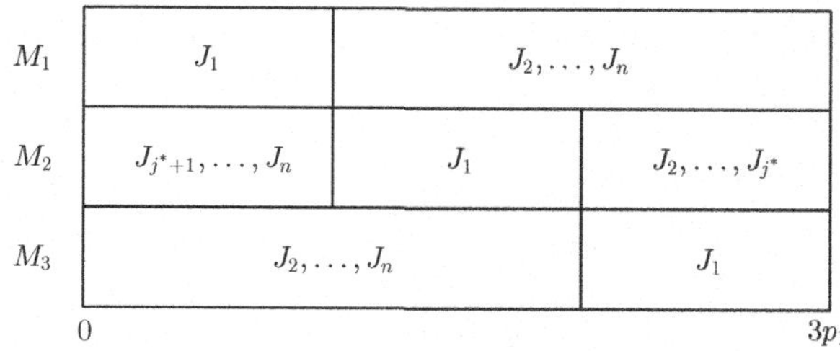


**Fig. 7.2** A schedule $S_1(j^*)$ for $p_2 + \cdots + p_{j*} \leq p_1$ and $p_{j^*+1} + \cdots + p_n \leq p_1$ for some $1 < j^* < n$

an optimal schedule; we show later that finding an optimal extension is $NP$-hard in the ordinary sense; however, it does provide a $\frac{7}{6}$-approximation of the optimal makespan. We now give details of the extension. For each $1 < j < n$ we have either

(1)

$$p_2 + \cdots + p_j \leq p_1 \quad \text{and} \quad p_{j+1} + \cdots + p_n \leq p_1, \text{ or} \tag{7.1}$$

(2)

$$p_2 + \cdots + p_j > p_1 \quad \text{and} \quad p_{j+1} + \cdots + p_n > p_1, \text{ or} \tag{7.2}$$

(3)

$$p_2 + \cdots + p_j > p_1 \quad \text{and} \quad p_{j+1} + \cdots + p_n \leq p_1, \text{ or} \tag{7.3}$$

(4)

$$p_2 + \cdots + p_j \leq p_1 \quad \text{and} \quad p_{j+1} + \cdots + p_n > p_1. \tag{7.4}$$

If either (7.1) or (7.2) holds for some $1 < j^* < n$, then optimal schedules $S_1(j^*)$ and $S_2(j^*)$ are shown in Figs. 7.2 and 7.3, respectively.

We have $C_{\max} = 3p_1$ for $S_1(j^*)$, also $C_{\max} = L$ for both equalities in (7.1). We have $C_{\max} = L$ for $S_2(j^*)$. We observe that $2p_1 > L$ implies (7.1) for each $1 < j < n$. Thus an optimal schedule in Fig. 7.2 is optimal for $2p_1 > L$, which means that the assumption $2p_1 \leq L$ is made without loss of generality.

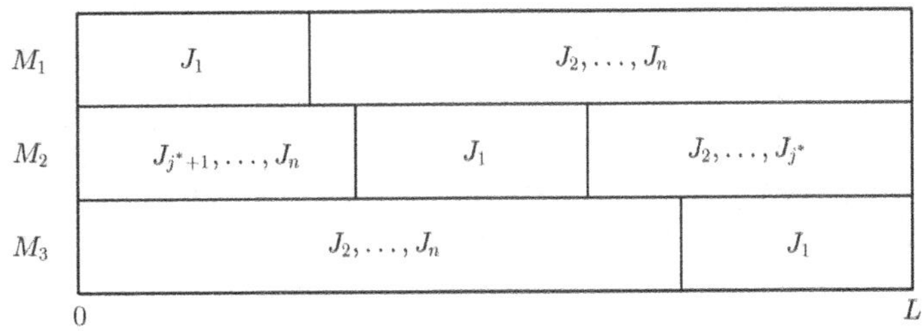


**Fig. 7.3** A schedule $S_2(j^*)$ for $p_2 + \cdots + p_{j*} > p_1$ and $p_{j^*+1} + \cdots + p_n > p_1$ for some $1 < j^* < n$

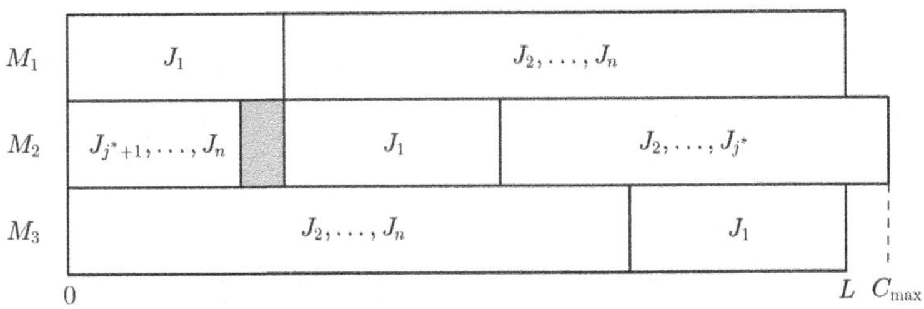


**Fig. 7.4** A schedule $S_3(j^*)$ for $p_2 + \cdots + p_{j*} > p_1$ and $p_{j^*+1} + \cdots + p_n \leq p_1$ for some $1 < j^* < n$

At this point it is worth observing that the existence of a partition of the multiset $\{p_2, \ldots, p_n\}$ into two disjoint subsets with processing times summing up to $p_1$ in either subset ensures an optimal schedule with $C_{\max} = 3p_2 = L$. The schedule is obtained in a similar fashion like $S_1(j^*)$. On the other hand the absence of such a partition may result in an optimal makespan being larger than $\max\{3p_1, L\}$, and also the makespan is intractable to compute. This observation naturally hints at the minimization of makespan for the three-machine proportionate open shop being an NP-hard problem. We return to the complexity of the three-machine problem later once we deal with the case where for *each* $1 < j < n$ we have either (7.3) or (7.4) satisfied.

Suppose (7.3) holds for some $1 < j^* < n$. For the *largest* such $j^*$ we have $C_{\max} - L < p_{j^*}$ and $j^* \geq 3$ in the schedule $S_3(j^*)$ in Fig. 7.4. Thus $C_{\max} - L < p_3$ for $S_3(j^*)$.

If (7.4) holds for some $1 < j^* < n$, then we have $C_{\max} - L \leq p_1 - (p_2 + \cdots + p_{j^*}) \leq p_1 - p_2$ in the schedule $S_4(j^*)$ in Fig. 7.5.

We now make a key observation for (7.3) that holds for some $1 < j^* < n$. Since $p_2 \leq p_1$, there is $1 < i^* < j^*$ such that $p_2 + \cdots + p_{i^*} \leq p_1$ and $p_{i^*+1} + \cdots + p_n > p_1$. Observe that $p_{i^*+1} + \cdots + p_n \leq p_1$ leads to contradiction since (7.1) would hold for $1 < i^* < n$. Therefore (7.4) holds for $i^*$, and we chose between shorter of the two schedules $S_3(j^*)$ and $S_4(i^*)$. Similarly we make a key observation for (7.4) that holds for some $1 < j^* < n$. Since $p_n \leq p_1$, there is $n > i^* > j^*$ such that $p_{i^*+1} + \cdots + p_n \leq p_1$ and $p_2 + \cdots + p_{i^*} > p_1$. Therefore (7.3) holds for

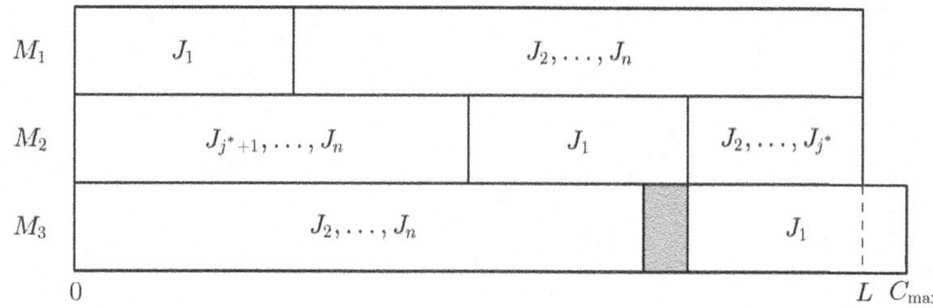


**Fig. 7.5** A schedule $S_4(j^*)$ for $p_2 + \cdots + p_{j*} \le p_1$ and $p_{j^*+1} + \cdots + p_n > p_1$ for some $1 < j^* < n$

$i^*$, and we chose between shorter of the two schedules $S_4(j^*)$ and $S_3(i^*)$. Thus we get $C_{\max} - L \le \min\{p_3, p_1 - p_2\}$ for the shorter of the two schedules in either of the two cases. If $p_2 \ge p_1/2$, then $p_1 - p_2 \le p_1/2$. If $p_2 < p_1/2$, then $p_3 < p_1/2$. Therefore, $C_{\max} - L \le \min\{p_3, p_1 - p_2\} \le p_1/2$ for the shorter of the two schedules in either case, and we get

$$\frac{C^*_{\max} - L}{C^*_{\max}} \le \frac{C_{\max} - L}{\max\{3p_1, L\}} \le \frac{1}{6}, \tag{7.5}$$

where $C^*_{\max}$ is an optimal makespan. Thus we obtain a $\frac{7}{6}$-approximation algorithm for the three-machine proportionate open shop. This proves the following theorem.

**Theorem 7.1** *The problem $O3|prpt|C_{\max}$ has $\frac{7}{6}$-approximation algorithm.*

We illustrate the algorithm for the following $5 \times 3$ job-proportionate open shop:

$$\mathbb{P} = \begin{bmatrix} 6\,6\,6 \\ 5\,5\,5 \\ 3\,3\,3 \\ 2\,2\,2 \\ 2\,2\,2 \end{bmatrix}.$$

Each machine load is $L = 18$, and $p_2 + p_3 = 8 > p_1 = 6$ and $p_4 + p_5 = 4 < p_1 = 6$. The $S_3(3)$ schedule with $C_{\max} = 20$ is shown in Fig. 7.6. However, $p_2 = 5 < p_1 = 6$ and $p_3 + p_4 + p_5 = 7 > p_1 = 6$. The $S_4(2)$ schedule with $C_{\max} = 19$ is shown in Fig. 7.7.

Recently, Ni and Chen [17] reported a $\frac{13}{12}$-approximation algorithm for the job-proportionate $O3|\text{prpt}|C_{\max}$ problem. It remains open whether there is an FPTAS for the job-proportionate open shop with fixed number of machines. Recall from Sect. 3.2.3 that PTAS for $Om||C_{\max}$, and thus for $Om|\text{prpt}|C_{\max}$, exists. Hence the factor $\frac{13}{12}$ can be improved by polynomial-time algorithms. We show a different PTAS for $Om|\text{prpt}|C_{\max}$ later in Theorem 7.5.

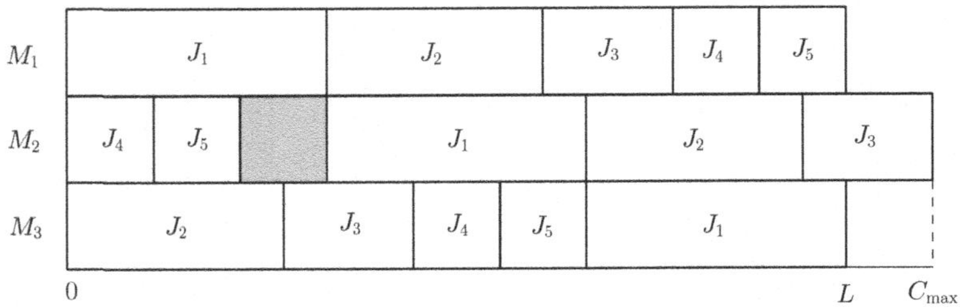


**Fig. 7.6** A schedule $S_3(3)$ with $C_{\max} - L = 2 \leq p_3 = 3$

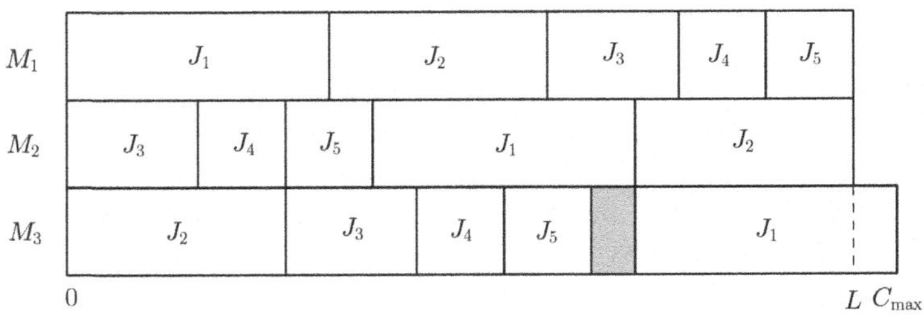


**Fig. 7.7** A schedule $S_4(2)$ with $C_{\max} - L = 1 \leq p_1 - p_2 = 1$

**Fig. 7.8** A schedule for partition $A_1$ and $A_2$

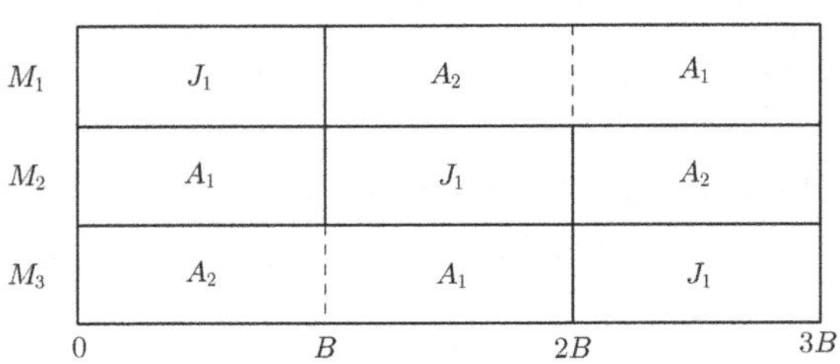


We now return to the computational complexity of $O3|\text{prpt}|C_{\max}$. The insights obtained through our analysis of the $\frac{7}{6}$-approximation should now easily lead to the NP-hardness proof. The proof is from the NP-complete PARTITION problem where for a multiset $\{a_1, \ldots, a_n\}$ of positive integers one asks whether there is a partition into disjoint sets $A_1$ and $A_2$, $A_1 \cup A_2 = \{1, \ldots, n\}$ such that $\sum_{i \in A_1} a_i = \sum_{i \in A_2} a_i = B$, see Garey and Johnson [6]. We need $n$ jobs, each with three operations, in the open shop instance. Job $J_i$ has processing time $a_i$ on each of the three machines. In addition there is a long job $J^*$ that has processing time $B$ on each of the three machines. Altogether $n + 1$ jobs. The $C_{\max} = 3B$.

Suppose $A_1$ and $A_2$ make up a partition, i.e., $\sum_{i \in A_1} a_i = \sum_{i \in A_2} a_i = B$. Then the schedule with $C_{\max} = 3B$ is shown in Fig. 7.8.

Suppose there is a schedule with $C_{\max} = 3B$. Clearly job $J^*$ is scheduled in the interval $[B, 2B]$ on one of the machines. Since the workload on each machine is exactly $3B$, and the operations are scheduled without preemption, the total processing time of all operations that complete *by* $B$ and *after* $2B$ on that machine equals $B$ and $B$, respectively. Thus clearly there exists a partition. This proves that $O3|\text{prpt}|C_{\max}$ is NP-hard. Sevastyanov [18] gives a pseudopolynomial-time algorithm for $O3|\text{prpt}|C_{\max}$. Therefore we have the following theorem.

**Theorem 7.2** *The problem $O3|prpt|C_{\max}$ is $NP$-hard in the ordinary sense.*

## 7.3 Machine-Proportionate Open Shops

### 7.3.1 The Job-Proportionate and Machine-Proportionate Open Shops Are Equivalent for Makespan

A *machine-proportionate* open shop is an open shop where $p_{i,h} = p_h > 0$ for each job $J_i$, $i = 1, \ldots, n$ and machine $M_h$, $h = 1, \ldots, m$. The machine-proportionate open shop is also referred to as open shop with machine-dependent processing times in the literature, Dror [5]. The $n \times m$ matrix of a machine-proportionate open shop is shown below

$$\mathbb{P} = \begin{bmatrix} p_1 & \cdots & p_m \\ p_1 & \cdots & p_m \\ \cdot & \cdots & \cdot \\ p_1 & \cdots & p_m \\ \cdot & \cdots & \cdot \\ p_1 & \cdots & p_m \end{bmatrix}.$$

All jobs of a machine-proportionate open shop are identical, thus all job lengths are equal, $P_1 = \cdots = P_n$, and equal to $P = \sum_{i=1}^{m} p_i$. The machine workloads are however different and equal $L_1 = np_1, \ldots, L_m = np_m$ for machines $M_1, \ldots, M_m$, respectively. The proportion $\frac{p_h}{P}$ of the total workload comes from machine $M_h$. We assume that the machines are ordered in non-increasing order of processing times $p_1 \geq \cdots \geq p_m$, which is the same as non-increasing order of the proportions. It is worth mentioning that the processing time matrix of machine-proportionate open shop admits a succinct encoding: One only needs the processing times $p_1, \ldots, p_m$ and the number of jobs $n$ to specify the input.

We observe that the transpose $\mathbb{P}^T$ of a processing time matrix $\mathbb{P}$ of a job-proportionate open shop is a processing time matrix of a machine-proportionate open shop. Thus a job-proportionate open shop with $n$ jobs on $m$ machines becomes a machine-proportionate open shop with $m$ jobs on $n$ machines.

Consider the makespan minimization in job-proportionate and machine-proportionate open shops. For a feasible schedule $S(\mathbb{P})$ of a job-proportionate open shop $\mathbb{P}$ we have a feasible schedule $S(\mathbb{P}^T)$ for a machine-proportionate shop $\mathbb{P}^T$ such that operation $O_{i,h}$ of job $J_i$ starts at $a$ and completes at $b$ on machine $M_h$ in $S$ if and only if operation $O_{h,i}$ of job $J_h$ starts at $a$ and completes at $b$ on machine $M_i$. This leads to the following observation.

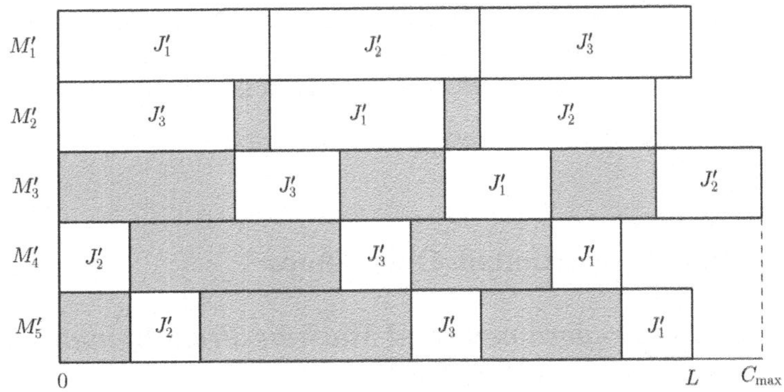


**Fig. 7.9** A schedule $S(\mathbb{P}^T)$ corresponding to the schedule in Fig. 7.7

**Observation 7.3** *Let $C_1, \ldots, C_m$ be completion times of machines $M_1, \ldots, M_m$, respectively, in schedule $S(\mathbb{P})$. Then $C_1, \ldots, C_m$ are completion times of jobs $J_1, \ldots, J_m$ respectively in schedule $S(\mathbb{P}^T)$. Thus the makespans of $S(\mathbb{P})$ and $S(\mathbb{P}^T)$ are equal.*

Consider a job-proportionate open shop from the previous section, see Fig. 7.7. We have

$$\mathbb{P}^T = \begin{bmatrix} 6 & 5 & 3 & 2 & 2 \\ 6 & 5 & 3 & 2 & 2 \\ 6 & 5 & 3 & 2 & 2 \end{bmatrix}$$

for the corresponding machine-proportionate open shop. The operation of job $J_i$, $i = 1, \ldots, 5$, on machine $M_h$, $h = 1, \ldots, 3$ in the job-proportionate open shop $\mathbb{P}$ becomes operation of job $J'_h$ on machine $M'_i$ in the machine-proportionate open shop $\mathbb{P}^T$. Thus we have three jobs $J'_1$, $J'_2$, and $J'_3$, and five machines $M'_1, \ldots, M'_5$ in the machine-proportionate open shop. Job $J'_i$ has five operations having processing times $p'_{i,1} = 6$, $p'_{i,2} = 5$, $p'_{i,3} = 3$, $p'_{i,4} = 2$, $p'_{i,5} = 2$. For the schedule $S(\mathbb{P})$ in Fig. 7.7 we get the corresponding schedule $S(\mathbb{P}^T)$ in Fig. 7.9. The operation $O_{2,3}$ is scheduled in the interval $[0, 5]$ on machine $M_3$ in $S(\mathbb{P})$, and the corresponding operation $O'_{3,2}$ is scheduled in the same interval on machine $M'_2$ in $S(\mathbb{P}^T)$. Similarly the operation $O_{5,1}$ is scheduled in the interval $[17, 18]$ on machine $M_1$ in $S(\mathbb{P})$, and the corresponding operation $O'_{1,5}$ is scheduled in the same interval on machine $M'_5$ in $S(\mathbb{P}^T)$. Both schedules $S(\mathbb{P})$ and $S(\mathbb{P}^T)$ have the same makespan $C_{\max} = 19$. However, total completion time of $S(\mathbb{P})$ in Fig. 7.7 equals $5(L - 1)$ whereas total completion time of $S(\mathbb{P}^T)$ in Fig. 7.9 equals $3L + 1$. Therefore, the job-proportionate open shop scheduling may *not* be equivalent to the machine-proportionate open shop scheduling for objective functions other than makespan.

An immediate consequence of Observation 7.3 is that the results obtained for the job-proportionate open shop makespan minimization in the previous section carry over to the machine-proportionate open shop makespan minimization considered in this section. In particular, the NP-hardness of *three-machine* makespan minimization for the job-proportionate open shops implies the NP-hardness of *three-job* makespan minimization for machine-proportionate open shops.

### 7.3.2 The $n \geq m$ Case Is Easy

Although it may seem somewhat surprising there exists a straightforward algorithm that produces optimal schedules as long as the number of machines does *not* exceed the number of jobs, i.e., $n \geq m$, see Dror [5] for instance. In this case the algorithm produces schedules with makespan equal to the workload of machine $M_1$ with the heaviest workload $C_{\max} = np_1$. The algorithm, we call it the cyclic permutation algorithm, starts with any permutation of jobs, say $1, 2, \ldots, n$, on machine $M_1$. Then it uses cyclic permutation $h, \ldots, n, 1, \ldots, h-1$ on machine $M_h$, $h = 2, \ldots, m$. The jobs $J_h, \ldots, J_n, J_1, \ldots, J_{h-1}$ are then scheduled in the intervals $[0, p_h], [p_1, p_1 + p_h], \ldots, [(n-1)p_1, (n-1)p_1 + p_h]$, respectively, on machine $M_h, h = 1, \ldots, m$. The schedule is feasible since job $J_i$ is scheduled in the intervals $[(i-1)p_1, ip_1]$ on $M_1$, $[ip_1, ip_1 + p_2]$ on $M_2, \ldots$, $[(i+h-1)p_1, (i+h-1)p_1 + p_h]$ on $M_h, \ldots$ , and $[(i+m-2)p_1, (i+m-2)p_1 + p_m]$ on $M_m$ for $i \leq n - m + 1$. Those intervals do not overlap. For $i > n - m + 1$, job $J_i$ is scheduled in the intervals $[(i-1)p_1, ip_1]$ on $M_1$, $[ip_1, ip_1 + p_2]$ on $M_2, \ldots$, $[(n-1)p_1, (n-1)p_1 + p_{n-i+1}]$ on $M_{n-i+1}$, in the interval $[0, p_{n-i+2}]$ on machine $M_{n-i+2}, \ldots, [(m-n+i-2)p_1, (m-n+i-2)p_1 + p_m]$ on machine $M_m$. Those intervals do not overlap either for $n \geq m$. Thus the algorithm produces a feasible schedule with $C_{\max} = np_1 = L_1$ which is clearly optimal.

The schedule in Fig. 7.10 illustrates the output of the algorithm for a machine-proportionate open shop with $n = 7$ jobs, $m = 4$ machines, and processing time matrix

$$\mathbb{P} = \begin{bmatrix} 3 & 2 & 2 & 1 \\ 3 & 2 & 2 & 1 \\ 3 & 2 & 2 & 1 \\ 3 & 2 & 2 & 1 \\ 3 & 2 & 2 & 1 \\ 3 & 2 & 2 & 1 \\ 3 & 2 & 2 & 1 \end{bmatrix}.$$

### 7.3.3 The $n < m$ Case

By Observation 7.3 and Theorem 7.2 the machine-proportionate open shop is NP-hard for $n < m$.

We can use the idea of cyclic permutations of $n$ jobs on $m$ machines, which ensures optimality for $n \geq m$, to the case $n < m$ with fewer jobs than machines to obtain approximate solutions. Just add $m - n$ dummy jobs and run the algorithm for the case $n \geq m$ with dummy jobs placed last in the cyclic permutation on $M_1$. Finally delete all dummy jobs from the resulting schedule to obtain a feasible schedule for the original instance. The makespan of the schedule equals

$$C_{\max} = np_1 + p_2 + \cdots + p_{m-n+1}, \tag{7.6}$$

which implies $(2-\frac{1}{n})\max\{np_1, p_1+\cdots+p_m\} \geq C_{\max} \geq \max\{np_1, p_1+\cdots+p_m\}$. Thus we get $(2-\frac{1}{n})$-approximation algorithm for the $n < m$ case, see Naderi et al. [16]. To illustrate this approach consider an example with $n = 3$ jobs and $m = 4$ where the jobs are as in the example of Fig. 7.10. The resulting schedule is shown in Fig. 7.11.

The schedule is not optimal since a shorter schedule is shown in Fig. 7.12. The idea behind the shorter schedule is to: first, reduce the original problem instance to a

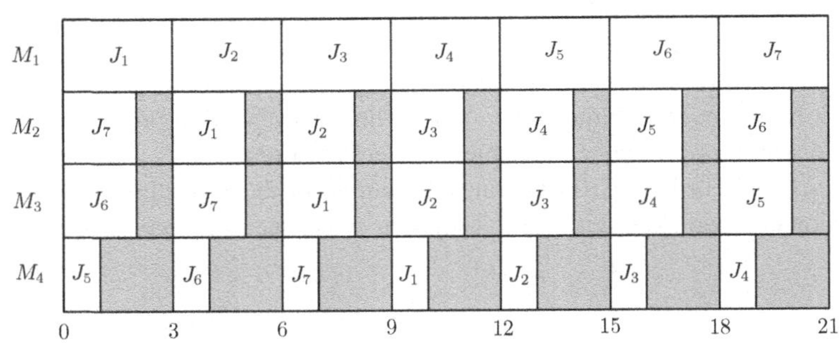


**Fig. 7.10** A schedule for machine-proportionate open shop with more jobs, $n = 7$, than machines, $m = 4$

**Fig. 7.11** A schedule for machine-proportionate open shop with fewer jobs, $n = 3$, than machines, $m = 4$

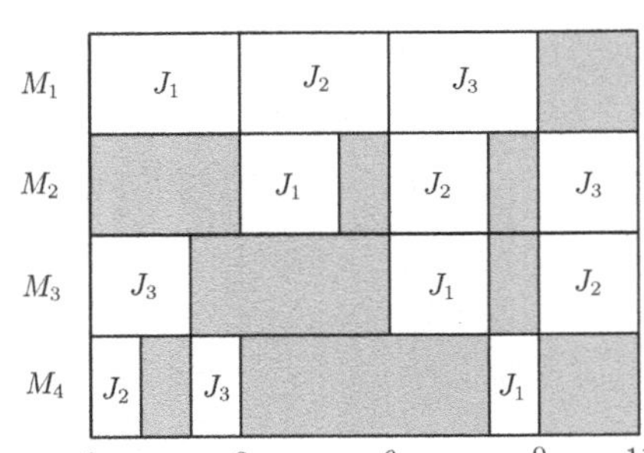

machine-proportionate open shop instance with possibly fewer machines $\ell$, $\ell \leq m$, and the same number of jobs $n$, then, to solve the latter, and finally to obtain a schedule for the former with the same makespan. Since this approach may reduce the number of machines, we may end up with $\ell \leq n$ which implies an optimal schedule by the cyclic permutation algorithm for the latter instance, and thus an optimal schedule for the former.

To be more specific we consider a BIN PACKING problem with each bin capacity equals $p_1$ and $m$ items $1, \ldots, m$ of sizes $p_1, \ldots, p_m$, respectively, to pack in as few bins as possible. Suppose that we can pack the items in $\ell \leq m$ bins $B_1, \ldots, B_\ell$. Let $A_h$ be the set of items in bin $B_h$, $h = 1, \ldots, \ell$. By definition the sets are pairwise disjoint, their union includes all $m$ items, and $q_h = \sum_{i \in A_h} p_i \leq p_1$ for $j = 1, \ldots, \ell$. We assume $q_1 \geq \cdots \geq q_\ell$. We thus get a machine-proportionate open shop with $\ell$ machines, $M'_1, \ldots, M'_\ell$, and $n$ jobs, $J'_1, \ldots, J'_n$, each having processing time $q_h$ on machine $M'_h$, $h = 1, \ldots, \ell$. To illustrate we have $\ell = 3$, $A_1 = \{1\}$, $A_2 = \{2, 4\}$ and $A_3 = \{3\}$, and $q_1 = 3$, $q_2 = 3$, and $q_3 = 2$, respectively, for the example in Figs. 7.11 and 7.12. Since $\ell = n$ we obtain an optimal schedule with $C_{\max} = 9$.

Now, let $S$ be a schedule with $C_{\max}$ obtained by the cyclic permutation algorithm for the reduced instance. We can then obtain a schedule for the original instance by unpacking the bins and moving the original operations from each bin to the original machines. Generally, consider bin $A_h = \{p_{h_1}, \ldots, p_{h_k}\}$ of job $J'_j$ scheduled on machine $M'_h$ in the interval $[a, b]$ in the schedule $S$. Unpack it, and schedule operation $O_{j,h_1}$ of $J_j$ in $[a, a+p_{h_1}]$ on machine $M_{h_1}, \ldots$ , and operation $O_{j,h_k}$ of $J_j$ in $[a+p_{h_1}+\cdots+p_{h_{k-1}}, a+p_{h_1}+\cdots+p_{h_k} = b]$ on machine $M_{h_k}$. The intervals are disjoint, thus no two operations of job $J_j$ are overlapping in the resulting schedule. Also the intervals occur on different machines thus after the unpacking no machine $M_h$, $h = 1, \ldots, m$, is required to process two or more operations of different jobs at the same time. Finally observe both $S$ and $S'$ have the same makespan. Therefore, if $\ell \leq n$, then makespan of $S$ is $np_1$ and so is of $S'$. Thus $S'$ is optimal. We observe that in the example in Fig. 7.12.

We refer to the algorithm, which is actually a scheme since various algorithms for BIN PACKING can be selected, just described as reduction-to-bin-packing algorithm. The algorithm gives the following approximation guarantee for the machine-proportionate open shop.

**Theorem 7.4** *The machine-proportionate open shop makespan minimization has a* $(1 + \frac{n-1}{\ell})$*-approximation algorithm.*

***Proof*** We observe that the reduction-to-bin-packing algorithm ensures that $q_1 - q_j < q_\ell$ for $j = 2, \ldots, \ell - 1$. Thus

$$C_{\max} = nq_1 + q_2 + \cdots + q_{\ell-n+1} \leq q_1 + (q_1 + \cdots + q_\ell) + (n-3)q_\ell.$$

Since $q_1 - q_\ell < q_j$ and $q_\ell \leq q_j$ for $j = 2, \ldots, \ell - 1$, we have $\max\{q_1 - q_\ell, q_\ell\} \leq q_j$. Thus

$$C^*_{\max} \geq q_1 + \cdots + q_\ell \geq q_1 + (\ell - 2)\max\{q_1 - q_\ell, q_\ell\} + q_\ell,$$

**Fig. 7.12** A schedule for machine-proportionate open shop with fewer jobs, $n = 3$, than machines, $m = 4$

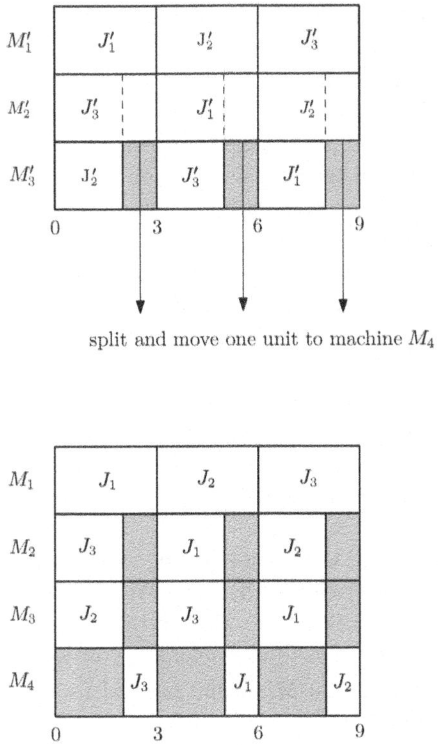


therefore

$$\frac{C_{\max} - C^*_{\max}}{C^*_{\max}} \leq 1 + \frac{q_1 + (n-3)q_\ell}{q_1 + (\ell - 2)\max\{q_1 - q_\ell, q_\ell\} + q_\ell}.$$

Assume $q_\ell \leq \frac{q_1}{2}$. Then $\max\{q_1 - q_\ell, q_\ell\} = q_1 - q_\ell$ and we get

$$\frac{q_1 + (n-3)q_\ell}{q_1 + (\ell - 2)\max\{q_1 - q_\ell, q_\ell\} + q_\ell} = \frac{q_1 + (n-3)q_\ell}{(\ell - 1)q_1 - (\ell - 3)q_\ell}.$$

Since $q_1 \geq 2q_\ell$, we have

$$\frac{q_1 + (n-3)q_\ell}{(\ell - 1)q_1 - (\ell - 3)q_\ell} \leq \frac{q_1 + (n-3)\frac{q_1}{2}}{(\ell - 1)q_1 - (\ell - 3)\frac{q_1}{2}} = \frac{n-1}{\ell + 1}.$$

Thus we get

$$\frac{C_{\max} - C^*_{\max}}{C^*_{\max}} \leq 1 + \frac{n-1}{\ell + 1}$$

for $q_\ell \leq \frac{q_1}{2}$. Now assume $q_\ell > \frac{q_1}{2}$. Then $\max\{q_1 - q_\ell, q_\ell\} = q_\ell$ and we get

$$\frac{q_1 + (n-3)q_\ell}{q_1 + (\ell - 2)\max\{q_1 - q_\ell, q_\ell\} + q_\ell} = \frac{q_1 + (n-3)q_\ell}{q_1 + (\ell - 1)q_\ell}.$$

Since $2q_\ell > q_1 \geq q_\ell$ we have

$$\frac{q_1 + (n-3)q_\ell}{q_1 + (\ell - 1)q_\ell} \leq \frac{n-1}{\ell}.$$

Thus we get

$$\frac{C_{\max} - C^*_{\max}}{C^*_{\max}} \leq 1 + \frac{n-1}{\ell}$$

for $q_\ell > \frac{q_1}{2}$. Therefore

$$\frac{C_{\max} - C^*_{\max}}{C^*_{\max}} \leq 1 + \frac{n-1}{\ell},$$

which proves the theorem. □

The advantage of the reduction-to-bin-packing algorithm will be illustrated on the following example given in Koulamas and Kyparisis [10] for job-proportionate open shop with $n = 6$ jobs and $m = 3$ machines

$$\mathbb{P} = \begin{bmatrix} 10 & 10 & 10 \\ 5 & 5 & 5 \\ 5 & 5 & 5 \\ 4 & 4 & 4 \\ 3 & 3 & 3 \\ 3 & 3 & 3 \end{bmatrix},$$

which becomes

$$\mathbb{P}^T = \begin{bmatrix} 10 & 5 & 5 & 4 & 3 & 3 \\ 10 & 5 & 5 & 4 & 3 & 3 \\ 10 & 5 & 5 & 4 & 3 & 3 \end{bmatrix},$$

for the corresponding machine-proportionate shop with $n' = 3$ jobs and $m' = 6$ machines. By applying a bin packing algorithm to the latter we get $A_1 = \{1\}$, $A_2 = \{2, 3\}$, and $A_3 = \{4, 5, 6\}$. This reduces the number of machines to $\ell = 3$ which

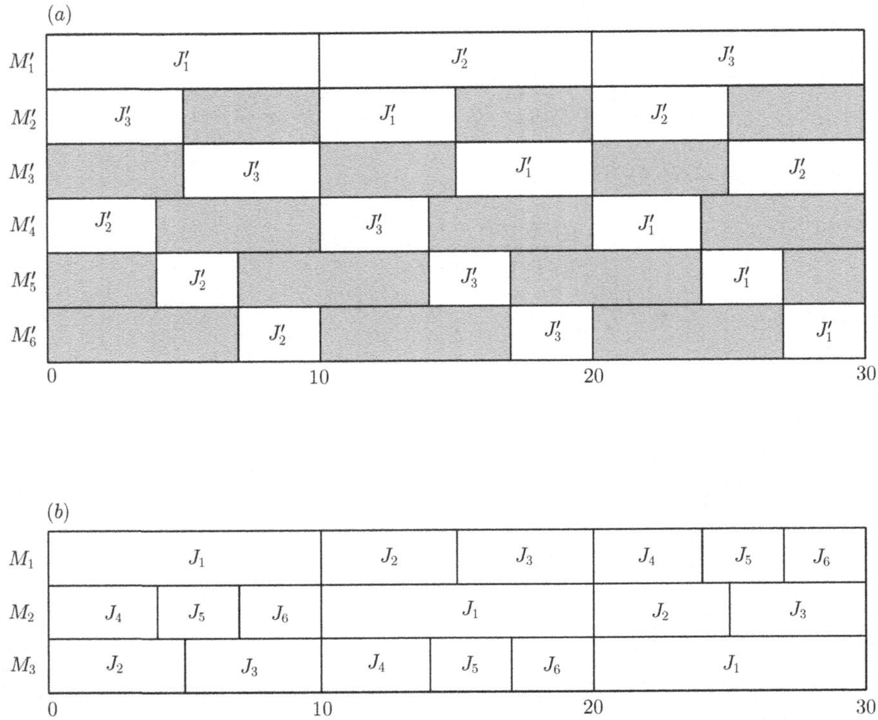


**Fig. 7.13** A schedule for machine-proportionate open shop with fewer jobs, $n = 3$, than machines, $m = 4$

allows to obtain an optimal schedule for the machine-proportionate open shop in Fig. 7.13a. By Observation 7.3, the original job-proportionate open shop has the same optimal makespan $C_{\max} = 30$. The optimal schedule for the job-proportionate open shop is shown in Fig. 7.13b.

We can use any approximation algorithm for the BIN PACKING problem, see Coffman et al. [2] for a survey, or a heuristic to efficiently solve the BIN PACKING problem which is NP-hard in the strong sense, Garey and Johnson [6].

The $\frac{7}{6}$-approximation of *three-machine* makespan minimization for job-proportionate open shops implies the $\frac{7}{6}$-approximation of *three-job* makespan minimization for machine-proportionate open shops.

Naderi et al. [16] give an MILP formulation of the problem for $n < m$. We close this section by presenting a PTAS for $Om|\text{prpt}|C_{\max}$.

**Theorem 7.5** *For a fixed positive integer k and a fixed number of machines m there is a* $(1 + \frac{1}{k})$*-approximation algorithm for the job-proportionate open shop problem* $Om|prpt|C_{\max}$. *The algorithm runs in polynomial time.*

***Proof*** Let $\mathbb{P}$ be an instance of the job-proportionate open shop problem $Om|\text{prpt}|C_{\max}$ with $n$ jobs and $m$ machines. Let $\mathbb{P}^T$ be the corresponding instance of the machine-proportionate open shop problem with $n' = m$ jobs and $m' = n$ machines. The reduction-to-bin-packing algorithm gives a $(1+\frac{n'-1}{\ell})$-approximation solution for $\mathbb{P}^T$, where $\ell \leq m'$. For $\frac{n'-1}{\ell} \geq \frac{1}{k}$, we have $k(m-1) \geq n$. Since both $k$ and $m$ are fixed, the number of jobs $n$ is fixed. The optimal schedules for fixed number of jobs can be computed in constant time. On the other hand if $\frac{n'-1}{\ell} < \frac{1}{k}$, we get $(1+\frac{1}{k})$- approximate solution by running the reduction-to-bin-packing algorithm. By Observation 7.3 this solution can be converted into $(1+\frac{1}{k})$-approximate solution for $\mathbb{P}$. □

## 7.4 Ordered Open Shops

Ordered open shops have been introduced by Liu and Bulfin [15]. Those open shops require the same order of jobs on each machine (job-ordered open shops)

$$p_{1h} \geq \cdots \geq p_{ih} \geq \cdots \geq p_{nh} \tag{7.7}$$

for each machine $M_h$, $h = 1, \ldots, m$, which means essentially the same descending order for each column of an instance $\mathbb{P}$ of the open shop. At the same time they require the same order of machines (machine-ordered open shops)

$$p_{i1} \geq \cdots \geq p_{ih} \geq \cdots \geq p_{im} \tag{7.8}$$

for each job $J_i$, $i = 1, \ldots, n$ which means essentially the same descending order for each row of an instance $\mathbb{P}$ of the open shop.

The following is an example of an instance of the ordered open shop:

$$\mathbb{P} = \begin{bmatrix} 10 & 8 & 7 \\ 7 & 6 & 5 \\ 6 & 5 & 5 \\ 5 & 4 & 4 \\ 3 & 2 & 2 \\ 2 & 1 & 1 \end{bmatrix}.$$

Clearly both the job-proportionate and the machine-proportionate open shops studied in Sects. 7.2 and 7.3 are special cases of the ordered open shops. Therefore the NP-hardness result obtained for the makespan minimization for the proportionate open shops applies to the ordered open shops as well, see also Liu and Bulfin [15] who prove that $O3|\mathrm{ord}|C_{\max}$ is NP-hard in ordered open shops.

Some special cases solvable in polynomial time have been reported for three-machine machine-ordered open shops. The idea is to order the jobs (the rows) in descending order of processing times on machine $M_1$ with the heaviest workload, and to consider column maximums

$$A_h = \max_i \{p_{ih}\} \tag{7.9}$$

for $h = 1, 2, 3$. For

$$p_{2,1} \geq A_2, \tag{7.10}$$

i.e., the *second* longest operation on $M_1$ is not shorter than the longest operation on $M_2$; Kyparisis and Koulamas [13] show an algorithm for makespan minimization that runs in $O(n)$ time.

Liu and Bulfin [15] propose a different sufficient condition for a linear-time algorithm to minimize makespan. The condition requires that $A_1$ and $A_2$ occur in different rows (belong to different jobs). Observe that their condition implies that of Kyparisis and Koulamas [13].

## 7.5 Maximal Machine

We say that open shop $\mathbb{P}$ has a *maximal machine* if there is a column in $\mathbb{P}$ that includes maximum value of each row. Without loss of generality we can assume that if such a column exists in $\mathbb{P}$, then it is column 1, or machine $M_1$. Also without loss of generality we can assume that jobs are ordered in descending order of their processing times on $M_1$, i.e.,

$$p_{11} \geq p_{21} \geq \cdots \geq p_{n1}. \tag{7.11}$$

Kyparisis and Koulamas [13] give a sufficient condition for $O||C_{\max}$ to be polynomially solvable for $m = n$. Let $q_i$, $i = 1, \ldots, m$, be maximum value in the $i \times (m - i + 1)$ upper-right corner submatrix of $\mathbb{P}$ made up of the entries in rows $1, \ldots, i$ and columns $i, \ldots, m$ of $\mathbb{P}$. The condition is

$$p_{i1} \geq q_i \tag{7.12}$$

for $i = 1, \ldots, m$. Kyparisis and Koulamas [13] show that the optimal makespan equals $C_{\max} = p_{11} + \cdots + p_{m1}$ ($n - m$ dummy machines with 0 processing time operations for all jobs is being added for the $m < n$ case). For $m > n$, we turn to the dual instance $\mathbb{P}^T$ with $m' = n < n' = m$. The optimal schedule is then obtained for $\mathbb{P}$ in polynomial time by Observation 7.3.

## 7.6 Dominated Machine

By definition of Adiri and Aizikowitz [1] $M_H$ dominates machine $M_h$ if the longest operation on $M_h$ is not longer than the shortest operation on $M_H$, formally

$$\max_j\{p_{j,h}\} \leq \min_j\{p_{j,H}\}. \tag{7.13}$$

For the instances of $O3||C_{\max}$ with a dominated machine Adiri and Aizikowitz [1] show how to start with an optimal schedule for $O2||C_{\max}$ without a dominated machine and extend this schedule to an optimal schedule for $O3||C_{\max}$ by adding a schedule on the dominated machine. Their algorithm runs in $O(n)$ time.

## 7.7 Bottleneck Machine

A bottleneck machine needs to process an operation of each job; in other words no job misses an operation on the bottleneck machine. The makespan minimization for open shops with bottleneck machines is $NP$-hard even for three machines since each machine in a job-proportionate open shop is a bottleneck machine. Drobouchevitch and Strusevich [4] study $O3||C_{\max}$ with at most two operations per job, and a *bottleneck* machine. They show an algorithm that runs in $O(n)$ time for the problem. Drobouchevitch [3] further simplifies the algorithm. For two bottleneck machines the problem reduces to $O2||C_{\max}$ since then each job has exactly two operations. By definition these two must be processed on the two bottleneck machines, and there is nothing to schedule on the machine that is non-bottleneck. The complexity status of the problem $O3||C_{\max}$ with at most two operations per job remains open; however, the problem $O4||C_{\max}$ with at most two operations per job is NP-hard in the ordinary sense, see Gonzalez and Sahni [7]. Kyparisis and Koulamas [14] show a polynomial-time algorithm for an arbitrary number of machines when the bottleneck machine is a maximal machine at the same time. Their algorithm runs in $O(n + m\log m)$ time.

## 7.8 Total Completion Time for Machine-Proportionate Open Shops

The total completion time minimization for the machine-proportionate open shops is NP-hard. This follows from the NP-hardness proof for the makespan minimization problem in Theorem 7.2. In the proof, the makespan equals $3B$ if and only if total completion time equals $9B$ since $P_1 = P_2 = P_3 = 3B$ for the three jobs in the instance of machine-proportionate open shop. Observe that this does not imply that the total completion time minimization problem is NP-hard for the

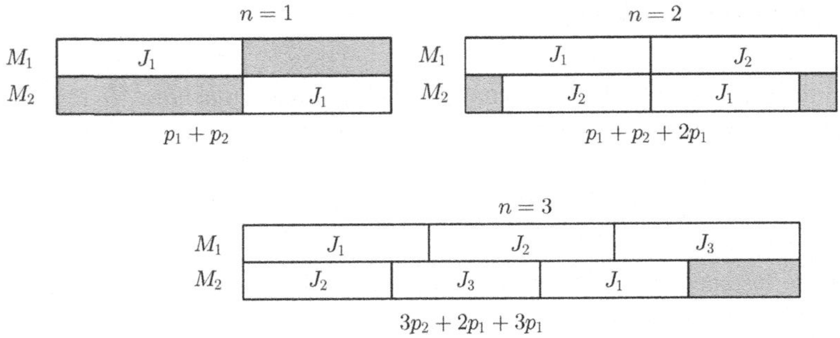


**Fig. 7.14** Optimal schedules for $n = 1, 2$, and 3

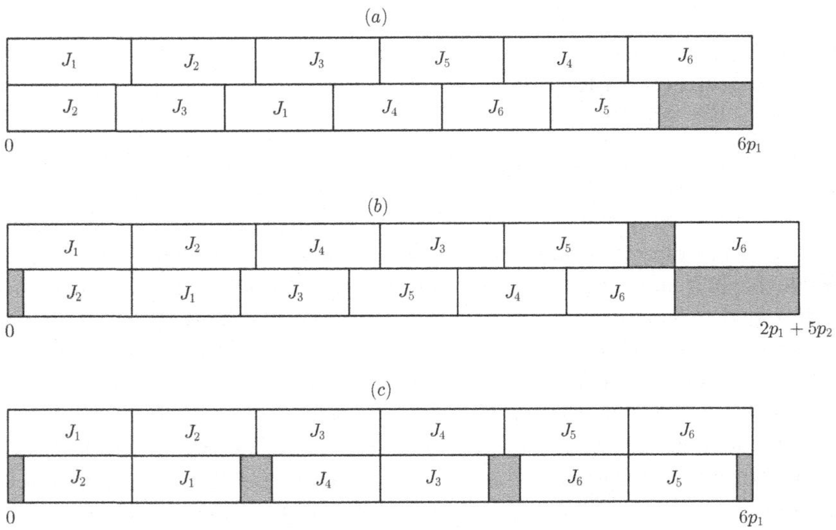


**Fig. 7.15** Schedules for an instance with $n = 6$, and $\frac{p_2}{p_1} = \frac{7}{8}$

job-proportionate open shops since Observation 7.3 may not hold for objective functions, total completion time objective in particular, other than makespan.

We now concentrate on two-machine machine-proportionate open shop. The problem instance is made up of $n$ jobs, each having processing time $p_1$ on $M_1$ and $p_2$ on $M_2$. Without loss of generality $p_1 \geq p_2$. Despite its simplicity the minimization of total completion time for two-machine machine-proportionate open shop poses some vexing open questions that we discuss in this section.

We assume $p_1 < 2p_2$ since the case $p_1 \geq 2p_2$ is easy to solve, see Problem 7.3. Optimal schedules for $n = 1, 2, 3$ are given in Fig. 7.14.

However, optimal schedules for $n > 3$ are not trivial to obtain generally. For example, consider an instance with $\frac{p_2}{p_1} = \frac{7}{8}$ and $n = 6$, see Vakharia and Çatay [19]. The schedule in Fig. 7.15a obtained by the greedy algorithm given in Dror [5]

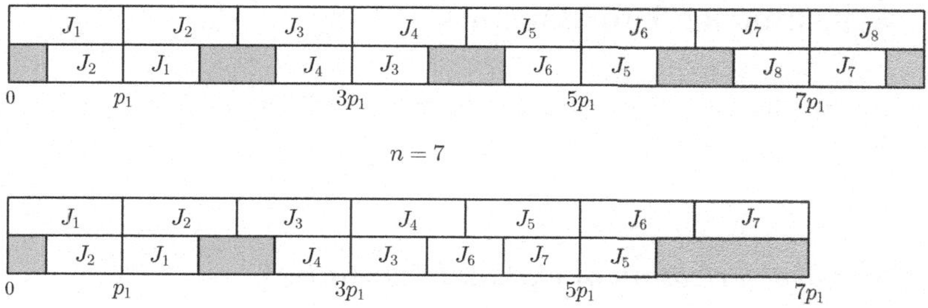


**Fig. 7.16** No-wait schedules for even and odd *n*

has total completion time $F_a = 16p_1 + 9p_2$ for the instance. The schedule avoids idle time on either machine. Figure 7.15b gives a schedule proposed by Vakharia and Çatay [19]. The schedule has total completion time $F_b = 15p_1 + 10p_2$ though it has idle time on both machines. Finally, the schedule in Fig. 7.15c has total completion time equal to $F_c = 21p_1 + 3p_2$. We have

$$F_c < F_b < F_a$$

since $F_b - F_c = -6p_1 + 7p_2 > 0$ for $\frac{p_2}{p_1} = \frac{7}{8}$, and $F_a - F_b = p_1 - p_2 > 0$. Thus neither schedule in Fig. 7.15a nor in Fig. 7.15b is optimal for the instance. The schedule in Fig. 7.15c is an example of a *no-wait* schedule for an even $n$. The no-wait schedules are defined in Fig. 7.16. Observe that the last three jobs for an odd $n$ are not per se scheduled in a no-wait fashion. The exception is made to reduce their total completion time. For the no-wait schedule $\mathcal{S}$ we have total completion time equal to

$$f(\mathcal{S}) = \frac{n(n+1)}{2}p_1 + \frac{n}{2}p_2,$$

for an even $n$, and

$$f(\mathcal{S}) = \left[\frac{n(n+1)}{2} - 2\right]p_1 + \frac{n+5}{2}p_2,$$

for an odd $n$. The no-wait schedules are interesting benchmark schedules. Besides, the start and completion time of any job in the no-wait schedule can be computed in $O(\max\{\log n, \log p_1\}$ time. However, it remains open under what conditions the no-wait schedules are optimal. In particular it remains open whether the schedules are optimal for the instances with $i^* > n$, where positive integer $i^*$ is defined as follows:

$$\left(1 + \frac{1}{i^*+1}\right)p_2 \le p_1 < \left(1 + \frac{1}{i^*}\right)p_2. \tag{7.14}$$

## 7.9 Algorithm for Two Machines

We begin by characterizing optimal schedules first. The algorithm is given in Sect. 7.9.2.

### 7.9.1 Characterization of Optimal Schedules

We give an algorithm for two machines in this section. We first introduce terminology needed to characterize optimal schedules. Let $\mathcal{S}$ be a feasible schedule. We say that job $J_i$ is in *$M_1$-configuration* in schedule $\mathcal{S}$ if the whole job $J_i$ completes on $M_1$ in $\mathcal{S}$. Thus, it is the operation of $J_i$ on $M_1$, i.e., $O_{j,1}$, that determines the job $J_i$ completion time in $\mathcal{S}$. Similarly, we say that job $J_i$ is in *$M_2$-configuration* in $\mathcal{S}$ if the whole job $J_i$ completes on $M_2$ in $\mathcal{S}$. Thus, it is the operation of $J_i$ on $M_2$, i.e., $O_{j,2}$, that determines the job $J_i$ completion time in $\mathcal{S}$.

For $i \geq 3$, let the operation in position $i-1$ on $M_1$ be scheduled in the interval $[S_{i-1}, C_{i-1}]$, and the operation in position $i$ on $M_1$ be scheduled in the interval $[s_i, c_i]$ on $M_2$. If $S_{i-1} \leq s_i < c_i \leq C_{i-1}$, then we say that there is a *skip* at $i$ in $\mathcal{S}$. We have the following classification of optimal schedules for total completion time.

**Lemma 7.1** *Let $\mathcal{S}$ be an optimal schedule. Then either*

- *$\mathcal{S}$ has idle time on $M_1$ or,*
- *$\mathcal{S}$ has no idle time on $M_1$ but has a skip or,*
- *$\mathcal{S}$ has no idle time on $M_1$ and $(i-1)p_1 < c_i < ip_1$, where $c_i$ is the completion time of the operation in position $i$ on $M_2$, for $i = 1, \ldots, n$.*

***Proof*** Let $\mathcal{S}$ be an optimal schedule without idle time on $M_1$, and $c_j \leq (j-1)p_1$ or $c_j \geq jp_1$ for some $j = 1, \ldots, n$. (Without loss of generality assume that $\mathcal{S}$ is active, i.e., each operation starts as early as other operations permit.) Take the smallest such $j$. Then $(k-1)p_1 < c_k < kp_1$ for $k = 1, \ldots, j-1$. If $c_j \leq (j-1)p_1$, then $S_{j-1} = (j-2)p_1 < c_{j-1} \leq s_j < c_j \leq (j-1)p_1 = C_{j-1}$. Moreover, since $2p_2 > p_1$, we have $j \geq 3$. Thus $j$ is a skip. Now, if $c_j \geq jp_1 = C_j$ for $j$, then since $(j-2)p_1 < c_{j-1} < (j-1)p_1$, we get $c_j - c_{j-1} > p_1$. Hence $s_j - \max\{S_j, c_{j-1}\} \geq p_1 - p_2 > 0$ and thus there is idle time in $[\max\{S_j, c_{j-1}\}, s_j]$ on $M_2$. Hence the schedule is not active since the operation in position $j$ on $M_2$ can start earlier than in $\mathcal{S}$. This leads to contradiction. □

If a skip at $i$ exists in $\mathcal{S}$, then the operations scheduled in the positions following the skip make up a *tail* which is easy to schedule to optimality. A tail is made up of three mutually disjoint sets: $Y$—a non-empty set of jobs with operations on $M_1$ only, $X$—a set of jobs with operations on $M_2$ only, $L$—a set of jobs with operations on $M_1$ and $M_2$, and two starting points $S$ on $M_1$ and $s$ on $M_2$, $s \leq S$. Moreover, $y = |Y| > |X| = x$. The schedule shown in Fig. 7.17 with each job in $L$ being in $M_1$-configuration is feasible for $Y \cup X \cup L$. The job from $L$ in position $y+i$ on

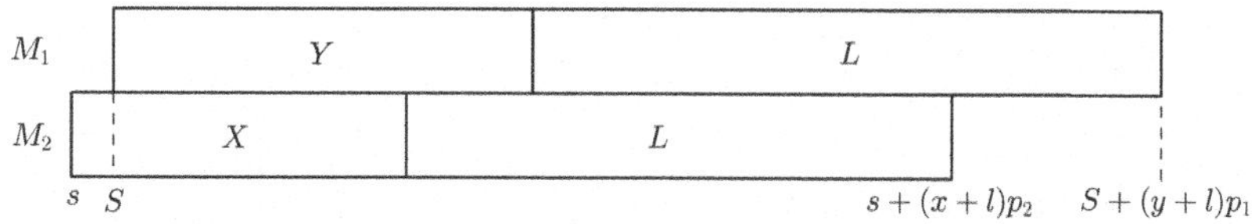


**Fig. 7.17** A tail that starts at $S$ on $M_1$ and at $s$ on $M_2$. Each job in $L$ is scheduled in $M_1$-configuration

$M_1$ starts at $S + (y + i - 1)p_1$ on $M_1$, the job in position $x + i$ on $M_2$ completes at $s + (x + i)p_2$ on $M_2$. We have $s + (x + i)p_2 \leq S + (y + i - 1)p_1$, which ensures feasibility since $y > x$ and $S \geq s$ by definition of the skip. Total completion time of the schedule equals

$$S(y + l) + \frac{(y + l)(y + l + 1)}{2} p_1 + s(x + l) + \frac{x(x + 1)}{2} p_2,$$

where $|L| = l$. No other schedule of $Y \cup X \cup L$ that starts at $S$ or later on $M_1$, and at $s$ or later on $M_2$ can have smaller total completion time. We have the following lemma.

**Lemma 7.2** *If $\mathcal{S}$ is an optimal schedule with a skip at $i$, then there is an optimal schedule $\mathcal{S}'$ with a skip at $i$, and a tail that starts at $S = C_{i-1}$ on $M_1$ and at $s = c_i$ on $M_2$.*

***Proof*** For a skip at $i$ in an optimal $S$, consider the earliest such skip. Let $x_{i-1}$ be the number of jobs in $M_2$-configuration that have one operation in some position in $\{1, \ldots, i - 1\}$ on $M_1$ (those operations complete by $C_{i-1}$ on $M_1$), and the other in some position in $\{i + 1, \ldots, n\}$ on $M_2$. Let $X_{i-1}$ be the set of those jobs. Let $y_i$ by the number of jobs in $M_1$-configuration that have one operation in some position in $\{1, \ldots, i\}$ on $M_2$ (those operations complete by $c_i$ on $M_2$), and the other in some position in $\{i + 1, \ldots, n\}$ on $M_1$. Let $Y_i$ be the set of those jobs. Let $c$ be the number of jobs that complete by $C_{i-1}$. Each such job has one of its operations in a position in $\{1, \ldots, i-1\}$ on $M_1$ and the other in a position in $\{1, \ldots, i\}$ on $M_2$. Since $p_1 < 2p_2$, no operation starts and completes in $[c_i, C_{i-1}]$ . We have $i - 1 = x_{i-1} + c$ and $i = y_i + c$. Thus $y_i = x_{i-1} + 1$. The sets $X_{i-1}$ and $Y_i$ are disjoint. Thus the sets $X_{i-1}$, $Y_i$, and the set of remaining jobs $L$ make up a tail with $S = C_{i-1}$ and $s = c_i$. Keep the schedule $\mathcal{S}$ in $[0, C_{i-1}]$ on $M_1$ and in $[0, c_i]$ on $M_2$ unchanged and schedule jobs in $Y_i$ on $M_1$ in $[C_{i-1}, C_{i-1} + y_i p_1]$ in any order and without idle time. Let the job $J_a$ be scheduled in position $i - 1$ on $M_1$. If $J_a \notin X_{i-1}$, then schedule the jobs in $X_{i-1}$ on $M_2$ in $[c_i, c_i + x_{i-1} p_2]$ in any order. If $J_a \in X_{i-1}$, then schedule the jobs in $X_{i-1} \setminus \{J_a\}$ in any order first and then job $J_a$ last on $M_2$ in $[c_i, c_i + x_{i-1} p_2]$. Since $C_{i-1} - c_i \geq 0$ and $p_1 - p_2 > 0$ we have $C_{i-1} + y_i p_1 - (c_i + x_{i-1} p_2) > p_1$. Therefore we can schedule each of the remaining $l$ jobs, $|L| = l$, in $M_1$-configuration. The total completion time of the tail is

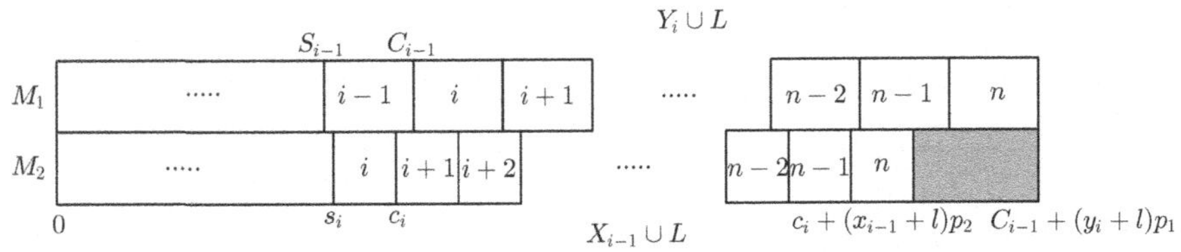


**Fig. 7.18** A schedule with a skip at $i$ and a tail $X_{i-1}$, $Y_i$, and $L$

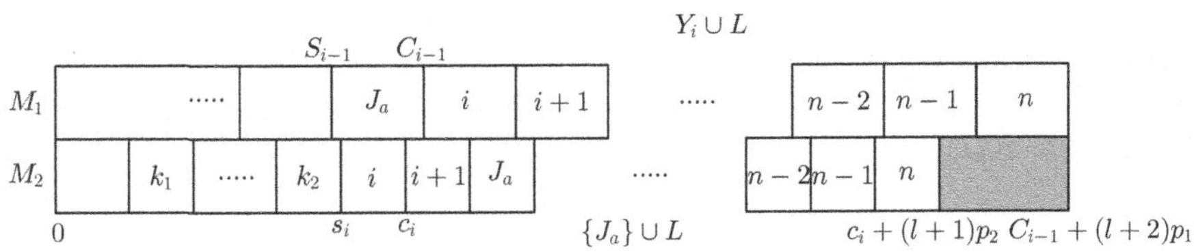


**Fig. 7.19** Exchange the operations of jobs $J_{k_1}$ and $J_a$ on $M_2$

$$yC_{i-1} + \frac{y(y+1)}{2}p_1 + x_{i-1}c_i + \frac{x_{i-1}(x_{i-1}+1)}{2}p_2,$$

where $y = y_i + l$. The resulting schedule, see Fig. 7.18, is feasible unless $X_{i-1} = \{J_a\}$ and $C_{i-1} > c_i$. To complete the proof assume this condition holds.

Let $k_1 < k_2 \in \{1, \ldots, i\}$ be the positions on $M_2$ with the two jobs in $Y_i$, recall that $y_i = x_{i-1} + 1$. Let $J_{k_1}$ and $J_{k_2}$ be the jobs in those positions on $M_2$. Those jobs are in $M_1$-configurations and occupy positions in $\{i+1, \ldots, n\}$ on $M_1$ in $\mathcal{S}$. If at least one of $J_{k_1}$ and $J_{k_2}$ is not done in parallel with $J_a$ on $M_1$ in $\mathcal{S}$, then exchange $J_a$ with that job on $M_2$, see Fig. 7.19. This exchange reduces the completion time of job $J_a$ by at least $p_2$. Then schedule each job in $Y_i \cup L$ in $M_1$-configuration. The resulting schedule is feasible since one of those jobs has an operation on $M_2$ scheduled in position $k_2 \in \{1, \ldots, i\}$ (i.e., by $c_i$) on $M_2$. Moreover total completion time of jobs in $Y_i \cup L$ is minimized by the resulting schedule and thus it does not exceed the one in $\mathcal{S}$. This however contradicts the optimality of $S$ since the job $J_a$ completes earlier than in $\mathcal{S}$.

If both jobs $J_{k_1}$ and $J_{k_2}$ on $M_2$ are done in parallel with $J_a$ on $M_1$, then $k_1 = i-1$ and $k_2 = i$ on $M_2$, see Fig. 7.20. Consider the job $J_b$ in position $i-2$ on $M_2$. Since the jobs in $Y_i = \{J_{k_1}, J_{k_2}\}$ are in $M_1$-configuration, job $J_b$ is in $M_2$-configuration. Move $J_b$ in position $i-2$ on $M_2$ to position $i-1$ on $M_2$, move $J_{k_1}$ in position $i-1$ on $M_2$ to position $i$ on $M_2$, and move $J_a$ to position $i-2$ on $M_2$. These exchanges delay the completion of job $J_b$ by $p_2$ but speed up the completion of $J_a$ by at least $p_2$. The sets $Y_i$, $X_{i-1} = \emptyset$, and $L$ make up a tail with $S = C_{i-1}$ and $s = c_i$. The schedule is optimal. □

A tail in a schedule may also be a result of idle time on $M_1$ even if there are no skips in the schedule. We have the following lemma.

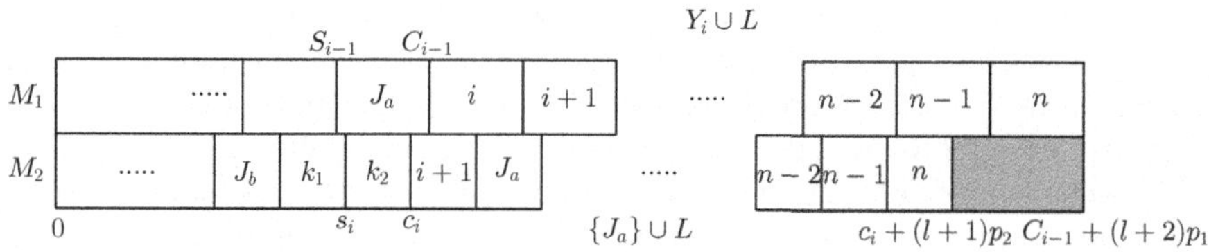


**Fig. 7.20** Exchange the operations of jobs $J_{k_1}$ and $J_a$ on $M_2$

**Lemma 7.3** *If $\mathcal{S}$ is an optimal schedule with idle time in the interval $[T, T + \Delta]$, $\Delta > 0$, on $M_1$, then there is a tail at $s = S = T + \Delta$ in $\mathcal{S}$.*

***Proof*** In $\mathcal{S}$, consider the earliest time $T$ such that machine $M_1$ is idle in the interval $[T, T + \Delta]$ for some $\Delta > 0$. Then there is a job $J_i$ such that it completes at $T + \Delta$ on $M_2$ and it starts at $T + \Delta$ on $M_1$. Let $J_i$ be in position $i$ on $M_2$. Then positions $1, \ldots, i - 1$ on $M_2$ are occupied by jobs that complete by $T$. Therefore there are at least $i - 1$ positions on $M_1$ by $T$. Suppose for contradiction that there are $i$ or more positions occupied by $T$ on $M_1$. Then $T \geq ip_1$ and $T + \Delta = xp_1 + yp_2$, where $x + y = i$. However, $T + \Delta = xp_1 + yp_2 \leq ip_1 \leq T$ which gives contradiction. Therefore, the sets $Y_i = \{J_i\}$, $X_{i-1} = \emptyset$, and $L$ make up a tail that starts at $s = S = T + \Delta$ in $\mathcal{S}$. □

The following lemma completes the characterization of optimal schedules.

**Lemma 7.4** *Let $\mathcal{S}$ be an optimal schedule with idle time on $M_1$. Then there is an optimal $\mathcal{S}'$ that has no idle time on $M_1$ but has a skip, or there is an optimal $\mathcal{S}'$ with a single idle time on $M_1$.*

***Proof*** Let $\mathcal{S}$ be an optimal schedule with a skip and idle time on $M_1$. Let $i$ be the earliest skip, i.e., let the operation in position $i - 1$ on $M_1$ be in the interval $[S_{i-1}, C_{i-1}]$, the operation in position $i$ on $M_1$ be in the interval $[s_i, c_i]$ on $M_2$, and $S_{i-1} \leq s_i < c_i \leq C_{i-1}$. Let $[T, T + \Delta]$ be the earliest idle interval on $M_1$ in $\mathcal{S}$. If $T < S_{i-1}$, then the lemma holds by Lemma 7.3. If $T \geq C_{i-1}$, then the lemma holds by Lemma 7.2. □

## 7.9.2 The Algorithm

We are now ready to present the algorithm. The algorithm relies on the characterization of optimal schedules given in Lemmas 7.1, 7.2, 7.3, and 7.4. It constructs a directed graph $G$ with weights on the arcs and then finds shortest paths in $G$. The node of $G$ is defined by a six-tuple $(m_1, m_2, j, c, t = xp_1 + yp_2, i)$ where

- $i$—position on $M_1$ and $M_2$, $i = 1, \ldots, n$. We assume $i = 0$ for the initial state.
- $m_1$—number of jobs in $M_1$-configuration that have one operation in some position in $\{1, \ldots, i\}$ on $M_2$, and the other in some position in $\{i+1, \ldots, n\}$ on $M_1$. We refer to these jobs as *open* jobs in $M_1$-configuration.
- $m_2$—number of jobs in $M_2$-configuration that have one operation in some position in $\{1, \ldots, i\}$ on $M_1$, and the other in some position in $\{i+1, \ldots, n\}$ on $M_2$. We refer to these jobs as *open* jobs in $M_2$-configuration.
- $j$—position of the last open job in $M_2$-configuration.
- $c$—number of jobs with both operations in positions in $\{1, \ldots, i\}$. We refer to these jobs as *complete* jobs.
- $t = xp_1 + yp_2$—completion time of the operation in position $i$ on $M_2$, $x + y = i$.

The starting node of in-degree 0 in $G$ is (0, 0, 0, 0, 0, 0). The shortest path from the starting note to a node $(m_1, m_2, j, c, t = xp_1 + yp_2, i)$ in $G$ gives an optimal schedule of positions $1, \ldots, i$ on $M_1$ and $M_2$ with $(i)$ $c$ jobs that complete in those positions, $(ii)$ $m_1$ jobs in $M_1$-configuration and $m_2$ jobs in $M_2$-configuration open in those positions, where $(iii)$ the last operation on $M_1$ completes at $ip_1$, and $(iv)$ the last operation on $M_2$ completes at $t$, and where $(v)$ the last open job in $M_2$-configuration is in position $j$. The path length equals total completion time of the schedule. The $G$ is constructed as follows. From the node

$$(m_1, m_2, j, c, t = xp_1 + yp_2, i)$$

with $(i-1)p_1 < t < ip_1$ for $n > i \geq 1$, $G$ has an arc to each of the following nodes:

- Completing an open job on $M_1$, and completing an open job on $M_2$:
  - For $m_1 \geq 1, m_2 \geq 2$

    $$(m_1 - 1, m_2 - 1, j, c + 2, t = xp_1 + (y+1)p_2, i + 1),$$

    the weight on the arc equals $w = (i+1)p_1 + t + p_2$.
  - For $m_1 \geq 1, m_2 = 1$, and $j < i$

    $$(m_1 - 1, m_2 - 1, 0, c + 2, t = xp_1 + (y+1)p_2, i + 1),$$

    the weight on the arc equals $w = (i+1)p_1 + t + p_2$.
  - For $m_1 \geq 1, m_2 = 1$, and $j = i$

    $$(m_1 - 1, m_2 - 1, 0, c + 2, t = ip_1 + p_2, i + 1),$$

    the weight on the arc equals $w = (i+1)p_1 + ip_1 + p_2$.
- Completing an open job on $M_1$, and opening a job in $M_1$-configuration on $M_2$ (we need $n - (c + m_1) \geq 1$ for such a job to exist):

– For $m_1 > 0$

$$(m_1, m_2, j, c+1, t = xp_1 + (y+1)p_2, i+1),$$

the weight on the arc equals $w = (i+1)p_1$.

- Opening a job in $M_2$-configuration on $M_1$ (we need $n - (c + m_2) \geq 1$ for such a job to exist), and completing an open job on $M_2$.

  – For $m_2 \geq 2$

$$(m_1, m_2, j, c+1, t = xp_1 + (y+1)p_2, i+1),$$

  the weight on the arc equals $w = t + p_2$.

  – For $m_2 = 1$, and $j < i$

$$(m_1, m_2, 0, c+1, t = xp_1 + (y+1)p_2, i+1),$$

  the weight on the arc equals $w = t + p_2$.

  – For $m_2 = 1$, and $j = i$

$$(m_1, m_2, 0, c+1, t = ip_1 + p_2, i+1),$$

  the weight on the arc equals $w = ip_1 + p_2$.

- Opening a job on $M_1$, and opening a job on $M_2$. We need $n - (c + m_1) \geq 2$.

  – For $m_2 \geq 1$

$$(m_1+1, m_2+1, j, c, t = xp_1 + (y+1)p_2, i+1),$$

  the weight on the arc equals $w = 0$.

  – For $m_2 = 0$

$$(m_1+1, m_2+1, i+1, c, t = xp_1 + (y+1)p_2, i+1),$$

  the weight on the arc equals $w = 0$.

Any node

$$(m_1, m_2, j, c, t = xp_1 + yp_2, i)$$

with $t \leq (i-1)p_1$ for $n \geq i \geq 3$ is terminal, i.e., has out-degree 0 in $G$. For terminal nodes to occur in $G$ we need

$$\left(1 + \frac{1}{n-1}\right)p_2 \leq p_1$$

which implies $i^* < n$, see (7.14) for $i^*$. Therefore $G$ for an instance with $i^* \geq n$ does not have terminal nodes. The shortest paths in $G$ are used to compute optimal schedules, by adding tails if need be.

For $i = 0, \ldots, n$, consider the nodes $n(i,t) = (0, 0, 0, i, t = xp_1 + yp_2, i)$ with $(i-1)p_1 < t < ip_1$ and $t + p_2 > ip_1$. Let $\alpha_{n(i,t)}$ be the shortest path from the starting node to $n(i,t)$. Let $f_{n(i,t)}$ be the length of $\alpha_{n(i,t)}$. Define

$$F_{n(i,t)} = f_{n(i,t)} + (t + p_2)(n - i) + \frac{(n-i)(n-i+1)}{2} p_1$$

to account for the tail. Let

$$F = \min_{n(i,t)} \{F_{n(i,t)}\}.$$

For each terminal node $n(m_2, i, t) = (m_1, m_2, j, c, t = xp_1 + yp_2, i)$ with $t \leq (i-1)p_1$ for $n \geq i \geq 3$. Let $\beta_{n(m_2,i,t)}$ be the shortest path from the starting node to $n(m_2, i, t)$. Let $h_{n(m_2,i,t)}$ be the length of $\beta_{n(m_2,i,t)}$. Define

$$H_{n(m_2,i,t)} = h_{n(m_2,i,t)} + ip_1(n-i) + \frac{(n-i)(n-i+1)}{2} p_1 + m_2 t + \frac{m_2(m_2+1)}{2} p_2$$

to account for the tail. Let

$$H = \min_{n(m_2,i,t)} \{H_{n(m_2,i,t)}\}.$$

Finally, consider the nodes $n(t) = (0, 0, 0, n, t = xp_1 + yp_2, n)$ with $(n-1)p_1 < t < np_1$. Let $\gamma_{n(t)}$ be the shortest path from the starting node to $n(t)$. Let $e_{n(t)}$ be the length of $\gamma_{n(t)}$ and

$$I = \min_{n(t)} \{e_{n(t)}\}.$$

The optimal solution has total completion time that equals $\min\{F, H, I\}$.

By induction, we can prove that $m_1 = m_2$, $m_1 + c = i$, $m_2 + c = i$, and $x + y = i$ for each node $(m_1, m_2, j, c, t = xp_1 + yp_2, i)$. The number of nodes in $G$ is bounded by a polynomial of $n$. Thus the shortest paths in $G$ can be computed in polynomial time, and so can be the optimal schedule after accounting for the possible tails. However, we need to keep in mind that the problem instances can have short inputs (succinct encoding) that require only $O(\max\{\log n, \log p_1\})$ bits, see Kubiak [11], Kubiak et al. [12], Hochbaum and Shamir [9], Grigoriev [8], and Dror [5] for similar type of problems. Thus the algorithm is not polynomial with the succinct encoding of the input. We have the following two open problems.

**Problem 7.1** Can the start and completion of any job in optimal schedules be computed in time polynomial in $O(\max\{\log n, \log p_1\})$? Can the value of minimum total completion time be computed in time polynomial in $O(\max\{\log n, \log p_1\})$?

## Problems

**7.1** Show that the job-proportionate open shop scheduling to minimize makespan, $O3|prpt|C_{\max}$, can be solved in polynomial time if $L \leq 2p_1 + p_2$.

**7.2** Show a schedule that minimizes total completion time for two-machine machine-proportionate open shop with $p_1 \geq 2p_2$.

**7.3** Find an optimal solution for the instance in Fig. 7.15.

## References


1. I. Adiri, N. Aizikowitz (Hefetz), Open-shop scheduling problems with dominated machines. Naval Res. Logist. **36**, 273–281 (1989)
2. E.G. Coffman Jr., J. Csirik, G. Galambos, S. Martello, D. Vigo, *Bin Packing Approximation Algorithms: Survey and Classification*, in eds. by P. Pardalos, D.Z. Du, R. Graham, *Handbook of Combinatorial Optimization* (Springer, New York, 2013), pp. 455–531
3. I.G. Drobouchevitch, Three-machine open shop with a bottleneck machine revisited. J. Schedul. **24**, 197–208 (2021)
4. I.G. Drobouchevitch, V.A. Strusevich, A polynomial algorithm for the three-machine open shop with bottleneck machine. Ann. Oper. Res. **92**, 185–210 (1999)
5. M. Dror, Openshop scheduling with machine dependent processing times. Discrete Appl. Math. **39**, 197–205 (1992)
6. M.R. Garey, D.S. Johnson, *Computers and Intractability: A Guide to the Theory of NP-Completeness* (W. H. Freeman, 1979)
7. T. Gonzalez, S. Sahni, Open shop scheduling to minimize finish time. J. ACM **23**, 665–679 (1976)
8. A. Grigoriev, *High multiplicity scheduling problems*. PhD thesis, Maastricht University, 2003
9. D.S. Hochbaum, R. Shamir, Strongly polynomial algorithms for the high multiplicity scheduling problem. Operations Research **39**, 648–653 (1991)
10. Ch. Koulamas, G.J. Kyparisis, The three-machine proprtionate open shop and mixed shop minimum makespan problems. Eur. J. Oper. Res. **243**, 70–74 (2015)
11. W. Kubiak, A pseudo-polynomial algorithm for a two-machine no-wait job-shop scheduling problem. Eur. J. Oper. Res. **43**, 267–270 (1989)
12. W. Kubiak, S. Sethi, C. Sriskandarajah, An efficient algorithm for a job shop problem. Ann. Oper. Res. **57**, 203–216 (1995)
13. G.J. Kyparisis, C. Koulamas, Open shop scheduling with maximal machines. Discrete Appl. Math. **78**, 175–187 (1997)
14. G.J. Kyparisis, C. Koulamas, Flow shop and open shop scheduling with a critical machine and two operations per job. Eur. J. Oper. Res. **127**, 120–125 (2000)
15. C.Y. Liu, R.L. Bulfin, Scheduling ordered open shops. Comput. Oper. Res. **14**, 257–264 (1987)
16. B. Naderi, M. Zandieh, M. Yazdani, Polynomial time approximation algorithms for proportionate open shop scheduling. Intl. Trans. Ops. Res. **21**, 1031–1044 (2014)

17. G. Ni, L. Chen, Improved scheduling for the three-machine proportionate open shop and mixed shop minimum makespan problems. IEEE Access **8**, 186805–186812 (2020)
18. S. Sevastyanov, Some positive news on the proportionate open shop problem. Sib. Èlektron. Mat. Izv. **16**, 406–426 (2019)
19. A.J. Vakharia, B. Çatay, Two machine open shop scheduling with machine-dependent processing times. Discrete Appl. Math. **73**, 283–288 (1997)

# Chapter 8
# Multiprocessor Open Shops

## 8.1 Preemptive Open Shops with Multiprocessors

The open shop scheduling has been extended to open shop scheduling with multiprocessors by replacing a single machine in each stage by a set of identical parallel machines or by having a single machine in each stage but allowing each operation of a job to be processed on a set of machines from different stages, see Lawler et al. [11] (similar models are also considered by Vairaktarakis and Sahni [20]). For these extensions machine workloads remain unknown until operations are allocated to machines, which is a main departure from the open shop scheduling studied in Chap. 3. The two extensions are particularly appealing for preemptive scheduling, since the allocation of operations to machines can then be obtained in those extensions so that the resulting maximum machine workload is a lower bound on the schedule makespan. Moreover, the allocation can be obtained in polynomial time. Once the allocation has been determined, one is left with an instance or a set of instances of an open shop with preemptions. By König's edge-coloring theorem the open shop has makespan that equals either the processing time of the longest job, which has not changed by the allocation, or the maximum machine workload, which due to the allocation, equals the lower bound on the schedule makespan. A schedule with this makespan can then be found in polynomial time by the algorithm of Gonzalez and Sahni [8] for instance. That is in a nutshell the approach taken in this chapter.

The open shop with multiprocessors is appealing also because of its applications. Matta [12] studies a multiprocessor open shop scheduling in a large oncology center where patients (jobs) need to be tested by multiple testing departments and each test can be performed by parallel identical machines. The applications of multiprocessor open shop scheduling are discussed in the last section.



W. Kubiak, *A Book of Open Shop Scheduling*, International Series in Operations Research & Management Science 325,
https://doi.org/10.1007/978-3-030-91025-9_8

### 8.1.1 Single-Operation Machines: McNaughton Meets König

In this section we consider the problem $O(P)|\text{pmtn}|C_{\max}$ with parallel identical machines in each stage, see Lawler et al. [11]. The notation $O(P)$ denotes open shop with an arbitrary number of stages and an arbitrary number of parallel identical machines in each stage. In other words, a single machine in each stage of the open shop $O$ is being replaced by a set of parallel identical machines to increase processing capacity of the stage and possibly reduce makespan in the $O(P)$ open shop. The operations $O_{1,h}, \ldots, O_{n,h}$ processed on machine $M_h$ in the open shop $O$ can now be processed on any parallel machine in the set $\mathcal{M}_h$ that replaces the machine (stage) $M_h$ in the open shop $O(M)$ with parallel machines. However, no operation $O_{i,h}$ is permitted to be processed on machines from $\mathcal{M}_{h'}$ for $h \neq h'$. The numbers of parallel machines in the sets $\mathcal{M}_h$ may differ. Each set $\mathcal{M}_h$ along with the operations $O_{1,h}, \ldots, O_{n,h}$ makes up an instance $I_h$ of the $P|\text{pmtn}|C_{\max}$ problem. The operation processing times $p_{1,h}, \ldots, p_{n,h}$ become the processing times of the jobs in the instance $I_h$. Without losing much generality let us assume integral processing times for the time being.

McNaughton [15] proposes a wrap-around rule to obtain an optimal schedule $S_h$ for $P|\text{pmtn}|C_{\max}$ with makespan

$$\max\left\{\max_i\{p_{i,h}\}, \frac{L_h}{|\mathcal{M}_h|}\right\} \tag{8.1}$$

for the $I_h$. The optimal schedules $S_1, \ldots, S_m$ for the instances $I_1, \ldots, I_m$, respectively, define an open shop with $|\mathcal{M}_1| + \cdots + |\mathcal{M}_m|$ single-machine stages. The number of jobs remains unchanged, yet each job may possibly have different operations that depend on the schedules $S_1, \ldots, S_m$. However, the construction ensures that each job $J_i$ has the same length $P_i$ as in the original instance. To illustrate this approach let us consider an open shop instance in Table 8.1 where $\mathcal{M}_1 = \{M_1, M_2\}$, $\mathcal{M}_2 = \{M_3, M_4, M_5\}$, and $\mathcal{M}_3 = \{M_6, M_7\}$. The operations $O_{i,1}, i = 1, \ldots, 5$ can be processed on two parallel identical machines in $\mathcal{M}_1$. The optimal preemptive schedule obtained by McNaughton wrap-around rule is given in Fig. 8.1. We have $L_1 = 23$ in the instance $I_1$; thus the optimal makespan equals $L_1/2 = 11\frac{1}{2}$ in the figure.

**Table 8.1** An instance $I$ of open shop with identical parallel machines in each stage

| $J_i$ | $p_{i,1}$ | $p_{i,2}$ | $p_{i,3}$ |
|---|---|---|---|
| $J_1$ | 4 | 2 | 1 |
| $J_2$ | 3 | 5 | 6 |
| $J_3$ | 7 | 2 | 0 |
| $J_4$ | 4 | 4 | 3 |
| $J_5$ | 5 | 8 | 2 |

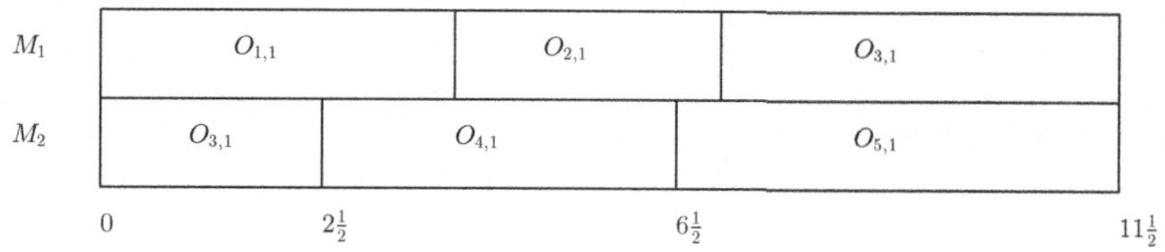


**Fig. 8.1** An optimal wrap-around schedule $S_1$ of $O_{1,1}, \ldots, O_{5,1}$ on parallel machines $M_1$ and $M_2$

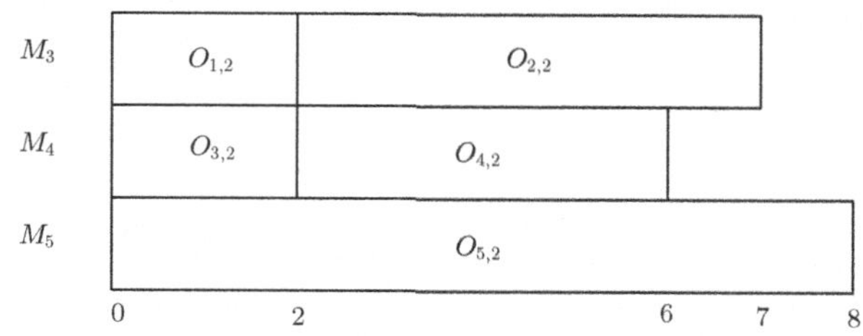


**Fig. 8.2** An optimal wrap-around schedule $S_2$ of $O_{1,2}, \ldots, O_{5,2}$ schedule on parallel machines $M_3$, $M_4$, and $M_5$

The operations $O_{i,2}$, $i = 1, \ldots, 5$ can be processed on three parallel identical machines in $\mathcal{M}_2$. The optimal preemptive schedule obtained by McNaughton wrap-around rule is given in Fig. 8.2, where we have $L_2 = 21$ and $L_2/3 = 7$, but $p_{5,2} = 8$. Finally, the operations $O_{i,3}, i = 1, \ldots, 5$ can be processed on two parallel machines in $\mathcal{M}_3$. The optimal preemptive schedule obtained by McNaughton wrap-around rule is given in Fig. 8.3, where $L_3 = 12$ and $L_3/2 = 6$. The new instance $I'$ obtained from the original instance $I$ and the schedules $S_1$, $S_2$, and $S_3$ is given in Table 8.2. Clearly, the new instance may have the operations of each job spread between more machines than the instance $I$, for example, job $J_3$ is processed by machines $M_1$ for $4\frac{1}{2}$ and $M_2$ for $2\frac{1}{2}$ and $M_4$ for 2 units of time in $I'$. This gives $P_1 = 9$, which is the same as in $I$, see Table 8.1. The instance $I'$ is an instance of $O|\text{pmtn}|C_{\max}$. By König's edge-coloring theorem the makespan of optimal preemptive open shop schedule for $I'$ equals

$$\max\left\{\max_i\{P_i\}, \max_h\{\frac{L_h}{|\mathcal{M}_h|}\}\right\}. \tag{8.2}$$

Compare this to the makespan of an optimal preemptive open shop with a single machine in each stage

$$\max\{\max_i\{P_i\}, \max_h\{L_h\}\}. \tag{8.3}$$

An optimal schedule for the instance $I'$ is given in Fig. 8.4. The schedule is optimal for the instance $I$ since $C_{\max} = P_5$.

**Fig. 8.3** An optimal wrap-around schedule $S_3$ of $O_{1,3}, \ldots, O_{5,3}$ schedule on parallel machines $M_6$ and $M_7$

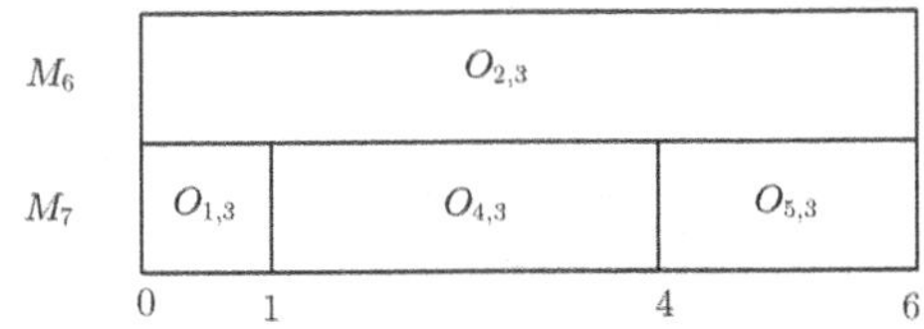


**Table 8.2** The instance $I'$ obtained by applying the wrap-around rule to each of the three stages

| $J_j$ | $M_1$ | $M_2$ | $M_3$ | $M_4$ | $M_5$ | $M_6$ | $M_7$ |
|---|---|---|---|---|---|---|---|
| $J_1$ | 4 | 0 | 2 | 0 | 0 | 1 | 0 |
| $J_2$ | 3 | 0 | 5 | 0 | 0 | 5 | 1 |
| $J_3$ | $4\frac{1}{2}$ | $2\frac{1}{2}$ | 0 | 2 | 0 | 0 | 0 |
| $J_4$ | 0 | 4 | 0 | 4 | 0 | 0 | 3 |
| $J_5$ | 0 | 5 | 0 | 0 | 8 | 0 | 2 |

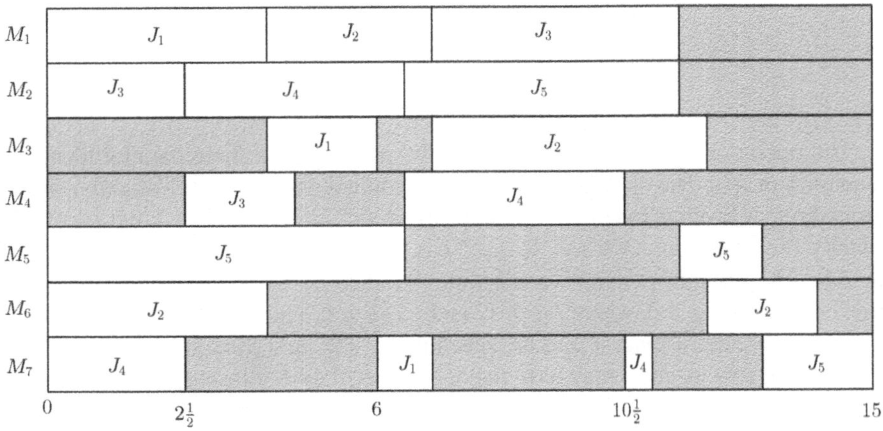


**Fig. 8.4** An optimal schedule for the instance $I'$ in Table 8.1 obtained from $I$ in Table 8.2

## 8.1.2 Multiple-Operation Machines

The model with multiple-operation machines was introduced by Lawler et al. [11]. Like in the traditional open shop job, $J_i$ consists of operations $O_{i,1}, \ldots, O_{i,m}$. However, the operation $O_{i,h}$ is no longer required to be processed on a single machine $M_h$, as in the traditional open shop, or on parallel machines in the set $\mathcal{M}_h$, as in the open shop with identical parallel machines, but on machines in the set $\mathcal{M}_{i,h}$. In contrast to parallel open shop some machines in $\mathcal{M}_{i,h}$ may also process operations $O_{j,k}$ where $h \neq k$, which happens when $\mathcal{M}_{i,h} \cap \mathcal{M}_{j,k} \neq \emptyset$. This makes it more difficult to determine the minimum maximum machine workload $L^*$ in an optimal solution. We use the network $\mathcal{N}$ in Fig. 8.5 to determine the smallest $L$, i.e., $L^*$, such that the $s - t$ flow in $\mathcal{N}$ equals

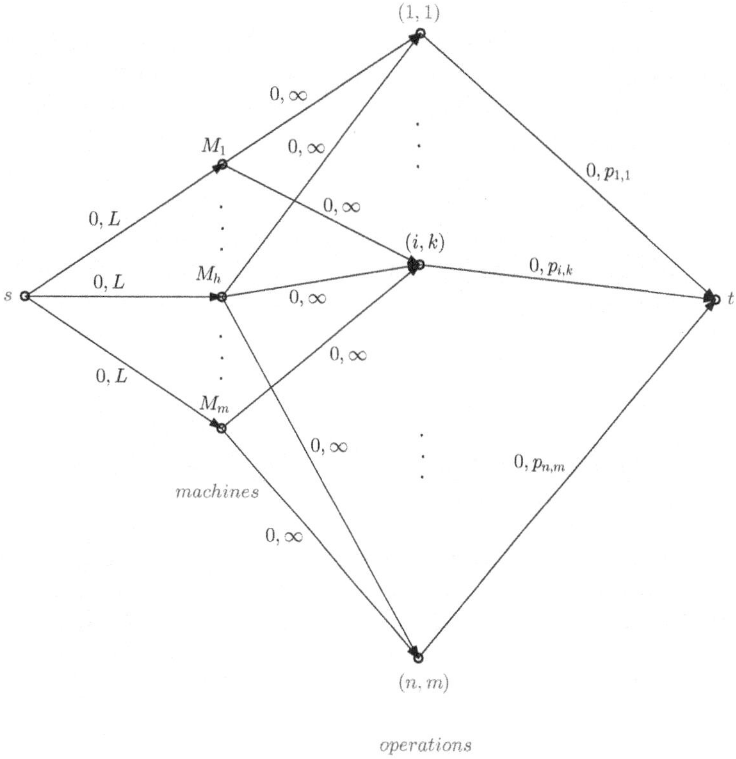


**Fig. 8.5** An arc from machine $M_h$ to operation $(i, k)$ exists if and only if $M_h \in \mathcal{M}_{i,k}$

$$F = \sum_{i=1}^{n} \sum_{h=1}^{m} p_{i,h}.$$

The search for $L^*$ is a binary search between the lower bound

$$LB = \left\lfloor \frac{\sum_{i,k} p_{i,k}}{m} \right\rfloor$$

and the upper bound

$$UB = \max_{h}\{L_h\}, \tag{8.4}$$

where

$$L_h = \sum_{O_{i,k} \in o_h} p_{i,k} \tag{8.5}$$

and $o_h = \{O_{i,k} : M_h \in \mathcal{M}_{i,k}\}$ be the set of all operations $O_{i,k}$ that can be processed on machine $M_h$. The binary search cannot however be limited to integers. Fortunately, the fractions are not too small in an optimal $L^*$, which makes the binary search to run in polynomial time. To see that, consider the set $\mathcal{L}$ of machines $M_h$ for which the capacity $L^*$ on the arc from $s$ to $M_h$ is attained by the $s - t$ flow in $\mathcal{N}$. Let $\mathcal{L}$ has the smallest cardinality among all such flows. Then no operation $O_{i,k}$ with positive flow from $M_h \in \mathcal{L}$ to $(i, k)$ has a positive flow $f > 0$ from machine $M \notin \mathcal{L}$ to $(i, k)$. If it does, then $f$ could be increased by some amount, and at the same time $M_h$ would no longer be saturated. Thus the contradiction. Let $O$ be the set of operations $O_{i,k}$ with positive flow from some $M_h \in \mathcal{L}$ to $(i, k)$. We have

$$|\mathcal{L}|L^* = \sum_{O_{i,k} \in O} p_{i,k}. \tag{8.6}$$

Therefore, if $L^*$ is fractional, then its fraction is $\frac{\alpha}{|\mathcal{L}|}$ for some $1 \leq \alpha < |\mathcal{L}| \leq m$. Thus the binary search runs in polynomial time. The whole algorithm however is not strongly polynomial, see Tardos [19], since the number of steps depends on $\log p_{\max}$. A strongly polynomial algorithm however can be obtained as follows. For each $h = 1, \ldots, m$ and $O_{i,k} \in o_h$ define a non-negative variable $x_{h,i,k}$ to denote the amount of operation $O_{i,k}$ processed on $M_h \in \mathcal{M}_{i,k}$. Let $L$ be a variable to denote the common bound on machine workloads. The following linear program LP determines the smallest possible bound $L$, i.e., $L^*$

$$\min L \tag{8.7}$$

subject to

$$\sum_{M_h \in \mathcal{M}_{i,k}} x_{h,i,k} = p_{i,k} \qquad \text{for each operation } O_{i,k}; \tag{8.8}$$

$$\sum_{O_{i,k} \in o_h} x_{h,i,k} \leq L \qquad \text{for each machine } M_h. \tag{8.9}$$

The linear program is *combinatorial* since each variable is multiplied by 1 in the constraints (8.8) and (8.9). Thus the numbers in the constraint matrix of the LP are bounded by a polynomial in $n$ and $m$. Tardos proposes a strongly polynomial algorithm for combinatorial linear programs that runs in time, which is polynomial in $n$ and $m$ but is independent of the length of binary encoding of $p_{i,k}$. The length of rational number $\frac{p}{q}$ is the number of bits in its binary encoding, $\lceil \log_2(|p| + 1) \rceil + \lceil \log_2(|q| + 1) \rceil$, see Tardos [19] for details and Dadush et al. [6] for recent developments.

The open shop instance corresponding to the solution is then obtained by taking processing time

$$q_{i,h} = \sum_{k:M_h \in \mathcal{M}_{i,k}} x_{h,i,k}$$

for the operation $O_{i,h}$. By König's edge-coloring theorem the makespan of an optimal preemptive open shop equals

$$\max\{\max_i\{Q_i\}, L^*\}, \tag{8.10}$$

where by (8.8)

$$Q_i = \sum_{h=1}^{m} q_{i,h} = \sum_{h=1}^{m} \sum_{k:M_h \in \mathcal{M}_{i,k}} x_{h,i,k} = \sum_{k=1}^{m} \sum_{M_h \in \mathcal{M}_{i,k}} x_{h,i,k} = \sum_{k=1}^{m} p_{i,k} = P_i. \tag{8.11}$$

To illustrate, consider processing times of the instance from Table 8.1. Let the operations in the set

$$\{O_{1,1}, O_{1.2}, O_{1,3}, O_{2,1}, O_{2,2}, O_{4,2}, O_{4,3}, O_{5,1}, O_{5,2}, O_{5,3}\}$$

be processed on $M_1$, the operations in the set

$$\{O_{1,1}, O_{1.2}, O_{1,3}, O_{2,1}, O_{2,2}, O_{2,3}, O_{3,1}, O_{3,2}, O_{3,3}, O_{4,1}\}$$

on $M_2$, and the operations in the set

$$\{O_{2,3}, O_{3,1}, O_{3,2}, O_{3,3}, O_{4,1}, O_{4,2}, O_{4.3}, O_{5,1}, O_{5,2}, O_{5,3}, \}$$

on $M_3$. Thus $\mathcal{M}_{1,1} = \{M_1, M_2\}$ and $\mathcal{M}_{3,1} = \{M_2, M_3\}$ for example. An optimal solution to the LP with $L^* = 18\frac{2}{3}$ is shown in Table 8.3. The open shop instance $I'$ corresponding to the solution is given in Table 8.4.

## 8.2 Non-Preemptive Open Shops with Parallel Machines

In this section we consider the problem $O(P)||C_{\max}$ with single-operation machines. The problem is NP-hard in the strong sense since it includes $P||C_{\max}$ as a subproblem for example. Schuurman and Woeginger [17] show a 2-approximation polynomial-time algorithm for the problem $O(P)||C_{\max}$ with an arbitrary number of stages and machines. The algorithm extends the idea of Bárány and Fiala [2].

**Theorem 8.1** *A greedy algorithm for $O(P)||C_{\max}$ produces schedules that are shorter than twice optimal makespan.*

***Proof*** Let

**Table 8.3** An optimal solution to the LP

| $O_{i,k}$ | $x_{1,i,k}$ | $x_{2,i,k}$ | $x_{3,i,k}$ |
|---|---|---|---|
| $O_{1,1}$ | 0 | 4 | 0 |
| $O_{1,2}$ | 0 | 2 | 0 |
| $O_{1,3}$ | 0 | 1 | 0 |
| $O_{2,1}$ | $2\frac{5}{6}$ | $\frac{1}{6}$ | 0 |
| $O_{2,2}$ | $4\frac{5}{6}$ | $\frac{1}{6}$ | 0 |
| $O_{2,3}$ | 0 | $1\frac{1}{6}$ | $4\frac{5}{6}$ |
| $O_{3,1}$ | 0 | 7 | 0 |
| $O_{3,2}$ | 0 | 2 | 0 |
| $O_{3,3}$ | 0 | 0 | 0 |
| $O_{4,1}$ | 0 | $1\frac{1}{6}$ | $2\frac{5}{6}$ |
| $O_{4,2}$ | 2 | 0 | 2 |
| $O_{4,3}$ | $1\frac{1}{2}$ | 0 | $1\frac{1}{2}$ |
| $O_{5,1}$ | $2\frac{1}{2}$ | 0 | $2\frac{1}{2}$ |
| $O_{5,2}$ | 4 | 0 | 4 |
| $O_{5,3}$ | 1 | 0 | 1 |

**Table 8.4** The open shop instance $I'$ obtained from the solution in Table 8.3

| $J_i$ | $p_{i,1}$ | $p_{i,2}$ | $p_{i,3}$ |
|---|---|---|---|
| $J_1$ | 0 | 7 | 0 |
| $J_2$ | $7\frac{2}{3}$ | $1\frac{1}{2}$ | $4\frac{5}{6}$ |
| $J_3$ | 0 | 9 | 0 |
| $J_4$ | $3\frac{1}{2}$ | $1\frac{1}{6}$ | $6\frac{1}{3}$ |
| $J_5$ | $7\frac{1}{2}$ | 0 | $7\frac{1}{2}$ |

$$L_h = \frac{1}{m_h} \sum_{i=1}^{n} p_{i,h} \tag{8.12}$$

be the average workload for stage $h = 1, \ldots, m$ where $m_h = |\mathcal{M}_h|$. Start with an empty schedule that has all machines available from $t = 0$ on. For each operation $O_{i,h}$ not yet scheduled, determine the earliest $t$ such that:

1. There is at least one machine $M$ in $\mathcal{M}_h$ with no job processed on $M$ at $t$.
2. Job $J_i$ is not processed on any machine in $\mathcal{M} \setminus \mathcal{M}_h$ at $t$.

The greedy algorithm proposed in Schuurman and Woeginger [17] works as follows. Select a not yet scheduled operation $O_{i,h}$ with the smallest $t$, break ties arbitrarily, and schedule it to start at $t$ on $M$. Continue until all operations are scheduled. The schedule $S$ so obtained is feasible since by (1) no operation can start on machine $M \in \mathcal{M}_h$, $h = 1, \ldots, m$ at $t$ when all machines in $\mathcal{M}_h$ are occupied by processing other operations at $t$, and by (2) no operation of job $J_i$

can start at $t$ if some other operation of job $J_i$ is being processed at $t$. Let $O_{i,h}$ be the last operation scheduled on machine $M \in \mathcal{M}_h$, $h = 1, \ldots, m$, in $S$. We have $C^h_{\max} \leq L_h + P_i - \frac{p_{i,h}}{m_h}$, $p_{i,h} > 0$. To see this we observe that $O_{ih}$ cannot start earlier in $S$ since either (1) or (2) does not hold for any $t$ earlier than the start time $s$ of $O_{i,h}$ in $S$, which means that for any $t < s$ either jobs are processed on all machines in $\mathcal{M}_h$ or job $J_i$ is processed on some machine in $\mathcal{M} \setminus \mathcal{M}_h$. Thus $s \leq L_h - \frac{p_{i,h}}{m_h} + P_i - p_{i,h}$, which proves $C^h_{\max} \leq L_h + P_i - \frac{p_{i,h}}{m_h}$. By (8.12) we have $C^*_{\max} \geq L_h$, and clearly $C^*_{\max} \geq P_i$. Hence, $C^h_{\max}/C^*_{\max} \leq 2 - \frac{p_{i,h}}{m_h C^*_{\max}} < 2$ for each $h = 1, \ldots, m$. Thus the greedy algorithm produces schedule $S$ with makespan $C_{\max}$ such that $C_{\max}/C^*_{\max} < 2$ for each instance of $O(P)||C_{\max}$. □

Sevastianov and Woeginger [18] propose a PTAS for a fixed number of stages and the number of processors in each stage bounded by a constant. Jansen and Sviridenko [9] strengthen this result by showing a PTAS for a fixed number of stages but allow the number of processors in each stage to be arbitrary. Kononow and Sviridenko [10] give a PTAS for a fixed number of stages and the number of processors in each stage bounded by a constant, but they allow operation release dates. Chen and Strusevich [4] give a $(2 - \frac{1}{\max\{m_1, m_2\}^2})$-approximation algorithm for two stages with $m_1$ machines in one stage and $m_2$ machines in the other. Naderi et al. [16] propose a mixed integer linear program and heuristics for the total completion time minimization problem, $O(P)||\sum C_i$. Adak et al. [1] review mathematical programming formulations, heuristic, metaheuristics, and test data for the open shop with parallel machines.

Chen et al. [5] and Dong et al. [7] propose a somewhat different model. The set of jobs $\mathcal{J}$ processed on $m$ machines $\mathcal{M}$ of an open shop needs to be partitioned into $k$ disjoint subsets $\mathcal{J}_1, \ldots, \mathcal{J}_k$ to be processed on sets of machines $\mathcal{M}_1, \ldots, \mathcal{M}_k$, respectively, in order to minimize the overall makespan. Each set $\mathcal{M}_h$, $h = 1, \ldots, k$, is a copy of $\mathcal{M}$. Thus we partition an open shop into $k$ open shops with fewer jobs each to minimize makespan. The problem is denoted by $P(O)||C_{\max}$ to underline the difference from the open shops with parallel machines discussed earlier. Chen et al. [5] show a 2-approximation algorithm for $Pk(O2)||C_{\max}$ and $\frac{3}{2}$-approximation algorithm for $P2(O2)||C_{\max}$. Observe that once a partition of jobs between $k$ two-machine open shops has been given in those problems, then an optimal schedule is easy to obtain since each of the $k$ two-machine open shops can be scheduled to optimality in linear time. Therefore, the key to the optimal solution for those problems is optimal partition of jobs between two-machine open shops. Dong et al. [7] give an FPTAS for $Pk(O2)||C_{\max}$, i.e., for partitioning a two-machine open shop into $k$ two-machine open shops. The $k$ is not part of the problem input. The partition of three-machine open shop into $k$ three-machine open shops, $Pk(O3)||C_{\max}$, looks more challenging since the existence of an $FPTAS$ for $Om||C_{\max}$, $m \geq 3$, is an open question, see Woeginger [21].

## 8.3 Applications

Matta and Patterson [14] and Matta [12] study a multiprocessor open shop scheduling in a large oncology center where patients (jobs) need to be tested by multiple testing departments all on a given day. The stages, i.e., types of tests carried out by various testing facilities, are as follows:

- Computer tomography (CT) scan
- Magnetic resonance imaging (MRI)
- Electrocardiogram (EKG)
- Echocardiogram (ECHO)
- Positron emission tomography (PET) scan
- Bone scan
- Pulmonary function test
- Barium swallow
- Barium enema
- X-ray
- Ultrasound imaging
- Mammogram
- Bone survey
- Blood draw

Matta [12] points out that the tests can be performed in any order on any patient, which makes her problem an instance of an open shop scheduling. Furthermore, each test can be performed by parallel identical machines, which makes it an instance of open shop with parallel single-operation machines. Finally, tests in any given stage take about the same time regardless of patient's gender, age, or illness, which makes the problem an instance of machine-proportionate open shop. This last claim holds since testing operations follow a well defined routine that results in little or no variance in their processing times. For instance an X-ray takes 30 min, while a bone scan takes 120 min regardless of the age, gender, or illness, see Matta and Elmaghraby [13]. Other applications given in Matta and Elmaghraby [13] include:

- A day spa
- Component tests before building a finished product
- College course registration

Vairaktarakis and Sahni [20] further expand the applications to include load balancing, see also Bondy and Murty [3], and preventive maintenance. Zhang et al. [22] apply a multiprocessor open shop scheduling with additional constraints to appointment scheduling for integrated practice units.

## References


1. Z. Adak, M.Ö. Arioğlu Akan, S. Bulkan, Multiprocessor open shop problem: literature review and future directions. J. Comb. Optim. **40**, 547–569 (2020)
2. I. Bárány, T. Fiala, Nearly optimum solution of multimachine scheduling problems (in Hungarian). Szigma **15**, 177–191 (1982)
3. J.A. Bondy, U.S.R. Murty, *Graph Theory with Applications* (North-Holland Publishing Co., 1976)
4. B. Chen, V.A. Strusevich, Worst-case analysis of heuristics for open shops with parallel machines. Eur. J. Oper. Res. **70**, 379–390 (1993)
5. Y. Chen, A. Zhang, G. Chen, J. Dong, Approximation algorithms for parallel open shop scheduling. Inf. Process. Lett. **113**, 220–224 (2013)
6. D. Dadush, B. Natura, L.A. Végh, Revisiting Tardos' Framework for Linear Programming: Faster Exact Solutions using Approximate Solvers. arXiv:2009.04942v1, September 2020
7. J. Dong, R. Jin, J. Hu, G. Lin, A fully polynomial time approximation scheme for scheduling on parallel identical two-stage openshops. J. Comb. Optim. **37**, 668–684 (2019)
8. T. Gonzalez, S. Sahni, Open shop scheduling to minimize finish time. J. ACM **23**, 665–679 (1976)
9. K. Jansen, M. Sviridenko, Polynomial time approximation schemes for the multiprocessor open and flow shop scheduling problems, in *Lecture Notes in Computer Science 1770* (STACS 2000, 2000), pp. 455–465
10. A. Kononov, M. Sviridenko, A linear time approximation scheme for makespan minimization in an open shop with release dates. O. R. Lett. **30**, 276–280 (2002)
11. E.L. Lawler, M.G. Luby, V.V. Vazirani, Scheduling open shops with parallel machines. Oper. Res. Lett. **1**, 161–164 (1982)
12. M.E. Matta, A genetic algorithm for the proportionate multiprocessor open shop. Comput. Oper. Res. **36**, 2601–2618 (2009)
13. M.E. Matta, S.E. Elmaghraby, Polynomial time algorithms for two special classes of the proportionate multiprocessor open shop. Eur. J. Oper. Res. **201**, 720–728 (2010)
14. M.E. Matta, S.S. Patterson, Evaluating multiple performance measures across several dimensions at a multi-facility outpatient center. Health Care Manag. Sci. **10**, 173–194 (2007)
15. R. McNaughton, Scheduling with deadlines and loss functions. Manag. Sci. **6**, 1–12 (1959)
16. B. Naderi, S.M.T. Fatemi Ghomi, M. Aminnayeri, M. Zandieh, Scheduling open shops with parallel machines to minimize total completion time. J. Comput. Appl. Math. **235**, 1275–1287 (2011)
17. P. Schuurman, G.J. Woeginger, Approximation algorithms for multiprocessor open shop scheduling problem. Oper. Res. Lett. **24**, 157–163 (1999)
18. S.V. Sevastianow, G.J. Woeginger, Linear time approximation scheme for multiprocessor open shop problem. Discrete Appl. Math. **114**, 273–288 (2001)
19. É. Tardos, A strongly polynomial algorithm to solve combinatorial linear programs. Opns. Res. **34**, 250–256 (1986)
20. G. Vairaktarakis, S. Sahni, Dual criteria preemptive open shop problems with minimum finish time. Naval Res. Logist. **42**, 103–121 (1995)
21. G.J. Woeginger, The open shop scheduling problem, in eds. by R. Niedermeier, B. Vallée, *35th Symposium on Theoretical Aspects of Computer Science (STACS 2018), Leibniz International Proceedings in Informatics* (Dagstuhl Publishing, 2018), pp. 4:1–4:12
22. P. Zhang, J.F. Bard, D.J. Morrice, K.M. Koenig, Extended open shop scheduling with resource constraints: Appointment scheduling for integrated practice units. IISE Trans. **51**(10), 1037–1060 (2019)

# Chapter 9
# Compact Scheduling of Open Shops

## 9.1 Compact Schedules

We discussed dense schedules for open shop in Chaps. 3 and 5. We now turn our attention to *compact schedules* where for each job $J_i$ there is a time interval $[S_i, S_i + P_i]$ where the job is processed—this implies no-waiting between operations of the job—and where for each machine $M_h$ there is a time interval $[s_h, s_h + L_h]$ where the machines process all its operations—this implies that compact schedules forbid idle time inserted *between* operations on machines. The compact schedules are motivated by compact timetabling where students do not wait for their next class to attend, and teachers do not wait for their next class to teach, see Asratian and Kamalian [9]. Another motivation comes from scheduling interviews for job fairs. Bartholdi and McCroan [10] give an example of The Southeastern Public Interest Job Fair held each year in a November weekend when 25–50 US law firms and 100–200 law students from the Southeast come to Atlanta for job interviews. To obtain a timetable of the meetings between the firms (machines) and the students (jobs) one can model this problem as an open shop or bipartite graph edge-coloring problem. However, students have normally fewer meetings than law firms, which may result in timetables that possibly force students to wait between consecutive interviews, and force law firms to stay idle. Thus the following question comes up naturally: Is it possible to obtain a compact timetable where students do not wait between interviews and law firms are not idle between interviews? If such compact timetables exist, find the one with minimum makespan. Hansen [29] provides yet another motivating example where a school looks for a schedule of parent–teacher meetings in time slots so that every person's meetings take place in consecutive time slots.

The compact schedules may not exist for some instances of open shop, for example, the following open shop with $m = 3$ machines and $n = 2$ jobs does not have a compact schedule:



W. Kubiak, *A Book of Open Shop Scheduling*, International Series in Operations Research & Management Science 325,
https://doi.org/10.1007/978-3-030-91025-9_9

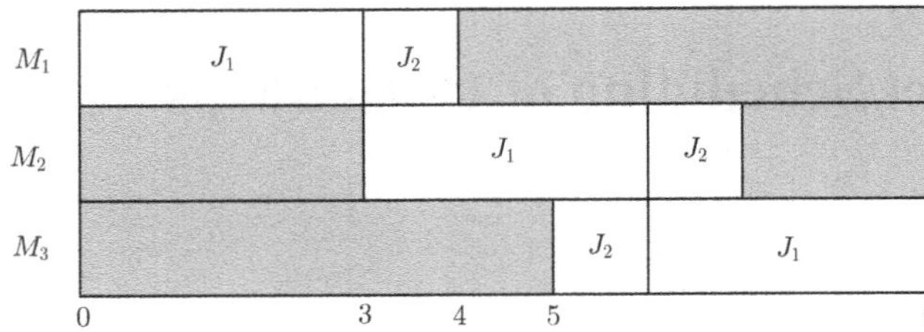


**Fig. 9.1** An open shop instance with no compact schedule

**Fig. 9.2** A compact schedule for $\mathbb{P}_{MSW}$ that permits preemptions

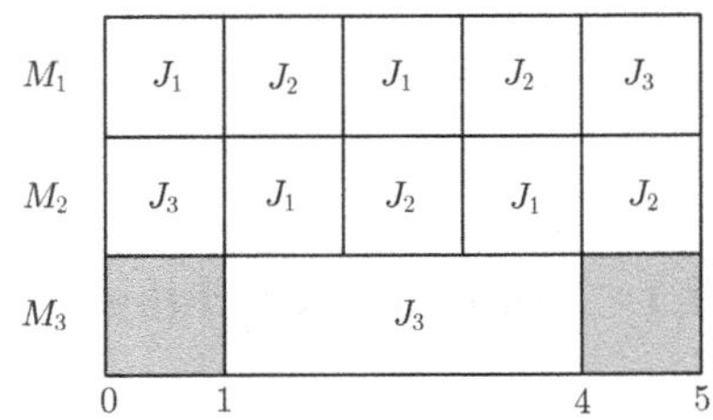


$$\mathbb{P}_{PK} = \begin{bmatrix} 3 & 3 & 3 \\ 1 & 1 & 1 \end{bmatrix},$$

see Petrosyan and Khachatrian [45]. The schedule for $\mathbb{P}_{PK}$ in Fig. 9.1 is compact on each machine, and also job $J_1$ is processed without waiting between its three operations. Job $J_2$, however, needs to wait between its operations on $M_1$ and $M_2$ in the schedule. No feasible schedule for $\mathbb{P}_{PK}$ is compact, see Problem 9.1.

Compact schedules permit preemptions as shown by the following open shop given by Mahadev et al. [40]:

$$\mathbb{P}_{MSW} = \begin{bmatrix} 2 & 2 & 0 \\ 2 & 2 & 0 \\ 1 & 1 & 3 \end{bmatrix}.$$

Figure 9.2 shows a compact schedule for $\mathbb{P}_{MSW}$, where $J_1$ is processed in $[0, 4]$, $J_2$ in $[1, 5]$, and $J_3$ in $[0, 5]$ and machines $M_1$ and $M_2$ are occupied in $[0, 5]$, and $M_3$ in $[1, 4]$. However, both $J_1$ and $J_2$ are preempted in the schedule.

Even for open shop instances with 0-1 operations, compact schedules may not exist, for example, the following instance $\mathbb{P}_M$ with $n + m = 19$ and $\Delta = 15$:

$$\mathbb{P}_M = \begin{bmatrix} 1\,1\,0\,1 \\ 1\,1\,0\,1 \\ 1\,1\,0\,1 \\ 1\,1\,0\,1 \\ 1\,1\,0\,1 \\ 1\,0\,1\,1 \\ 1\,0\,1\,1 \\ 1\,0\,1\,1 \\ 1\,0\,1\,1 \\ 1\,0\,1\,1 \\ 0\,1\,1\,1 \\ 0\,1\,1\,1 \\ 0\,1\,1\,1 \\ 0\,1\,1\,1 \\ 0\,1\,1\,1 \end{bmatrix}$$

was given by Małafiejski [24]. Other instances without compact schedules were given by Mirumyan ($n + m = 19$ and $\Delta = 15$), Sevast'janow ($n + m = 28$ and $\Delta = 21$), Erdős ($n + m = 27$ and $\Delta = 13$), Hertz and de Werra ($n + m = 21$ and $\Delta = 14$), see Petrosyan and Khachatrian [45], and Jensen and Toft [32] for further details. Petrosyan and Khachatrian [45] give examples of open shops with $n + m = 20$ and $\Delta = 11$, and $n + m = 19$ and $\Delta = 12$ that have no compact schedules thus leaving the existence of such open shops for $\Delta = 4, \ldots, 10$ open, see Jensen and Toft [32]. On a positive side, Giaro [22] uses computer search to prove that each open shop with $n + m \leq 14$ has a compact schedule, and recently Khachatrian and Mamikonyan [35] extended the computer search to prove the result for $n + m \leq 15$.

### *9.1.1 Interval Edge Coloring*

Unless explicitly stated otherwise our discussion in this chapter concentrates on an open shop scheduling problem with $0 - 1$ operations and makespan minimization. We denote this problem as $O|p_{i,j} \in \{0, 1\}, \text{compact}|C_{\max}$. Thus, the instances of the open shops can be recast as *simple* bipartite graphs, and the search for compact schedules can be recast as the search for *interval* edge coloring. The interval edge coloring of $G$ was introduced by Asratian and Kamalian [9] who define it as an edge coloring of $G$ with colors $1, \ldots, t$ where each color is used for at least one edge, and for each vertex the edges incident with the vertex are colored with colors making up an interval in $1, \ldots, t$.

Many results that prove the existence of interval edge coloring in bipartite graph $G$ have been obtained by first identifying a subgraph of $G$, then coloring its edges, and finally extending the edge coloring to all edges so that the resulting edge coloring is an interval edge coloring of $G$, see Theorems 9.5 and 9.4 for instance. The main challenge in this approach is to identify the right subgraph that is included in as many as possible bipartite graphs and ideally to prove that there is a polynomial-time test to check for a given bipartite graph whether it includes the required subgraph or not. The subgraphs in question have been searched for among factors of bipartite graphs. Thus we now briefly review main terminology and definitions concerning factors of graphs. We follow a survey of graph factors by Plummer [46], see also Akiyama and Kano [1].

We denote the set of vertices and the set of edges of graph $G$ by $V(G)$ and $E(G)$, respectively. Graph $H$ is a *subgraph* of $G$ if $V(H) \subseteq V(G)$ and $E(H) \subseteq E(G)$. A subgraph $H$ of $G$ *spans* $G$ if $V(H) = V(G)$. Subgraph $H$ of $G$ that spans $G$ is called a *factor* of $G$, see Plummer [46]. A factor defined in terms of its vertex degrees will be referred to as a *degree factor*. Hence, if a factor has all of its degrees equal to 1, it will be referred to as 1-factor or a *perfect matching*. A factor defined in terms of desired features of a subgraph other than vertex degree alone will be referred to as a *component factor*. For instance, a factor is made up of paths having endpoints in a given set of vertices.

## 9.2 Complexity of Short Compact Schedules

We study the complexity of the compact open shop scheduling with respect to the value of maximum vertex degree $\Delta$. That is, for a given bipartite graph $G = (V = X \cup Y, E)$, one asks whether there is an interval edge coloring of $G$ that uses $\Delta$ colors or not. Giaro [21] proved that the test can be done in polynomial time for $\Delta \leq 4$; however, the problem becomes $NP$-complete for any $\Delta \geq 5$.

For a bipartite $G$ with $\Delta \leq 2$, every connected component of $G$ is either a path or an even cycle. Thus, each component can be edge colored alternately with two colors. This gives the required interval edge coloring of $G$.

Let us now consider $\Delta = 3$.

### 9.2.1 *Test for* $\Delta = 3$

Giaro [21] shows the following lemma for bipartite graphs with maximum degree $\Delta = 3$ and minimum degree $d$ at least 2.

**Lemma 9.1** *A bipartite graph $G$ with $\Delta = 3$ and $d \geq 2$ has an interval $\Delta$-edge coloring if and only if it has a perfect matching.*

***Proof*** Suppose $I$ is a perfect matching in $G = (V, E)$. Color each edge in $I$ with color 2. The graph $H = (V, E \setminus I)$ has vertices with degree 1 or 2. Hence, $H$ is a collection of paths and even cycles; thus, each component of $H$ can be colored alternately with two colors 1 and 3. Therefore, each vertex of degree 2 in $G$ is colored with either 1 and 2 or 2 and 3, and each vertex of degree 3 in $G$ is colored with 1, 2, and 3. This gives an interval 3-edge coloring of $G$. Now suppose there is an interval 3-edge coloring of $G$ with 1, 2, and 3. Since $\Delta = 3$ and $d \geq 2$, for each vertex there is an edge incident with that vertex and colored with 2. All such edges make up a perfect matching. This proves the lemma. □

It remains to consider bipartite graphs $G = (V, E)$ with vertices of degree 1. Let $U \subseteq V$ be the set of all vertices of degree 1 in $V$. Take a copy $G^c = (V^c, E^c)$ of $G$ to create a graph $H = (V \cup V^c, E \cup E^c \cup \{(u, u^c) : u \in U\})$. The new graph $H$ is bipartite, and informally, machines in $G$ become jobs in $G^c$ and jobs in $G$ become machines in $G^c$, with $\Delta = 3$. If $H$ has an interval $\Delta$-edge coloring, then so does its subgraph $G$. On the other hand, if $G$ has an interval $\Delta$-edge coloring, then so does $G^c$. This interval edge coloring of $G$ and $G^c$ can be extended to an interval edge coloring of $H$ by coloring $(u, u^c)$ with color 2 if the edges incident with $u$ and $u^c$ are colored with either 1 or 3, and with color 1 if the edges incident with $u$ and $u^c$ are colored with 2. Therefore, an interval $\Delta$-edge coloring exists in $G$ if and only if it exists in $H$. Since $H$ meets the condition of Lemma 9.1, we can test the existence of interval $\Delta$-edge coloring in $G$ by computing a maximum cardinality matching in $H$ that can be done in $O(|E|\sqrt{|V|})$ time, see Micali and Vazirani [41].

### 9.2.2 Makespan Minimization for Compact Open Shops with $\Delta \leq 3$

The following result is obtained by Hansen [29], see also Giaro [21].

**Lemma 9.2** *A bipartite graph with $\Delta = 3$ has an interval edge coloring with at most 4 colors.*

***Proof*** Let $c$ be an edge coloring of a bipartite graph $G = (V, E)$ having $\Delta(G) = 3$ with a minimum number of colors. By König's edge-coloring theorem $c$ uses 3 colors, say 1, 2, and 3. Let $F \subset E$ be the subset of edges colored with 2 or 3 by $c$. We now show how to obtain interval edge coloring with at most 4 colors from $c$. Let $W \subseteq V$ be the subset of vertices incident with at least one edge in $F$. The graph $H = (W, F)$ is bipartite with $\Delta(H) = 2$. Hence, $H$ is a collection of paths and even cycles. Each connected component of $H$ is colored alternately with two colors 2 and 3. Moreover, each vertex of degree 1 in $H$ has degree 1 or 2 in $G$. We now show how to extend the edge coloring of $H$ to an interval edge coloring of $G$. Let $e = (v, u) \in E \setminus F$:

- $u, v \in W$, and $\deg(u) = \deg(v) = 2$ in $H$. Color $e$ with color 1. Thus, the three edges incident with $u$ and $v$ in $G$ become colored with 1, 2, and 3.
- $u, v \in W$, and $\deg(u) = 1$ and $\deg(v) = 2$ in $H$. If the edge incident with $u$ in $H$ is colored with 2, then color $e$ with 1. Otherwise, if the edge incident with $u$

in $H$ is colored with 3, then color $e$ with 4. Thus the three edges incident with $v$ in $G$ become colored with 1, 2, and 3 or 2, 3, and 4, and the two edges incident with $u$ in $G$ become colored with 1, and 2 or 3 and 4.

- $u, v \in W$, and $\deg(u) = \deg(v) = 1$ in $H$. Then $u$ is an endpoint of a path $C$ in $H$, and $v$ is an endpoint of a path $D$ in $H$. If $C \neq D$, then the path $C \cup D \cup \{u, v\}$ can be colored alternately with two colors 2 and 3. Hence, the two edges incident with $u$ and $v$ in $G$ become colored with 2 and 3. If $C = D$, then $C \cup \{u, v\}$ is an even cycle, which can be alternately colored with 2 and 3.
- $u \in V \setminus W$ and $v \in W$, and $\deg(v) = 2$ in $H$. Then $e$ is colored with 1 in $c$ so that the three edges incident with $v$ in $G$ remain colored with 1, 2, and 3.
- $u \in V \setminus W$ and $v \in W$, and $\deg(v) = 1$ in $H$. Then $v$ is an endpoint of a path $C$ in $H$. The path $C \cup \{u, v\}$ can be colored alternately with two colors 2 and 3. Hence, the two edges incident with $v$ in $G$ become colored with 2 and 3.
- $u, v \in V \setminus W$. Then, $\deg(u) = \deg(v) = 1$ in $G$, and $e$ is colored with 1.

Therefore, we obtain an interval edge coloring of $G$ with at most 4 colors. □

Lemmas 9.1 and 9.2 give a polynomial-time algorithm for the shortest compact open shop schedule for open shops $G$ with $\Delta \leq 3$. First, we observe that compact schedules for $\Delta \leq 2$ always exist. For $\Delta = 1$ or $\Delta = 2$ the shortest schedule has makespan 1 or 2, respectively, and it is easy to obtain in linear time as pointed out at the beginning of this section. For $\Delta = 3$, check whether $G$ has a perfect matching. This can be done in $O(|E|\sqrt{|V|})$ time. If $G$ has a perfect matching, then by Lemma 9.1 the shortest schedule has makespan 3, and the schedule can easily be obtained as in the proof of that lemma. Otherwise, by Lemma 9.2, the shortest schedule has makespan 4. The schedule can be obtained as in the proof of Lemma 9.2. First, we obtain a proper edge coloring of $G$ in $O(|E|)$ time by the algorithm of Cole et al. [17]. This gives us $H$ and its interval edge coloring. Second, we extend the interval edge coloring of $H$ to the interval edge coloring of $G$. The overall complexity is $O(|E|\sqrt{|V|})$.

### 9.2.3 Test for $\Delta = 4$

Giaro [21] proposes the following test for bipartite graphs with maximum degree $\Delta = 4$ and minimum degree $d$ at least 3.

**Lemma 9.3** *A bipartite graph $G$ with $\Delta = 4$ and $d \geq 3$ has an interval $\Delta$-edge coloring if and only if it has a 2-factor.*

***Proof*** Suppose $H = (V, F)$ is a 2-factor in a bipartite $G = (V, E)$. Thus $H$ is made up of even cycles. Color each even cycle of $H$ alternately with colors 2 and 3. The graph $G' = (V, E \setminus F)$ has vertices with degree 1 or 2. Hence, $G'$ is a collection of paths and even cycles; thus, each component of $G'$ can be colored alternately with two colors 1 and 4. Therefore, each vertex of degree 3 in $G$ is colored with either 1, 2, and 3 or 2, 3, and 4, and each vertex of degree 4 in $G$ is colored with 1, 2, 3, and 4. This gives an interval 4-edge coloring of $G$. Now suppose there is an interval 4-edge coloring of $G$. Each vertex of degree 4 is colored with 1, 2, 3, and 4, and each vertex of degree 3 is colored either with 1, 2, and 3, or 2, 3, and 4. Thus each vertex in $G$ is incident with an edge colored with 2 and with an edge colored with 3. Therefore, the edges colored with 2 or 3 make up a 2-factor of $G$. □

It remains to consider bipartite graphs $G = (V, E)$ with vertices of degree 1 or 2. Let $U \subseteq V$ be the set of all vertices of degree 1 in $V$. Take a copy $G^c = (V^c, E^c)$ of $G$ to create a graph $H = (W = V \cup V^c, F = E \cup E^c \cup \{(u, u^c) : u \in U\})$. The new graph $H$ is bipartite with $\Delta = 4$ and $d = 2$. If $H$ has an interval $\Delta$-edge coloring, then so does its subgraph $G$. On the other hand, if $G$ has an interval $\Delta$-edge coloring, then so does $G^c$. This interval edge coloring of $G$ and $G^c$ can be extended to an interval edge coloring of $H$ by coloring $(u, u^c)$ with color 2 if the edges incident with $u$ and $u^c$ are colored with either 1 or 3, with color 1 if the edges incident with $u$ and $u^c$ are colored with 2, and with color 3 if the edges incident with $u$ and $u^c$ are colored with 4. Therefore, an interval $\Delta$-edge coloring exists in $G$ if and only if it exists in $H$. We now deal with vertices of degree 2 in $H$. Let $T \subseteq W$ be the set of all vertices of degree 2 in $H$. Take a copy $H^c = (W^c, F^c)$ of $H$ to create a graph $I = (W \cup W^c, F \cup F^c \cup \{(u, u^c) : u \in T\})$. The new graph $I$ is bipartite with $\Delta = 4$ and $d = 3$. Unfortunately, it is no longer straightforward to show that if $I$ has an interval $\Delta$-edge coloring, then so does its subgraph $H$. Observe that the color on some edges $(u, u^c)$ can be right in the middle of the interval for $u$ and $u^c$ in the $\Delta$-edge coloring of $I$. Thus removing the edges from $\{(u, u^c) : u \in T\}$ results in the two copies $H$ and $H^c$, but their $\Delta$-edge colorings may no longer be interval for the vertices of degree 2. However, for either copy the resulting edge colorings meet the conditions of the following lemma proved by Giaro [21]:

**Lemma 9.4** *Let $G$ be a bipartite graph with $\Delta \leq 4$. If $G$ has a 4-edge coloring with colors 1, 2, 3, and 4 that is interval for each vertex of degree 3 or 4, and for each vertex of degree 2 the difference between colors of edges incident with the vertex does not exceed 2, then $G$ has an interval 4-edge coloring with colors 1, 2, 3, and 4.*

This proves that the existence of an interval $\Delta$-edge coloring of $I$ implies the existence of an interval $\Delta$-edge coloring of $H$.

On the other hand, if $H$ has an interval $\Delta$-edge coloring, then so does $H^c$. This interval edge coloring of $H$ and $H^c$ can be extended to an interval edge coloring of $I$ by coloring $(u, u^c)$ with color 3 if the two edges incident with $u$ and $u^c$ are colored with 1 and 2, with color 1 if the two edges incident with $u$ and $u^c$ are colored with 2 and 3, and with color 2 if the two edges incident with $u$ and $u^c$ are colored with 3 and 4. Therefore, an interval $\Delta$-edge coloring exists in $G$ if and only if exists in $I$. Since $I$ meets the condition of Lemma 9.3, we can test the existence of $\Delta$-edge coloring in $G$ by checking whether $G$ has a 2-factor. The test can be done in $O(|E|\sqrt{|V|})$ time whenever $\Delta$ is constant by reduction to the maximum cardinality matching problem.

### 9.2.4 Makespan Minimization for Compact Open Shops with $\Delta \leq 4$

Let us begin with bipartite graph $G = (X, Y, E)$ having each vertex of degree 4 or 2. The $G = (X, Y, E)$ can be partitioned into bipartite subgraphs $G_1 = (X, Y, E_1)$ and $G_2 = (X, Y, E_2)$ such that $E_1 \cap E_2 = \emptyset$, $E_1 \cup E_2 = E$, and $\deg_G(v) = 2\deg_{G_1}(v) = 2\deg_{G_2}(v)$, see Giaro [21]. We leave the proof as an exercise in Problem 9.3. Then the $G_1 = (X, Y, E_1)$ and $G_2 = (X, Y, E_2)$ are collections of even cycles and paths. Thus the edges of $G_1$ can be colored with colors 2 or 3, and the edges of $G_2$ with colors 1 or 4. This gives an interval edge coloring with 1, 2, 3, and 4 of the edges incident with each vertex of degree 4 in $G$. Each vertex of degree 2 in $G$ has the edges incident with the vertex colored with 2 and 1, or 2 and 4, or 3 and 1, or 3 and 4. Thus, the difference between the colors does not exceed 2. Therefore, the edge coloring meets the assumptions of Lemma 9.4, which guarantees an interval 4-edge coloring of $G$. This result can be extended to bipartite graphs with each vertex of degree 4 or 2 or 1 by using the construction described in Sect. 9.2.3. Therefore, the following lemma, see Giaro [21], holds.

**Lemma 9.5** *A bipartite graph $G$ with $\Delta = 4$ and no vertex of degree 3 has an interval $\Delta$-edge coloring.*

Observe that Lemmas 9.3 and 9.5 do *not* give a polynomial-time algorithm for the shortest compact open shop schedules for open shops $G$ with $\Delta \leq 4$.

Lemma 9.5 indicates that the presence of vertices of degree 3 in $G$ may be the cause of the difficulty in getting such an algorithm. Considerable efforts have been made to understand the difficulty, in particular for (3, 4)-biregular bipartite graphs. A bipartite graph $G = (X, Y, E)$ is (3, 4)-biregular if each vertex in $X$ has degree 3 and each vertex in $Y$ has degree 4. In other words, each job has exactly three unit-time operations, and each machine workload equals 4 (equivalently, for dual open shop, each job has exactly 4 unit-time operations, and each machine workload equals 3). We shall leave details of those efforts for Sect. 9.3. It now suffices to say that lower bound on the makespan for such graphs is 6, see Theorem 9.6. Moreover, Theorems 9.3, 9.4, and 9.5 show sufficient conditions for (3, 4)-biregular bipartite graphs to admit makespan 6. However, some of those conditions remain $NP$-complete to check. The computational complexity status of the following problem remains open.

**Problem 9.1** Given a bipartite graph $G$ with $\Delta = 4$. Find a compact schedule, if any, with minimum makespan or conclude that no compact schedule for $G$ exists.

Actually, the following problem posed by Jensen and Toft [32], see also Petrosyan and Khachatrian [45], remains open.

**Problem 9.2** Is there a bipartite graph $G$ with $4 \leq \Delta \leq 10$ that does not have interval edge coloring?

### 9.2.5 Test for $\Delta \geq 5$

The problem to determine whether a bipartite graph has an interval edge coloring is shown $NP$-complete by Sevast'janow [49]. However, we can observe that the $NP$-completeness of the problem to decide whether there is an interval $\Delta(G)$-edge coloring of a bipartite graph $G$ follows from the Gonzalez's [26] proof of $NP$-hardness of $O|p_{i,j} \in \{0, 1\}|\sum C_i$, see Sect. 3.6.2 for the proof details. Actually the proof in Sect. 3.6.2 simplifies significantly when feasible schedules are limited to compact schedules. Moreover, the optimal compact schedules for the instances in the proof are ideal, that is a schedule for the instance that minimizes makespan minimizes the total completion time at the same time, and the other way round, a schedule for the instance that minimizes the total completion time minimizes makespan at the same time. Thus, we get $NP$-completeness for compact open shop scheduling for both the makespan and the total completion time. Observe, however, that each machine workload in the proof equals 15 and length of each job equals 5; hence, $\Delta = 15$ in the bipartite graph of the open shop. Now we strengthen this result by showing that the open shop compact scheduling remains $NP$-complete for bipartite graphs with maximum degree $\Delta = 5$, see Giaro [21].

**Theorem 9.1** *The problem to decide whether there is an interval $\Delta(G)$-edge coloring of a bipartite graph $G$ with $\Delta(G) = 5$ is $NP$-complete.*

***Proof*** We follow the proof given by Giaro, see also Giaro et al. [23]. The point of departure in that proof is the bipartite edge pre-coloring extension problem where an instance is a bipartite graph $H$ with each vertex of degree 1 or 3, and some pendant edges (an edge $(u, e)$ is pendant if $\deg(u) = 1$ or $\deg(v) = 1$) pre-colored with 0 or 1. Can we extend this pre-coloring to an edge coloring of $H$ using colors 0, 1, or 2? The problem is $NP$-complete, see Even et al. [20], Giaro [21], and Kubale [37].

Let us first select a list of five colors $L = \{-2, -1, 0, 1, 2\}$. Though the choice of the interval of five colors is arbitrary, it will become later clear why this particular choice fits the proof best. Second let us consider a series of simple bipartite graphs that will be appended to $H$ in order to enforce the required pre-coloring on the pendant edges of $H$ and to enforce a 3-edge coloring of $H$. Let us start with a bipartite graph $G$ in Fig. 9.3. Any interval edge coloring of $G$ using colors from $L$ uses color 0 on one edge incident with $d$, on one edge incident with $e$, and on one edge incident with $f$. Thus any interval edge coloring of $G$ using colors from $L$ forbids color 0 on the edge $(a, u)$. This is the main feature of $G$ that will be used in the proof.

The graph $G$ will be used as a main building block in the construction of bipartite graphs $G_0$, $G_{\pm 1}$, and $G_{\pm 2}$. The required features of those graphs will immediately follow from the main feature of $G$. Any interval edge coloring of $G_0$ in Fig. 9.4 using colors from $L$ uses 0 on one edge incident with $\gamma$; moreover, this edge must be $(\gamma, \theta)$.

It then follows that any interval edge coloring of $G_{\pm 1}$ in Fig. 9.5 using colors from $L$ uses either $-1$ or 1 on the edge $(\sigma, \omega)$.

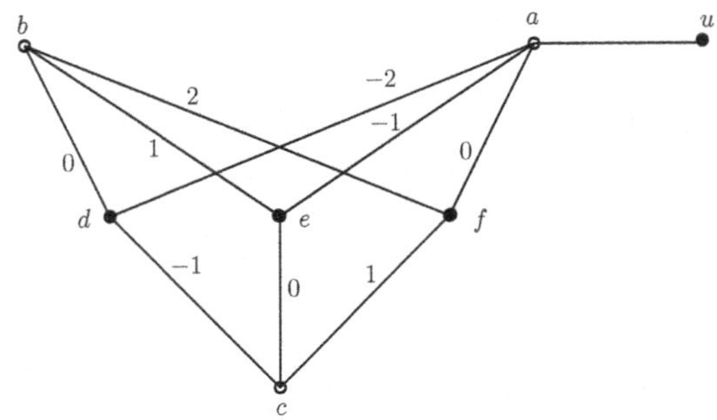


**Fig. 9.3** A bipartite graph $G$ that forbids color 0 on the edge $(a, u)$ in any interval edge coloring using colors in $L$

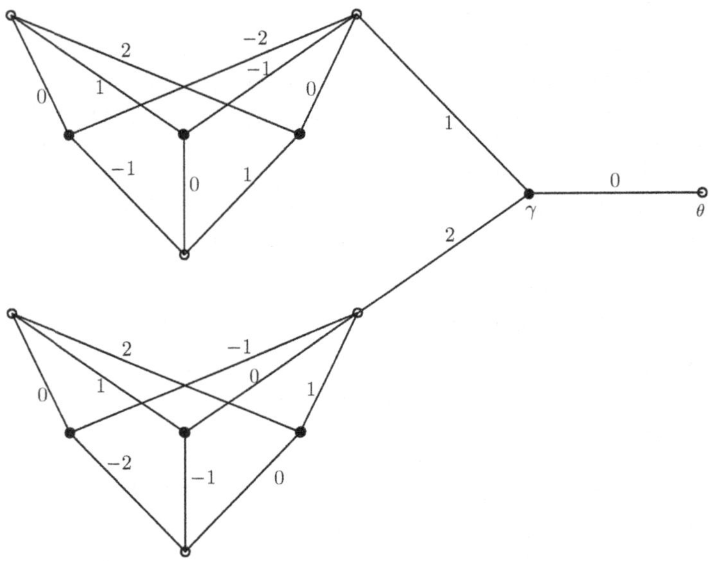


**Fig. 9.4** A bipartite graph $G_0$ that enforces color 0 on the edge $(\gamma, \theta)$ in any interval edge coloring using colors in $L$

Finally, a copy of $G_0$ and two copies of $G_{\pm 1}$ can be used to construct $G_{\pm 2}$ in Fig. 9.6 that enforces colors $-2$ or 2 on the edge $(\alpha, \beta)$ in any interval edge coloring using colors in $L$.

Let us now return to graph $H$, to each vertex $v$ of degree 3 in $H$ append a copy of $G_{\pm 1}$ by making $\omega = v$ and a copy of $G_{\pm 2}$ by making $\beta = v$. Thus each vertex of degree 3 in $H$ becomes a vertex of degree 5. That is not all, for each pendant edge $(v, w)$ in $H$, where $v$ is of degree 1, which is pre-colored with 0 append a copy of $G_0$ by making $\gamma = v$ and $\theta = w$. Finally, for each pendant edge $(v, w)$ in $H$, where $v$ is of degree 1, which is pre-colored with 1 append a copy of $G_{\pm 1}$ by making $\sigma = v$ and $\omega = w$. This defines graph $H'$ that is bipartite and with maximum degree $\Delta = 5$.

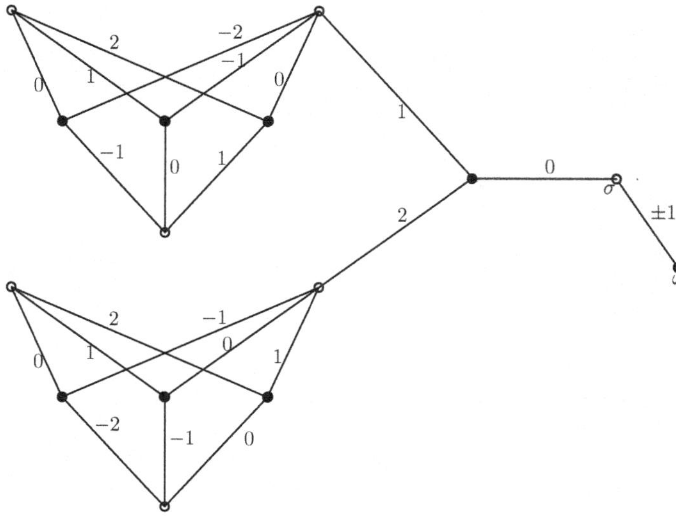


**Fig. 9.5** A bipartite graph $G_{\pm 1}$ that enforces colors $-1$ or 1 on the edge $(\sigma, \omega)$ in any interval edge coloring using colors in $L$

Suppose there is an edge coloring of $H$ with colors 0, 1, or 2 that respects the pre-coloring of the pre-colored pendant edges. Then the three edges incident with any vertex $v$ of degree 3 in $H$ are colored by 0, 1, and 2. The remaining two edges incident with $v$ in $H'$ come from a copy of $G_{\pm 1}$ and a copy of $G_{\pm 2}$. They can then be colored with $-1$ and $-2$, respectively, to make an interval edge coloring of $v$ using the interval $L$. The remaining edges in the copies of $G_{\pm 1}$ and $G_{\pm 2}$ can be colored as in Figs. 9.5 and 9.6. Any pendant edge $(v, w)$ in $H$ pre-colored with 0 is replaced by a copy $G_0$ for which the interval edge coloring is shown in Fig. 9.4, and any pendant edge $(v, w)$ in $H$ pre-colored with 1 is replaced by a copy $G_{\pm 1}$ for which the interval edge coloring is shown in Fig. 9.5 with color 1 on the edge $(\sigma, \omega)$. This clearly gives the required interval edge coloring of $H'$ using colors from $L$.

Now suppose there is an interval edge coloring of $H'$ using colors from $L$. Consider a vertex $v$ with degree 3 in $H$, thus with degree 5 in $H'$. There are five different colors on the five edges incident with $v$. Two of these edges come from the copies $G_{\pm 1}$ and $G_{\pm 2}$: one with the color 1 or $-1$ and the other with the color 2 or $-2$. The remaining three belong to graph $H$: one is colored with 0, one with $-1$ or 1, and one with 2 or $-2$. Thus taking the absolute value of the color gives the edge coloring of the edges incident with $v$ with three colors 0, 1, and 2. By definition of $H$ each its vertex is either degree 3 or degree 1; thus we get a proper edge coloring of $H$ using colors 0, 1, or 2. Finally, the interval edge coloring of the copy of $G_0$ that replaced a pendant edge of $H$ pre-colored with 0 gives that edge color 0, and the interval edge coloring of the copy of $G_{\pm 1}$ that replaced a pendant edge of $H$ pre-colored with 1 gives that edge color 1 or $-1$. Thus, the absolute value of the colors in the interval edge coloring preserves the pre-coloring of $H$. This proves

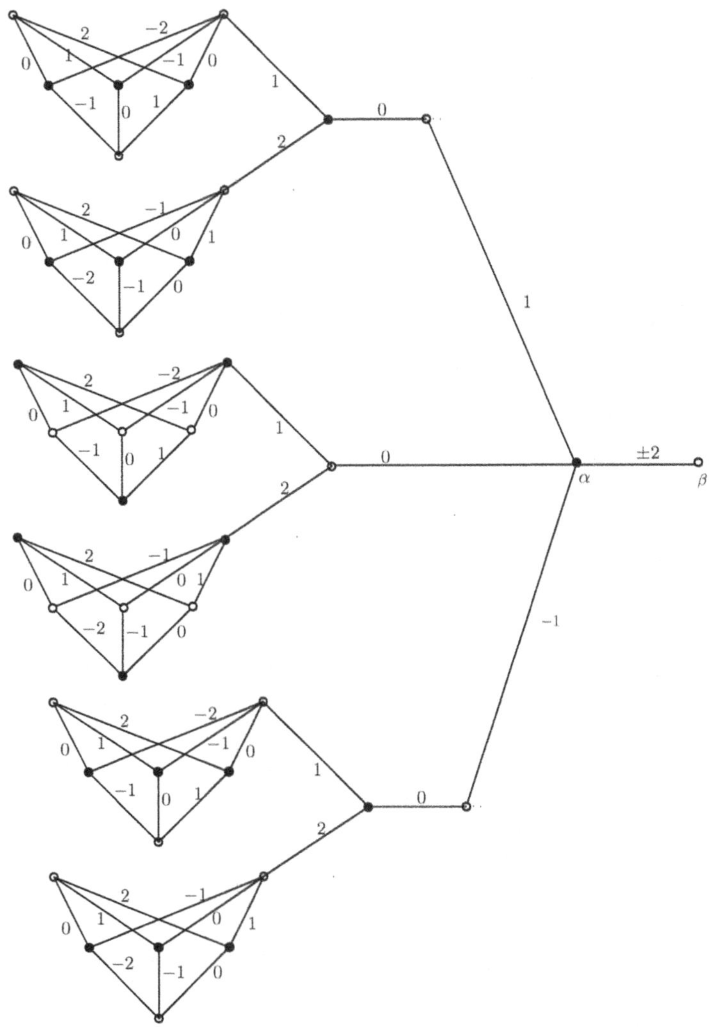


**Fig. 9.6** A bipartite graph $G_{\pm 2}$ that enforces colors $-2$ or 2 on the edge $(\alpha, \beta)$ in any interval edge coloring using colors in $L$

that it is possible to extend the pre-coloring to an edge coloring of $H$ using colors 0, 1, or 2. Thus we proved that the problem in $NP$-complete. □

Theorem 9.1 implies that the problem $O|p_{i,j} \in \{0, 1\}, \text{compact}|C_{\max}$ is $NP$-hard in the strong sense. A polynomial-time approximation algorithm for $O|p_{i,j} \in \{0, 1\}, \text{compact}|C_{\max}$ does not exist unless $P = NP$ since the decision problem to determine whether a bipartite graph has an interval edge coloring is $NP$-complete, see Sevast'janow [49].

## 9.3 Biregular Open Shops

We have seen in Sect. 9.1 that compact open shop schedules may not always exist, even for quite regular open shops like $\mathbb{P}_M$ with 15 nodes of degree 3, 3 nodes of degree 10, and 1 node of degree 15. Actually each job in $\mathbb{P}_M$ has exactly three unit-time operations, three machines have workload 10 each, and one has workload 15. Thus it comes only naturally to ask whether by making open shop even more regular we can ensure that compact schedules always exist. This has turned out a very challenging question that despite substantial efforts remains still wide open. We discuss this question for some *biregular bipartite graphs* (or biregular open shops) in this section. A bipartite graph $G = (X, Y, E)$ is $(a, b)$-biregular if each vertex in $X$ has degree $a$ and each vertex in $Y$ has degree $b$. An example of (2, 6)-biregular bipartite graph is given in Fig. 9.7. We focus on $(2, \Delta)$- and (3, 4)-biregular bipartite graphs in this section and assume the set of machines $X = \mathcal{M}$ and the set of jobs $Y = \mathcal{J}$.

### 9.3.1 (2, Δ)-Biregular Open Shops

In the $(2, \Delta)$-biregular open shop, each machine processes exactly two unit-time operations, and each job has the same length $\Delta$. The open shop has always a compact schedule with makespan $\Delta$ or $\Delta + 1$. The solution for an even $\Delta$ has makespan $\Delta$ and thus it is optimal. The solution is given by Hansen [29], and it is based on the following seminal result of Petersen [44], see Akiyama and Kano [1] and Plummer [46].

**Theorem 9.2** *For every integer $k \geq 1$, every $2k$-regular multigraph is 2-factorable.*

Let $G = (\mathcal{M}, \mathcal{J}, E)$ be a $(2, \Delta)$-biregular bipartite graph. By the convention, each vertex in $\mathcal{M}$ has degree 2, and each vertex in $\mathcal{J}$ has degree $\Delta$. For $G$, let $H = (J, E')$ be a job-multigraph where $(J_i, J_j) \in E'$ if and only if there is $M_h \in M$ such that $(J_i, M_h) \in E$ and $(J_j, M_h) \in E$. Observe that there may be multiple edges $(J_i, J_j)$ since the two jobs $J_i$ and $J_j$ may be processed together on more than one machine. Therefore, we label each edge $(J_i, J_j)$ with the machine where $J_i$ and $J_j$ are processed together. The labeling is unique since each machine processes exactly two jobs in the $(2, \Delta)$-regular open shop. There are exactly $|M| = \frac{\Delta|J|}{2}$ edges in the job-multigraph $H$, and $H$ is $\Delta$-regular. For $\Delta = 2k$ for some integer $k \geq 1$, Theorem 9.2 ensures that there are 2-factors $F_1, \ldots, F_k$ of $H$. The 2-factors are edge-disjoint and $F_1 \cup \cdots \cup F_k = H$. Consider 2-factor $F_\ell$, $\ell = 1, \ldots k$, and replace each edge $(J_i, J_j)$ in $F_\ell$ by two edges $(J_i, M_h)$ and $(M_h, J_j)$, where $M_h$ is the label on $(J_i, J_j)$. The resulting graph $G_\ell$ is an even cycle in $G$. Thus $G_\ell$ can be edge colored with two colors $2\ell - 1$ and $2\ell$. The two edges incident with each machine $M_h$ in $G_\ell$ are colored with $2\ell - 1$ and $2\ell$, and the two edges incident with each job $J_j$, $j = 1, \ldots, n$ are colored with $2\ell - 1$ and $2\ell$ in $G_\ell$. Thus $\Delta$ edges incident

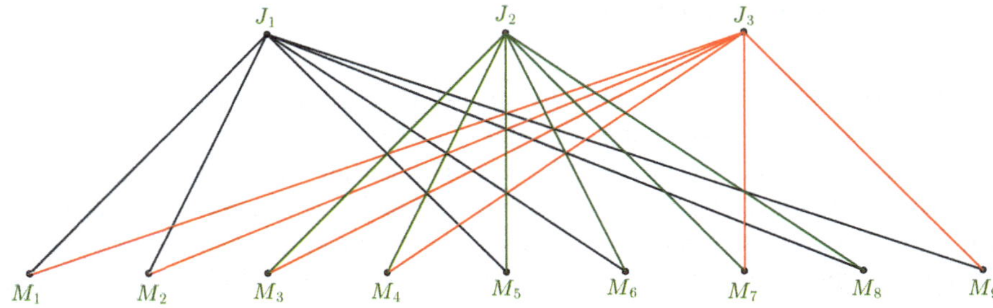


**Fig. 9.7** An instance $G$ of $(2, \Delta)$-biregular open shop with $\Delta = 6$, $m = 9$, and $n = 3$

**Fig. 9.8** The job $\Delta = 6$-regular graph $H$ for $G$

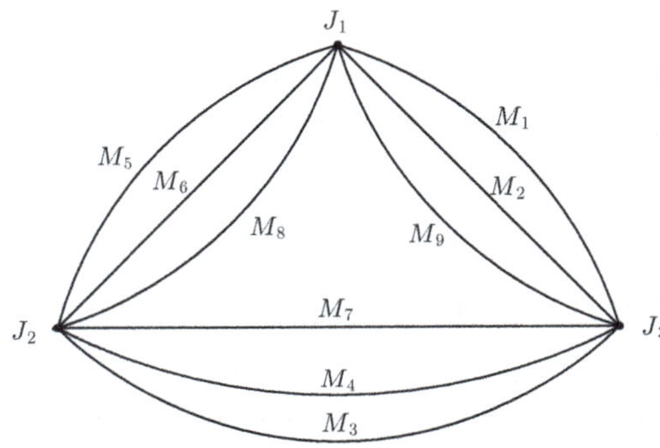


with each job in $G$ are colored with colors in the interval $[1, \Delta]$. This proves that $(2, \Delta)$-biregular bipartite graph $G$ has a $\Delta$-interval edge coloring for an even $\Delta$.

We illustrate this approach for a $(2, \Delta)$-biregular open shop $G$ in Fig. 9.7. The labeled $\Delta$-regular job-multigraph $H$ for $G$ is given in Fig. 9.8. The factorization of $H$ results in the following three job–machine cycles: $G_1 = (J_1, M_1, J_3, M_3, J_2, M_5, J_1)$; $G_2 = (J_1, M_2, J_3, M_4, J_2, M_6, J_1)$; and $G_3 = (J_1, M_9, J_3, M_7, J_2, M_8, J_1)$. Finally, the compact schedule for $G$ is given in Fig. 9.9.

For an odd $\Delta$, Hanson et al. [31] and Kostochka [36] show that there always is an interval edge coloring using no more than $\Delta + 1$ colors. The case $\Delta = 3$ is considered in Lemma 9.2. Somewhat surprisingly, the proof for an odd $\Delta \geq 5$ is much more involved than this for an even $\Delta$ presented earlier. The proof will not be presented here, see Hanson et al. [31] for the proof.

### 9.3.2 (3, 4)-*Biregular Open Shops*

The $(3, 4)$-biregular open shops are the simplest still unsolved cases of compact open shop scheduling. There have however been a number of sufficient conditions that guarantee the existence of compact schedules for those open shops. Pyatkin

Fig. 9.9 A compact schedule for $G$

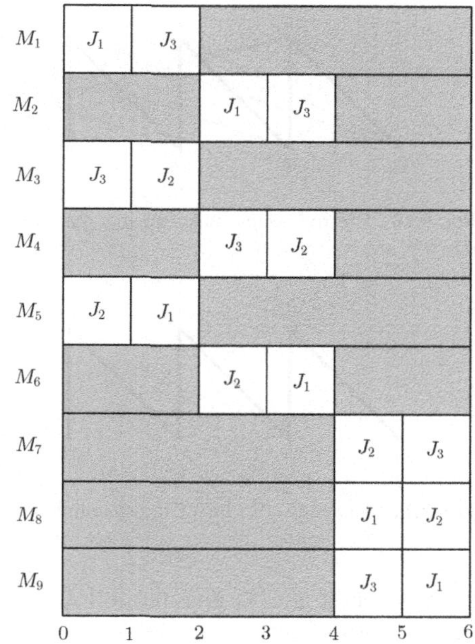


[47] gives the following sufficient condition for a (3, 4)-biregular bipartite graph to have interval edge coloring.

**Theorem 9.3** *Let $G = (X, Y, E)$ be a $(3, 4)$-biregular bipartite graph. If $G$ has a 3-regular subgraph $H$ such that $Y \subseteq V(H)$, then $G$ has an interval* 6*-coloring.*

Unfortunately, the problem to determine whether a $(3, 4)$-biregular bipartite graph $G$ has a 3-regular subgraph $H$ such that $Y \subseteq V(H)$ is $NP$-complete, see Pyatkin [47]. Thus, the sufficient condition is intractable to check for a given $G$. Asratian et al. [8] show that the condition proposed by Pyatkin may not be satisfied by some $(3, 4)$-biregular bipartite graphs.

Yang and Li [51] give another sufficient condition for a $(3, 4)$-biregular bipartite graph to have interval edge coloring.

**Theorem 9.4** *Let $G = (X, Y, E)$ be a $(3, 4)$-biregular bipartite graph. If $G$ has two $(2, 3)$-biregular bipartite components: $G_1 = (Y, X_1, E_1)$ where all vertices in $Y$ are of degree* 2 *and all vertices in $X_1$ are of degree* 3*, and $G_2 = (Y, X_2, E_2)$ where all vertices in $Y$ are of degree* 2 *and all vertices in $X_2$ are of degree* 3 *such that $X_1 \cup X_2 = X$, $X_1 \cap X_2 = \emptyset$, $E_1 \cup E_2 = E$, and $E_1 \cap E_2 = \emptyset$. Then $G$ has an interval* 6*-edge coloring.*

Yang and Li [51] show that for some graphs the sufficient condition in Theorem 9.4 can be satisfied but not the sufficient condition in Theorem 9.3, and the other way round. The complexity of the decomposition into the required two $(2, 3)$-biregular bipartite components seems open.

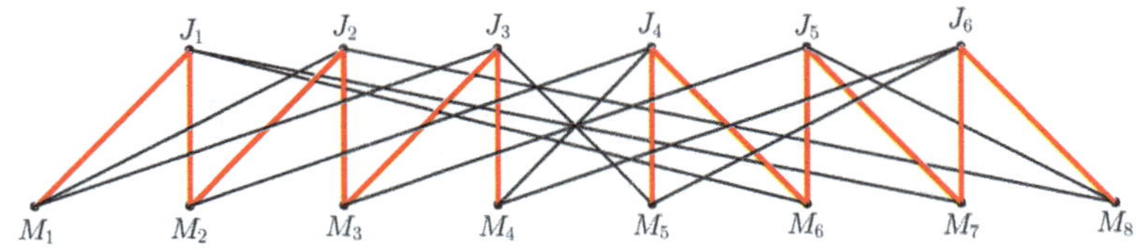


**Fig. 9.10** The red edges make up the spanning subgraph $P$ that meets the condition of Theorem 9.5, see Asratian et al. [8]

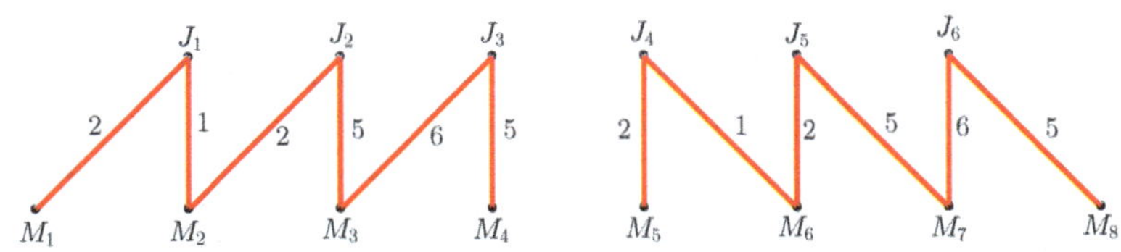


**Fig. 9.11** The edge coloring of the spanning subgraph $P$, see Asratian et al. [8]

Asratian et al. [8] give the following sufficient condition for a $(3, 4)$-biregular bipartite graph to have interval edge coloring.

**Theorem 9.5** *Let $G = (X, Y, E)$ be a $(3, 4)$-biregular bipartite graph. If $G$ has a spanning subgraph made up of paths with endpoints in $X$ and lengths (the number of edges on the path) in $\{2, 4, 6, 8\}$, then $G$ has an interval 6-edge coloring.*

Furthermore, Asratian et al. [8] conjecture that the condition is always satisfied for $(3, 4)$-biregular bipartite graphs. In an attempt to prove the conjecture, Asratian and Casselgren [6] show that each $(3, 4)$-biregular bipartite graph has a spanning subgraph made up of paths with endpoints in $X$; however, they do not establish any bound on the lengths of the paths in the subgraph. Casselgren [13] further shows that each path in the subgraph is not longer than 22. Both Asratian and Casselgren [6] and Casselgren [13] give polynomial-time algorithms to find a spanning subgraph made up of paths with endpoints in $X$.

To illustrate the application of Theorem 9.5 consider the $(3, 4)$-biregular bipartite graph $G$ in Fig. 9.10. The graph comes from Asratian et al. [8]. It satisfies the condition of Theorem 9.5, and the spanning subgraph $P$ is made up of two paths of length 6 with endpoints in $X = \mathcal{M}$. Figures 9.11 and 9.12 show how to obtain an interval 6-edge coloring for $G$. Finally, Fig. 9.13 shows the compact schedule for $G$ with makespan 6. The graph $G$ in Fig. 9.10 does not satisfy the condition of Theorem 9.3, see Asratian et al. [8].

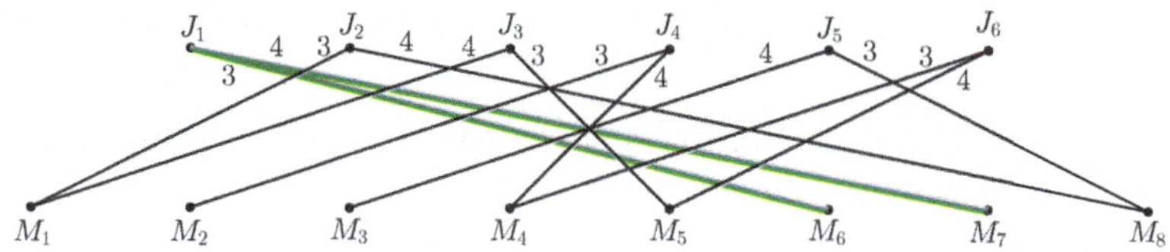


**Fig. 9.12** The edge coloring of the graph $G - P$ made up of two paths, the black and the green

**Fig. 9.13** The compact schedule with $C_{\max} = 6$ for the open shop in Fig. 9.10

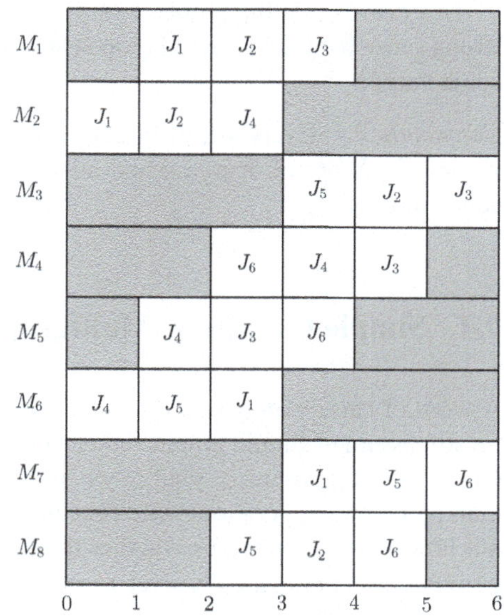


### 9.3.3 $(a, b)$-Biregular Open Shops

Hanson and Loten [30] prove the following lower bound on the number of colors required by an interval edge coloring, if any exists, of $(a, b)$-biregular bipartite graph.

**Theorem 9.6** *No $(a, b)$-biregular bipartite graph has an interval edge coloring with fewer than $a + b - \gcd(a, b)$ colors, where $\gcd(a, b)$ is the greatest common divisor of $a$ and $b$.*

Observe that the interval edge colorings in Theorems 9.3, 9.4, and 9.5 use 6 colors, which is the lower bound for (3, 4)-biregular bipartite graphs. Therefore, the schedules obtained in those theorems, if sufficient conditions are met, minimize makespan.

Not much is known about interval edge coloring of $(a, b)$-biregular open shops with $a > 3$. It has been observed at the beginning of Sect. 9.2.5 that the interval 15-edge coloring of (5, 15)-biregular bipartite graph is $NP$-complete. In the same vein Asratian and Casselgren [5] show:

**Theorem 9.7** *The interval* 6*-edge coloring of a* (3, 6)*-biregular bipartite graph is* $NP$*-complete.*

However, Casselgren and Toft [15] prove that (3, 6)-biregular bipartite graphs always have interval edge colorings using 7 colors.

**Theorem 9.8** *Every* (3, 6)*-biregular bipartite graph has an interval* 7*-edge coloring.*

This proves that the problem $O|p_{i,j} \in \{0, 1\}, \text{compact}|C_{\max}$ is $NP$-hard in the strong sense for (3, 6)-biregular open shops. Hansen [29] and Jensen and Toft [32] conjectured.

*Conjecture 9.1* Every $(a, b)$-biregular bipartite graph $G = (X, Y, E)$ where all vertices in $X$ are of degree $a$ and all vertices in $Y$ are of degree $b$ has an interval edge coloring.

## 9.4 Simple Graphs or Multigraphs

A word of caution may be in order at this point. The discussion in this chapter has mostly assumed simple graphs rather than multigraphs. This assumption is made in Giaro [21] for instance. That is we assume 0-1 operations for open shops, rather than preemptive operations with preemptions permitted at integer points only. It is not however often obvious whether the results reported in the literature for simple graphs carry automatically over to multigraphs or not. Casselgren and Toft [15] point out that Theorem 9.8 holds for multigraphs, and also the algorithm for $(2, \Delta)$-biregular bipartite graphs can be extended to multigraphs, see Hansen [29]. That is those results hold for operations with processing times equal $0, 1, \ldots, \Delta$. Similarly, Asratian et al. [8] assume multigraphs in their paper. At the same time, however, those authors show that (3, 4)-biregular bipartite multigraphs $G = (X, Y, E)$ need not have spanning subgraphs made up of paths with endpoints in $X$ and lengths in $\{2, 4, 6, 8\}$ to have interval 6-edge coloring. Their example is as follows:

$$\mathbb{P}_A = \begin{bmatrix} 1\,3\,0\,0 \\ 1\,0\,3\,0 \\ 1\,0\,0\,3 \end{bmatrix},$$

and a compact schedule with makespan $C_{\max} = 6$ is shown in Fig. 9.14. All this despite the fact that no spanning subgraphs made up of paths with endpoints in $X$ and lengths in $\{2, 4, 6, 8\}$ exists in $\mathbb{P}_A$. On the other hand Asratian et al. [8] conjecture that every simple (3, 4)-biregular bipartite graph has spanning subgraphs made up of paths with endpoints in $X$ and with lengths in $\{2, 4, 6, 8\}$. Petrosyan and Khachatrian [45] prove Lemma 9.2 for multigraphs.

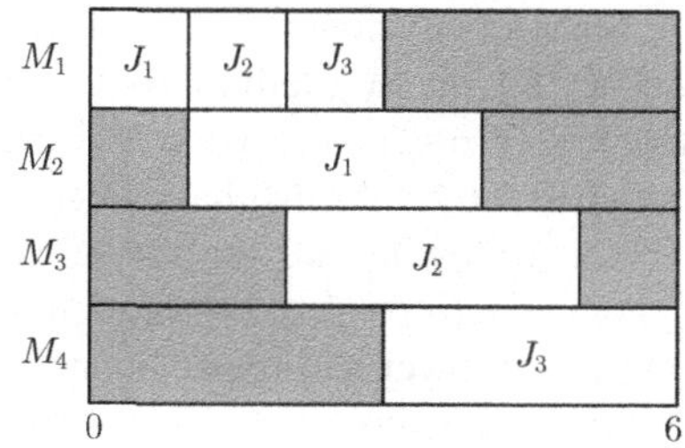


**Fig. 9.14** The compact schedule with $C_{\max} = 6$ for the open shop $\mathbb{P}_A$

## 9.5 Deficiency and Spans

In a schedule $\mathcal{S}$, for each job $J_i$, there is a time interval $[S_i, C_i]$ where the job is processed, and for each machine $M_h$ there is a time interval $[s_h, c_h]$ where the machine processes all its operations. Recall that a compact schedule requires $C_i = S_i + P_i$ for each job $J_i$ and $c_h = s_h + L_h$ for each machine $M_h$. However, compact schedules do not always exist, see for instance open shop instances $\mathbb{P}_{PK}$ and $\mathbb{P}_M$ given at the beginning of this chapter. This means that $C_i - S_i > P_i$ for some job $J_i$ or $c_h - s_h > L_h$ for some machine $M_h$ in $\mathcal{S}$. We refer to the difference $d(J_i) = (C_i - S_i) - P_i$ as *job deficiency*, and to the difference $d(M_h) = (c_h - s_h) - L_h$ as *machine deficiency*. The job deficiency of job $J_i$ in $\mathcal{S}$ is the total *waiting* time of the job in $[S_i, C_i]$ from its start at $S_i$ until its completion at $C_i$. The machine deficiency of machine $M_h$ in $\mathcal{S}$ is the total *idle* time of the machine in $[s_h, c_h]$ from the start of its first operation at $s_h$ until the completion of its last operation at $c_h$. The schedule $\mathcal{S}$ *deficiency* is then defined as the total job and machine deficiency

$$d(\mathcal{S}) = \sum_i d(J_i) + \sum_h d(M_h). \tag{9.1}$$

For an instance $\mathbb{P}$ of an open shop, the deficiency $d(\mathbb{P})$ equals deficiency of a feasible schedule $\mathcal{S}$ of $\mathbb{P}$ with minimum deficiency

$$d(\mathbb{P}) = \min_{\mathcal{S}} d(\mathcal{S}). \tag{9.2}$$

The concept of deficiency in the context of interval graph edge coloring was first introduced by Giaro et al. [24]. For a given graph $G = (V, E)$ and an edge coloring $c$ of $G$ they define deficiency $d(v)$ of vertex $v \in V$ to be equal to the deficiency of the set of colors used by $c$ to color the edges incident with $v$ (for a subset $A$ of the set of colors $\{1, \ldots, t\}$, the deficiency of $A$ is the number of integers missing from $A$ between the minimum and maximum of $A$, see Giaro et al. [24]). The deficiency $d(G, c)$ of $c$ is then the total efficiency of all vertices in $V$. Finally, the deficiency of the graph $G$ equals

$$d(G) = \min_c d(G, c). \tag{9.3}$$

Since open shops $\mathbb{P}$ with 0-1 operations are equivalent to bipartite graphs $G = (\mathcal{J}, \mathcal{P}, E)$, the two definitions are equivalent for those open shops. The test whether $d(\mathbb{P}) = 0$ (or $d(G) = 0$) is $NP$-complete, see Sevast'janow [49]. The deficiency equals 0 for graphs that have interval edge coloring, and it is positive otherwise.

The following are the results on deficiency of some special open shops and graphs. Giaro et al. [24] show that $1 \leq d(\mathbb{P}_M) \leq 3$, recall $\mathbb{P}_M$ from Sect. 9.1, and prove lower and upper bounds on deficiency of some bipartite graphs without interval edge coloring. They also show that there exist bipartite graphs with deficiency approaching the number of vertices. For the sake of completeness we also mention that in a similar vein Giaro et al. [25] compute deficiency of odd cycles, wheels, broken wheels, and complete graphs. Schwartz [48] does this computation for $k$-regular graphs with $k = 1, 2, 3, 4$ and shows upper bounds on the deficiency of $k$-regular graphs for higher values of $k$. The deficiency of the multigraph $\mathbb{P}_{PK}$ equals 1. Section 9.4 briefly reviews some bipartite multigraphs with deficiency 0. Petrosyan and Khachatrian [45] show that for any $\Delta \geq 9$, among bipartite multigraphs $G$ with $\Delta(G) = \Delta$, there is a connected bipartite multigraph with positive deficiency.

For a graph $G$ with $d(G) > 0$, Giaro et al. [25] observe that the deficiency $d(G)$ is the number of *pendant edges* whose attachment to $G$ makes it interval edge colorable, i.e., reduces the deficiency to 0. For open shops, a pendant edge appended to a job-vertex is equivalent to adding one more unit-time dummy operation to the job and a unique dummy machine where this operation is processed on. A pendant edge appended to a machine-vertex is equivalent to adding one more unique dummy single-operation job to be processed on the machine. To our knowledge, not much research has been conducted on the deficiency of compact schedules for bipartite multigraphs.

### *9.5.1 Spans*

For a graph $G$ with deficiency $d(G) = 0$, there may generally be more than one interval edge coloring $c$ with $d(G, c) = 0$. For a bipartite $G$ that means that there may be compact open shop schedules with different values of makespan. Those schedules with minimum makespan are then clearly desirable since they minimize makespan and deficiency at the same time. Thus they naturally extend the concept of ideal schedules to makespan and schedule deficiency, see Coffman et al. [16] for the concept of ideal schedules.

Following Giaro et al. [25], we define the minimum span of graph $G$, $s(G)$, and the maximum span of $G$, $S(G)$, as the minimum and maximum number of colors among all edge colorings $c$ of $G$ with deficiency $d(G, c) = 0$, if any, respectively. Giaro et al. [25] show $\Delta(G) \leq s(G) \leq S(G) \leq 2|V| - 4$ for graphs $G = (V, E)$ with at least three vertices and $d(G) = 0$, and $\Delta(G) \leq s(G) \leq S(G) \leq |V| - 1$ if in addition $G$ is bipartite. The computation of $s(G)$ and $S(G)$ for a given graph $G$ appears a challenging and still open problem even for bipartite $G$, i.e., for open shops. The computation is reported for special graphs in Giaro et al. [25] and Altinakar et al. [2]. Some insight into the difficulty is given by Sevast'janov [49] who presents a bipartite $G$ with $s(G) = 100$ and $S(G) = 173$ but no interval $k$-edge coloring with $k = 101, \ldots, 172$.

Altinakar et al. [2] extend the definition of spans to graphs with positive deficiency by defining the minimum span of $G$, $s(G)$, and the maximum span of $G$, $S(G)$, as the minimum and maximum number of colors among all edge colorings $c$ with deficiency $d(G, c) = d(G)$, respectively. They show $\Delta(G) \leq s(G) \leq S(G) \leq 2|V| - 4 + d(G)$ for graphs $G = (V, E)$ with at least three vertices. To the best of our knowledge, no algorithms and computational experiments have been reported for the computation of $s(G)$ or $S(G)$, specifically for bipartite graphs $G$, in the literature.

### 9.5.2 Computation of Deficiency

Bouchard et al. [12] present a tabu search algorithm to compute graph deficiency $d(G)$ of $G$. Their computational experiments lead them to a conclusion that graphs with odd number of vertices pose a greater challenge for their algorithm, and so do the graphs with large distance $s(G) - \Delta(G)$.

Altinakar et al. [2] propose edge-based integer programming (IP) models and a constraint programming (CP) model to calculate $d(G)$. Their IPs could calculate $d(G)$ of graphs with up to 7 vertices, and their CP could calculate $d(G)$ of graphs with up to 8 vertices in an hour or so in their experiments. The CP outperforms the IP in the experiments. They also point out that though often the true deficiency value can be found quickly for a graph $G$, the proof that the value is indeed true deficiency of $G$ can take much longer. To speed up their CP model computations, Altinakar et al. [3] propose a set of symmetry braking constraints to supplement the model. While the extended CP model can find and prove deficiency of some graphs in a matter of seconds, which took hours for the original CP, the symmetry braking constraints seem still insufficient to efficiently find $d(G)$ of graphs with 10 or more vertices.

Bodur and Luedtke [11] propose a matching-based integer programming model to minimize $d(G, c)$ over all edge colorings $c$ with $k \geq \Delta(G)$ colors. By starting with $k = \Delta(G)$ and systematically increasing $k$ by 1, one can find $s(G)$ for which the IP gives $d(G)$. By running computational experiments, Bodur and Luedtke [11] compare their integer programming model with the CP model of Altinakar et al. [2]. Both models use some additional symmetry breaking constraints. Those experiments show that the CP model is significantly faster than the proposed IP model for low density graphs; however, the IP provides significant reduction in running time for high density graphs, where the density of $G = (V, E)$ equals $\frac{2|E|}{|V|(|V|+1)}$. It seems that further improvements to the IP model are needed to insure that the IP model can solve instances with 10 or more vertices in reasonable time.

Those models and computational experiments presented in the literature have focused mainly on general graphs, and hardly any computational results have been reported explicitly for *bipartite* graphs. Bouchard et al.'s study [12] is an exception that it presents computational results for complete bipartite graphs. However for those graphs, Kamalian [33, 34] proves that complete bipartite graph $K_{a,b}$ has interval $k$-edge coloring, and thus its deficiency equals 0, if and only if $a + b - \gcd(a, b) \leq k \leq a + b - 1$, where $\gcd(a, b)$ is the greatest common divisor of $a$ and $b$. To the best of our knowledge no algorithms have been developed specifically for bipartite graphs or bipartite multigraphs thus far. Such algorithms and computational experiments seem a novel avenue for further research.

Shao et al. [50] propose integer programming formulation for the interval edge coloring and use it on instances of complete $k$-partite graphs to disprove some conjectures posed by Grzesik and Khachatrian [27].

## 9.6 Cyclic Compact Open Shop Scheduling

The *cyclic interval edge coloring* was introduced by de Werra and Solot [18] who define it as an edge coloring of $G$ with colors $1, \ldots, t$ where each color is used for at least one edge, and for each vertex the edges incident with the vertex are colored with colors making up an interval in $1, \ldots, t$, or the colors from $\{1, \ldots, t\}$ *not* used to color the edges incident with the vertex make up an interval in $1, \ldots, t$.

For bipartite $G$, we refer to cyclic interval edge coloring as *cyclic compact open shop schedule* (also called *cylindrical open shop schedule* by de Werra and Solot [18]). Thus in a compact open shop schedule, for each job, there is a *single* interval between 0 and $C_{\max}$ where the job is being processed; similarly for each machine there is a *single* interval between 0 and $C_{\max}$ where the machine is occupied. On the other hand in a cyclic compact open shop schedule, for each job, there is a *single* interval between 0 and $C_{\max}$ where the job is *not* being processed, or there is a *single* interval between 0 and $C_{\max}$ where it *is*; similarly for each machine there is a *single* interval between 0 and $C_{\max}$ where the machine is *idle*, or there is a *single* interval between 0 and $C_{\max}$ where it is *occupied*.

Clearly, if an interval edge coloring exists for a graph $G$, then so does a cyclic interval edge coloring. The opposite however does not hold. For instance, the open shop $\mathbb{P}_M$ presented in Sect. 9.1 does not have an interval edge coloring, yet it has cyclic interval edge coloring, see Nadolski [43]. A cyclic compact open shop schedule for $\mathbb{P}_M$ is shown in Fig. 9.15. Observe that colors used for machine-vertex $M_1$ make up an interval $2, \ldots, 11$, and those used by machine-vertex $M_4$ make up an interval $1, \ldots, 15$. The machine-vertex $M_2$ does *not* use colors $2, 3, 4, 5, 6$ that make up an interval of integers, and the machine-vertex $M_3$ does not use colors $7, 8, 9, 10, 11$ that also make up an interval of integers. We can also easily verify that the definition of cyclic compact colorings is satisfied by the schedule for every job-vertex.

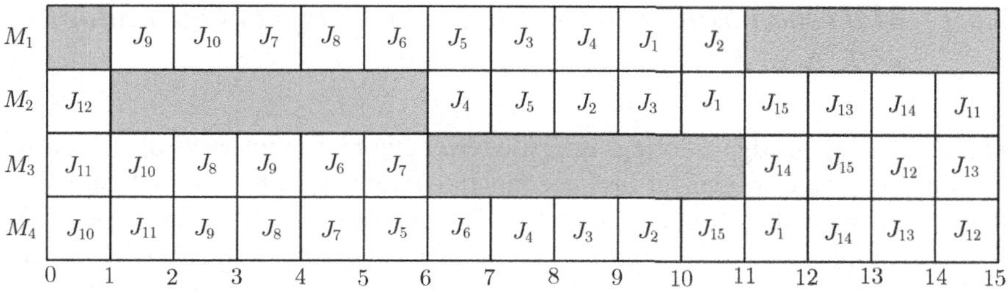


**Fig. 9.15** A cyclic compact schedule for $\mathbb{P}_M$

Even so, cyclic compact schedules may not exist for some open shops. The following open shop $\mathbb{P}_{ACP}$ with $m = 6$ machines, $n = 15$ jobs, and $\Delta = 14$ is shown by Asratian et al. [4] not to have cyclic interval edge coloring:

$$\mathbb{P}_{ACP} = \begin{bmatrix} 1&1&1&1&1&0 \\ 1&0&0&0&0&1 \\ 1&0&0&0&0&1 \\ 1&0&0&0&0&1 \\ 0&1&0&0&0&1 \\ 0&1&0&0&0&1 \\ 0&1&0&0&0&1 \\ 0&0&1&0&0&1 \\ 0&0&1&0&0&1 \\ 0&0&0&1&0&1 \\ 0&0&0&1&0&1 \\ 0&0&0&1&0&1 \\ 0&0&0&0&1&1 \\ 0&0&0&0&1&1 \\ 0&0&0&0&1&1 \end{bmatrix}.$$

Another example is the open shop $\mathbb{P}_{PK}$ presented in Sect. 9.1, see also Asratian et al. [4].

The motivation to study cyclic interval edge coloring comes from the just-in-time systems where it is a common practice to repeat relatively short sequences to build a sequence for a longer time horizon, see Monden [42], Hall [28], de Werra and Solot [19], and Kubiak [39], to obtain a closer match with demand and improve quality in just-in-time production.

### 9.6.1 *Makespan Minimization for Cyclic Compact Open Shops with* $\Delta = 3$

The cyclic interval edge colorings may use fewer colors than interval edge colorings for some graphs. Consider for instance bipartite graphs $G$ with $\Delta = 3$. Lemma 9.2 shows that there always is an interval 4-edge coloring of such graphs. However, we observe that any 3-edge coloring $c$ of a bipartite graph $G = (X, Y, E)$ having $\Delta(G) = 3$ is actually a cyclic interval edge coloring of $G$. Namely, by König's edge-coloring theorem, the edge coloring $c$ uses 3 colors, say 1, 2, and 3. Thus, the colors the coloring $c$ uses to color the edges incident with each vertex of degree 3 or 1 make up intervals. For each vertex of degree 2, the $c$ colors the two edges incident with the vertex either with 1 and 2, or 2 and 3, or 1 and 3. The first two are intervals in $1, 2, 3$. The third does *not* use color 2 that makes up an interval in $1, 2, 3$. Thus, $c$ is a cyclic interval 3-edge coloring of $G$. Nadolski [43] proves that each connected *graph* $G$ with $\Delta(G) = 3$ has cyclic interval edge coloring with at most 4 colors.

### 9.6.2 *Makespan Minimization for Cyclic Compact Open Shops with* $\Delta = 4$

For bipartite graphs $G$ with $\Delta = 4$, Lemma 9.5 guarantees the existence of interval 4-edge coloring for any $G$ *without* vertices of degree 3. It remains to consider bipartite graphs $G = (V, E)$ with $\Delta = 4$ and vertices of degree 3. Let $U \subseteq V$ be the set of all vertices of degree 3 in $V$. Take a copy $G^c = (V^c, E^c)$ of $G$ to create a graph $H = (W = V \cup V^c, F = E \cup E^c \cup \{(u, u^c) : u \in U\})$. The graph $H$ is bipartite with $\Delta = 4$ and no vertices of degree 3.

If $H$ has a cyclic interval $\Delta$-edge coloring, then so does its subgraph $G$. Namely, for any vertex $u \in U$ the edges incident with $u$ are colored with $1, 2, 3, 4$ in any cyclic interval $\Delta$-edge coloring $c$ of $H$. Thus the edge $(u, u^c)$ is colored with either 1, or 2, or 3, or 4 in $c$, and the remaining three edges with colors $2, 3, 4$, or $1, 3, 4$, or $1, 2, 4$, or $1, 2, 3$, respectively. Therefore, the cyclic interval edge coloring for $u \in U$ exists, which implies that $G$ has a cyclic interval $\Delta$-edge coloring. On the other hand, if $G$ has a cyclic interval $\Delta$-edge coloring, then so does $G^c$. The edge coloring of $G$ (and $G^c$) can be extended to a cyclic interval edge coloring of $H$ by coloring $(u, u^c)$ with color 1 if the edges incident with $u$ and $u^c$ are colored with either $2, 3, 4$, with color 2 if the edges incident with $u$ and $u^c$ are colored with $1, 3, 4$, with color 3 if the edges incident with $u$ and $u^c$ are colored with $1, 2, 4$, and with color 4 if the edges incident with $u$ and $u^c$ are colored with $1, 2, 3$. Therefore, a cyclic interval $\Delta$-edge coloring exists for $G$ if and only if it exists for $H$. The latter exists by Lemma 9.5 since $H$ has no vertices of degree 3. This proves that cyclic interval $\Delta$-edge coloring always exists for bipartite graphs with $\Delta = 4$.

Theorems 9.6 and 9.5 show that if an interval edge coloring of a graph bipartite with $\Delta = 4$ exists, then the coloring may require more than 4 colors, for instance,

6 colors. Therefore, the difference between minimum makespans of compact and cyclic compact schedules may be generally greater than 1, even for open shops with $\Delta = 4$.

### 9.6.3 Makespan Minimization for Other Cyclic Compact Open Shops

Asratian et al. [4] generalize the algorithms from the previous Sects. 9.6.1 and 9.6.2 in the following theorem.

**Theorem 9.9** *For a bipartite graph* $G = (X, Y, E)$*:*

- *If* $\Delta(G) = 2k$, $k \geq 2$, *and each vertex* $v \in V$ *has degree* 1 *or* 2 *or* $2k - 2$ *or* $2k - 1$ *or* $2k$, *then* $G$ *has a cyclic interval* $2k$*-edge coloring.*
- *If* $\Delta(G) = 2k - 1$, $k \geq 2$, *and each vertex* $v \in V$ *has degree* 1 *or* 2 *or* $2k - 2$ *or* $2k - 1$, *then* $G$ *has a cyclic interval* $(2k - 1)$*-edge coloring or a cyclic interval* $2k$*-edge coloring.*

They further show:

**Theorem 9.10** *Every bipartite graph* $G$ *with each vertex having degree* 1 *or* 2 *or* 4 *or* 6 *or* 7 *or* 8 *has a cyclic interval edge coloring.*

Kubale and Nadolski [38] prove that the problem to determine whether a bipartite graph $G(X, Y, E)$ has a cyclic interval edge coloring is $NP$-complete. Their proof however requires $G$ to have a large $\Delta(G)$, actually larger than $|E|/2$. Thus the complexity of the problem remains open for bipartite graphs $G$ with $\Delta = 5$. However, by Theorem 9.9 if $G$ with $\Delta(G) = 5$ misses vertices of degree 3, then it has a cyclic interval 5-edge coloring or 6-edge coloring. Thus again vertices with degree 3 would be key for graphs used in an $NP$-completeness proof, if any. Actually, the following problem posed by Asratian et al. [4] remains open.

**Problem 9.3** Is there a bipartite graph $G$ with $5 \leq \Delta(G) \leq 8$ that does not have cyclic interval edge coloring?

For bipartite graphs with $\Delta = 5$, Casselgren et al. [14] prove the following theorem:

**Theorem 9.11** *Every* $(3, 5)$*-biregular bipartite graph has a cyclic interval* 6*-edge coloring.*

Casselgren and Toft [15] show:

**Theorem 9.12** *Every* $(4, 8)$*-biregular bipartite graph has a cyclic interval* 8*-edge coloring.*

Asratian et al. [7] study deficiency of cyclic compact scheduling.

## Problems

**9.1** Prove that there is no compact schedule for the instance $\mathbb{P}_{PK}$.

**9.2** Prove Lemma 9.4.

**9.3** Let $G = (X, Y, E)$ be a bipartite graph having each vertex of degree 4 or 2. Show that $G = (X, Y, E)$ can be partitioned into bipartite subgraphs $G_1 = (X, Y, E_1)$ and $G_2 = (X, Y, E_2)$ such that $E_1 \cap E_2 = \emptyset$, $E_1 \cup E_2 = E$, and $\deg_G(v) = 2\deg_{G_1}(v) = 2\deg_{G_2}(v)$. Hit: Use the fact that $G$ is Eulerian.

**9.4** Prove that an interval edge coloring exists for any $(2, \Delta)$-biregular bipartite graph with an odd $\Delta$. Show that the number of colors used in the interval edge coloring does not exceed $\Delta + 1$.

## References

1. J. Akiyama, M. Kano, *Factors and Factorizations of Graphs: Proof Techniques in Factor Theory*, vol. 2031 of *Lecture Notes in Mathematics* (Springer Berlin/Heidelberg, 2011)
2. S. Altinakar, G. Caporossi, A. Hertz, A comparison of integer and constraint programming models for the deficiency problem. Comput. Oper. Res. **68**, 89–96 (2016)
3. S. Altinakar, G. Caporossi, A. Hertz, Symmetry breaking constraints for the minimum deficiency problem. J. Graph Algorithms Appl. **21**, 195–218 (2017)
4. A. Asratian, C.J. Casselgren, P.A. Petrosyan, Some results on cyclic interval edge colorings of graphs. J. Graph Theory **87**, 239–252 (2018)
5. A.S. Asratian, C.J. Casselgren, On interval edge colorings of $(\alpha, \beta)$-biregular bipartite graphs. Discrete Mathematics **307**, 1951–1956 (2007)
6. A.S. Asratian, C.J. Casselgren, On path factors of (3,4)-biregular bigraphs. Graphs Comb. **24**, 405–411 (2008)
7. A.S. Asratian, C.J. Casselgren, P.A. Petrosyan, Cyclic deficiency of graphs. Discrete Appl. Math. **266**, 171–185 (2019)
8. A.S. Asratian, C.J. Casselgren, J. Vandenbussche, D.B. West, Proper path-factors and interval edge-coloring of (3, 4)-biregular bigraphs. J. Graph Theory **61**, 88–97 (2009)
9. A.S. Asratian, R.R. Kamalian, Interval colorings of edges of a multigraph (in Russian). Appl. Math. (also arXiv:1401.8079v1) **5**, 25–34 (1987)
10. J.J. Bartholdi, K.L. McCroan, Scheduling interviews for a job fair. Operations Research **38**, 951–960 (1990)
11. M. Bodur, J.R. Luedtke, Integer programming formulations for minimum deficiency interval coloring. Networks **72**, 249–271 (2018)
12. M. Bouchard, A. Hertz, G. Desaulniers, Lower bounds and a tabu search algorithm for the minimum deficiency problem. J. Comb. Optim. **17**, 168–191 (2009)
13. C.J. Casselgren, A note on path factors of (3, 4)-biregular bigraphs. Electron. J. Comb. **18**, 1–12 (2011)
14. C.J. Casselgren, P.A. Petrosyan, B. Toft, On interval and cyclic interval edge colorings of (3,5)-biregular graphs. Discrete Mathematics **340**, 2678–2687 (2017)
15. C.J. Casselgren, B. Toft, On interval edge colorings of biregular bipartite graphs with small vertex degrees. J. Graph Theory **80**, 83–97 (2015)
16. E.G. Coffman Jr., D. Dereniowski, W. Kubiak, An efficient algorithm for finding ideal schedules. Acta Informatica **49**, 1–14 (2012)

17. E. Cole, K. Ost, S. Schirra, Edge-coloring Bipartite Multigraphs in $O(E \log D)$. Combinatorica **21**, 5–12 (2001)
18. D. de Werra, Ph. Solot, Compact cylindrical chromatic scheduling. SIAM J. Disc. Math. **4**, 528–534 (1991)
19. D. de Werra, Ph. Solot, Some graph-theoretical models for scheduling in automated production systems. Networks **23**, 651–660 (1993)
20. S. Even, A. Itai, A. Shamir, On the complexity of timetable and multicommodity flow problems. SIAM J. Comput. **5**, 691–703 (1976)
21. K. Giaro, The complexity of consecutive $\Delta$-coloring of bipartite graphs: 4 is easy, 5 is hard. Ars Combin. **47**, 287–298 (1997)
22. K. Giaro, *Compact task scheduling on dedicated processors with no waiting periods*. PhD thesis, Gdańsk University of Technology, EIT Faculty, 1999
23. K. Giaro, M. Kubale, M. Małafiejski, Compact scheduling in open shop with zero-one time operations. INFOR **37**, 37–47 (1999)
24. K. Giaro, M. Kubale, M. Małafiejski, On the deficiency of bipartite graphs. Discrete Appl. Math. **94**, 193–203 (1999)
25. K. Giaro, M. Kubale, M. Małafiejski, Consecutive colorings of the edges of general graphs. Discrete Mathematics **236**, 131–143 (2001)
26. T. Gonzalez, Unit execution time shop problems. Math. O. R **7**, 57–66 (1982)
27. A. Grzesik, H. Khachatrian, Interval edge-coloring of $K_{1,m,n}$. Discrete Appl. Math. **174**, 140–145 (2014)
28. R.W. Hall, Cyclic scheduling for improvement. Int. J. Prod. Res. **26**, 457–472 (1988)
29. H.M. Hansen, Scheduling with minimum waiting periods (in Danish). Master's thesis, Odense University, Denmark, 1992
30. D. Hanson, C.O.M. Loten, A lower bound for interval coloring bi-regular bipartite graphs. Bull. Inst. Comb. Appl. **18**, 69–74 (1996)
31. D. Hanson, C.O.M. Loten, B. Toft, On interval colourings of bi-regular bipartite graphs. Ars Comb. **50**, 23–32 (1998)
32. T.R. Jensen, B. Toft, *Graph Coloring Problems* (Wiley, 1995)
33. R.R. Kamalian, Interval colorings of complete bipartite graphs and trees. Comput. Cen. of Acad. Sci. of Armenian SSR, Erevan, 1989 (in Russian)
34. R.R. Kamalian, *Interval edge-colorings*. PhD thesis, Novosibirsk, 1990
35. H.H. Khachatrian, T. Mamikonyan, On interval edge-colorings of bipartite graphs of small order. arXiv:1508.02851v1, 2015
36. A.V. Kostochka, Unpublished manuscript (1995)
37. M. Kubale, Some results concerning the complexity of restricted coloring of graphs. Discrete Appl. Math. **36**, 35–46 (1992)
38. M. Kubale, A. Nadolski, Chromatic scheduling in a cyclic open shop. Eur. J. Oper. Res. **164**, 585–591 (2005)
39. W. Kubiak, *Proportional Optimization and Fairness* (Springer, 2009)
40. N.V.R. Mahadev, Ph. Solot, D. de Werra, The cyclic compact open-shop scheduling problem. Discrete Mathematics **111**, 361–366 (1993)
41. S. Micali, V.V. Vazirani, An $O(m\sqrt{n})$ algorithm for finding maximum matching in general graphs, in *Proc. 21st Ann. IEEE Symp. on Foundations of Computer Science*, pp. 17–27 (1980)
42. Y. Monden, *Toyota Production System: An Integrated Approach to Just–In–Time* (Engineering and Management Press, Norcross, GA, 1998)
43. A. Nadolski, Compact cyclic edge-colorings of graphs. Discrete Mathematics **308**, 2407–2417 (2008)
44. J. Petersen, Die teorie der regulären graphs. Acta Math., 193–220 (1891)
45. P.A. Petrosyan, H.H. Khachatrian, Interval non-edge-colorable bipartite graphs and multigraphs. J. Graph Theory **76**, 200–216 (2014)
46. M.D. Plummer, Graph factors and factorization:1985 - 2003: A survey. Discrete Mathematics **307**, 791–821 (2007)

47. A.V. Pyatkin, Interval coloring of (3,4)-biregular bipartite graphs having large cubic subgraphs. J. Graph Theory **47**, 122–128 (2004)
48. A. Schwartz, The deficiency of a regular graph. Discrete Mathematics **306**, 1947–1954 (2006)
49. S.V. Sevast'janov, Interval colorability of the edges of a bipartite graph (in Russian). Met. Diskret Analiz. **50**, 61–72 (1990)
50. Z. Shao, Z. Li, B. Wang, S. Wang, X. Zhang, Interval edge-coloring: A model of curriculum scheduling. AKCE Int. J. Graphs Comb. **17**, 725–729 (2020)
51. F. Yang, X. Li, Interval coloring of (3,4)-biregular graphs having two (2,3)-biregular subgraphs. Appl. Math. Lett. **24**, 1574–1577 (2011)

# Chapter 10
# No-Wait Open Shop Scheduling

## 10.1 No-Wait Open Shop Schedules Can Be Preemptive

We discussed *compact* schedules for open shop in Chap. 9. We now turn our attention to *no-wait schedules* where for each job $J_i$ there is a time interval $[S_i, S_i + P_i]$ where the job is processed—this implies no-waiting between operations of the job. The no-wait schedule is motivated by applications where there is:

- Limited or no intermediate storage between machines or on machines to store jobs between consecutive operations, see Papadimitriou and Kanellakis [17]
- In hot-potato routing where a packet that arrives at a node is immediately forwarded to another node, the routs may not be fixed as in open shop. This occurs in optical networks because it may be difficult to buffer optical messages, see Acampora and Shah [1] and Szymanski [21].

A list of other no-wait scheduling applications can be found in Hall and Sriskandarajah [11].

To begin, we need to observe that no-wait requirement does not rule out preemptions for open shops, which is first observed in Shani and Cho [18]. To see this consider the following two-machine open shop with $n = 4$ jobs and $m = 2$ machines

$$\mathbb{P} = \begin{bmatrix} 4 & 4 \\ 3 & 3 \\ 2 & 2 \\ 1 & 1 \end{bmatrix}.$$

Its optimal no-wait non-preemptive and preemptive schedules are given in Fig. 10.1a and b, respectively.



W. Kubiak, *A Book of Open Shop Scheduling*, International Series in Operations Research & Management Science 325,
https://doi.org/10.1007/978-3-030-91025-9_10

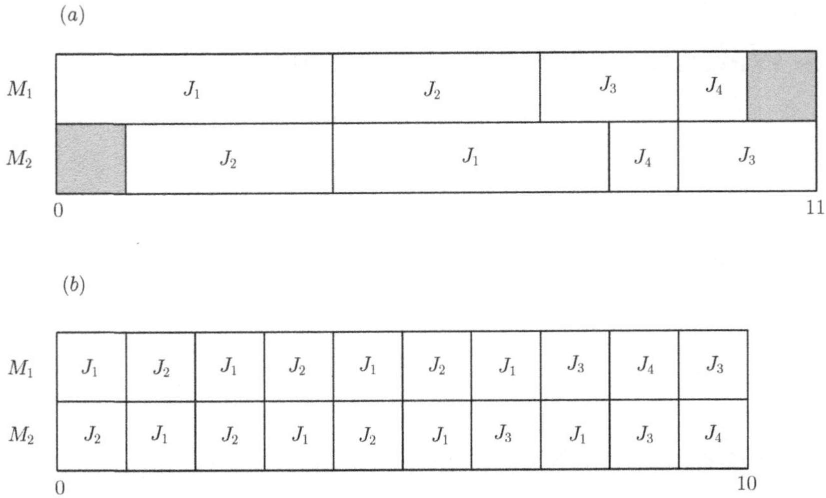


**Fig. 10.1** An optimal non-preemptive (**a**), and preemptive (**b**) no-wait schedule for the open shop in $\mathbb{P}$

The minimization of makespan for two-machine non-preemptive open shop is proved NP-hard in the strong sense by Shani and Cho [18]. They present proofs for the case where no job operation is missing and for the case where some operations of jobs may be missing. Somewhat surprisingly the proof for the latter seems simpler. It is worth noticing that a job with one operation missing in a two-machine no-wait open shop cannot be preempted. Thus there is no difference between preemptive and non-preemptive schedules for such jobs. Therefore the NP-hardness proof for no-wait non-preemptive case works for the preemptive case if we allow missing operations, see Giaro [8]. This is however not the case for the instances where none operation is missing, i.e., each job has exactly two operations with positive processing times in the two-machine open shop. We prove that the no-wait *preemptive* case is NP-hard in the strong sense for two-machine open shop in the next section.

The optimal no-wait preemptive and non-preemptive schedules in Fig. 10.1 indicate that optimal schedules for these two cases may look quite different in general. An intriguing question arises about the makespan reduction offered by optimal no-wait preemptive schedules in comparison to optimal no-wait non-preemptive schedules. The reduction equals 1 for the schedules in Fig. 10.1.

In general, let $C_{\max}(I)$ be the makespan of an optimal no-wait non-preemptive schedule for an $I$ instance of $O|\text{no-wait}|C_{\max}$ and $C^{p}_{\max}(I)$ be the makespan of an optimal no-wait preemptive schedule for the same instance $I$ of the preemptive problem $O|\text{pmtn, no-wait}|C_{\max}$. The following problem remains open.

**Problem 10.1** What is the upper bound on the difference $C_{\max}(I) - C^{p}_{\max}(I)$ that holds for all instances $I$?

In this chapter an instance of an open shop with $n$ jobs and $m$ machines will be represented by an $n \times m$ matrix of non-negative integers $\mathbb{P}$, or equivalently by a bipartite multigraph $G = (\mathcal{J}, \mathcal{M}, E)$ with the job-vertices in the set $\mathcal{J}$ and the machine-vertices in the set $\mathcal{M}$. Recall that the two representations are equivalent by taking $\mathrm{mp}(J_i, M_h) = p_{i,h}$ for the edge $e = (J_i, M_h) \in E$, $|\mathcal{J}| = n$, and $|\mathcal{M}| = m$. We have $\deg(J_i) = P_i$ for the job $J_i$ length, and $\deg(M_h) = L_h$ for the machine $M_h$ workload. Finally, $\Delta(G) = \max\{\max_j\{P_j\}, \max_h\{L_h\}\}$. For the open shops with unit-time operations, the matrix $\mathbb{P}$ has each entry either 0 or 1, and $G = (\mathcal{J}, \mathcal{M}, E)$ is bipartite simple graph.

### *10.1.1 Strong NP-Hardness for Two Machines*

We proof the following lemma in this section.

**Lemma 10.1** *The problem $O2|pmtn, no-wait|C_{\max}$ is NP-hard in the strong sense.*

***Proof*** The reduction is from 3-partition problem, see Garey and Johnson [6]. Let non-negative integer sizes $a_1, \ldots, a_{3n}$ and $B$ make up an instance of the 3-partition problem. Without loss of generality we assume that all $a_i$ and $B$ are even and divisible by 11, and

$$\frac{B}{4} < a_i < \frac{B}{2} \tag{10.1}$$

for $i = 1, \ldots, 3n$. The instance $\mathbb{P}$ of the open shop is defined as follows. We have $(n-2)$ *half-B* jobs $\mathcal{E} = \{E_1, \ldots, E_{n-2}\}$, each with $p_{i,1} = \frac{B}{2} + 4$ and $p_{i,2} = 5$, and $(n-2)$ *half-B* jobs $\mathcal{F} = \{F_1, \ldots, F_{n-2}\}$, each with $p_{i,1} = \frac{B}{2} + 4$ and $p_{i,2} = 6$. We have two *full-B* jobs $D_1$ and $D_2$, each with $p_{i,1} = B + 4$ and $p_{i,2} = 7$. The jobs $D_1, D_2, E_1, \ldots, E_{n-2}, F_1, \ldots, F_{n-2}$ are called *long* jobs. Finally, there are $3n$ *short* jobs $\omega_1, \ldots, \omega_{3n}$ with $p_{i,1} = 1$ and $p_{i,2} = a_i$. The makespan is set to $C_{\max} = n(B + 11) - 8$.

Let $A_i = \{\alpha_i, \beta_i, \gamma_i\}$, for $i = 1, \ldots, n$, be a partition such that $s(\alpha_i) = a_{3(i-1)+1}$, $s(\beta_i) = a_{3(i-1)+2}$, and $s(\gamma_i) = a_{3i}$ and $a_{3(i-1)+1} + a_{3(i-1)+2} + a_{3i} = B$, where $s(\alpha)$ is the size of $\alpha$. The required schedule for $\mathbb{P}$ is shown in Fig. 10.2. Observe that by (10.1), $a + b > \frac{B}{2}$ and $a < \frac{B}{2}$ for any sizes $a$ and $b$ in the 3-partition instance. Thus the job $\beta_i$ on $M_2$ *straddles* the start of job $F_{i-1}$ on $M_1$ for $i = 2, \ldots n-1$. Therefore, the $M_1$ operation of job $\beta_i$ and the $M_1$ operation of job $\gamma_i$ can be done after the start of job $F_{i-1}$ in parallel with $M_2$ operation of $F_{i-1}$, and the $M_1$ operation of $\alpha_i$ can be done before the start of job $F_{i-1}$ in parallel with the $M_2$ operation of $E_{i-1}$, see Fig. 10.2. Thus the schedule is feasible, no-wait, and with makespan $C_{\max} = n(B + 11) - 8$.

Now, let $\mathcal{S}$ be a no-wait schedule for $\mathbb{P}$ with makespan $C_{\max} = n(B + 11) - 8$. Observe that there is no idle time in $[0, C_{\max}]$ on either machine in $\mathcal{S}$. Let $\mathcal{S}^R$ be a

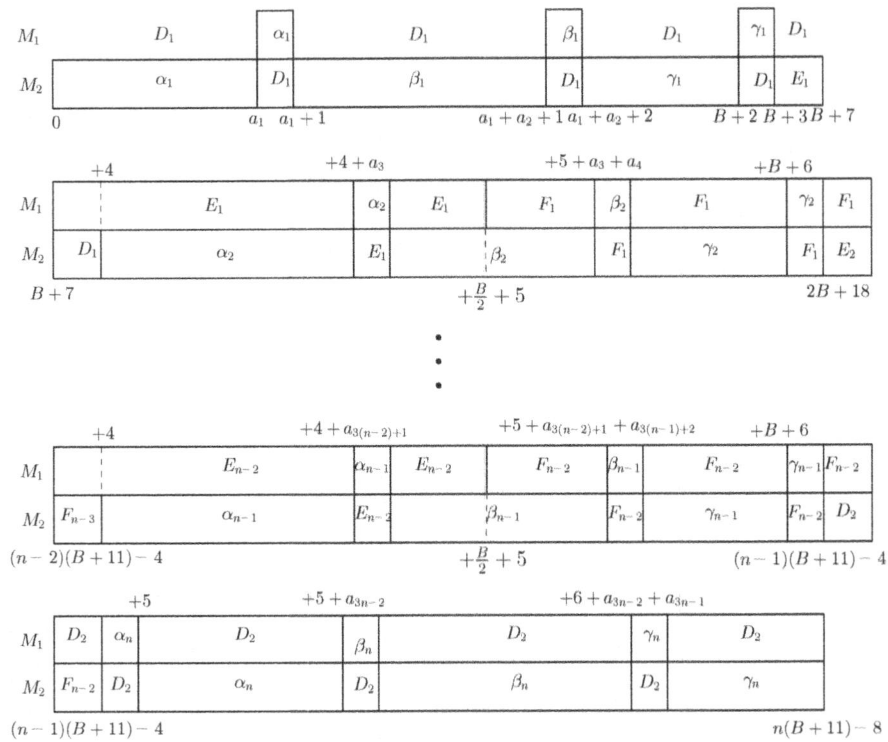

**Fig. 10.2** An optimal no-wait schedule for the three partition $A_i = \{\alpha_i, \beta_i, \gamma_i\}$, $i = 1, \ldots, n$

schedule obtained by the transformation $C_{\max} - t$, where $t$ is a start or completion of an operation in $\mathcal{S}$. The schedule $\mathcal{S}^R$ is feasible for $\mathbb{P}$, no-wait, and it has the same makespan as $\mathcal{S}$. Let jobs $J_1, J_2, \ldots, J_{2(n-1)}$ be the first, the second, ..., and the $2(n-1)$-th long job to start in $\mathcal{S}$. That is, $S_1 \leq S_2 \leq \cdots \leq S_{2(n-1)}$. We show that $\mathcal{S}$ needs to look like the schedule in Fig. 10.2.

We first show that $J_1$ and $J_n$ are *full-B* jobs. Let $t_1$ be the *earliest* moment when $J_1$ and $J_2$ are processed in parallel in $\mathcal{S}$. We prove now that $t_1$ exists, $t_1 > 0$, and $J_1$ is a *full-B* job. Let $J_1$ complete at $C_1$ in $\mathcal{S}$. Suppose for contradiction that $J_1$ and $J_2$ are never done in parallel, i.e., $C_1 \leq S_2$. Assume $J_1$ is a *full-B* job. Then the job $J_1$ is processed for $\ell = 7$ time units on $M_2$ in $[0, C_1]$ in parallel with short jobs being done on $M_1$ to avoid idle time in $\mathcal{S}$. However, each *short* job has only *one* unit-time operation on $M_1$. Thus at least $\ell \geq 7$ short jobs must start before $C_1$. Therefore, at least $\ell - 1 \geq 6$ short jobs must complete before $C_1$. Observe that no two short jobs are being processed in parallel by $C_1$. Because if they were, then there would be $t < C_1$ where two short jobs are processed in parallel. Take the earliest such $t$. Then $S_1 > t$ since otherwise $J_1$ would not be no-wait. Because of no-wait execution, one of those short jobs finishes at $T > t \geq a_{\min} \geq 22$, the other starts at $t$, but then $M_1$ is idle in $[0, t]$ in $\mathcal{S}$, which gives contradiction. Hence, for *full-B* job $J_1$, we have

$$a_1 + \dots a_{\ell-1} + \ell - 1 < C_1 = B + 4 + \ell, \tag{10.2}$$

where $a_1, \dots, a_{\ell-1}$ are the processing times of $M_2$ operations of short jobs that finish before $C_1$. On the other hand by (10.1) we have

$$a_1 + \dots a_{\ell-1} > \frac{(\ell-1)B}{4}, \tag{10.3}$$

which leads to contradiction since

$$\frac{(\ell-1)B}{4} > B + 5 \tag{10.4}$$

holds for $\ell - 1 \geq 6$ and $B \geq 33$. This proves that $J_1$ and $J_2$ must be done in parallel and $t_1$ exists. Observe that $t_1 > 0$. Otherwise, $J_1$ and $J_2$ would be done in parallel from time $t_1 = 0$ for at least $\frac{B}{2}$ units of time, and thus one of them needs its operation on $M_2$ to be at least that long, which leads to contradiction, since the long jobs have their operations on $M_2$ not longer than 7.

We now prove that $J_1$ cannot be a *half-B* job. Suppose for contradiction that $J_1$ is a *half-B* job. We first prove that $J_1$ and $J_2$ must be done in parallel. We use similar arguments as we did assuming that $J_1$ is a *full-B* job. Some details are different however. Suppose for contradiction that $J_1$ and $J_2$ are never done in parallel. A *half-B* job $J_1$ is processed on $M_2$ for at least $\ell = 5$ time units in parallel with short jobs being done on $M_1$ to avoid idle time in $\mathcal{S}$. However, each *short* job has only *one* unit-time operation on $M_1$. Thus at least $\ell \geq 5$ short jobs must start before $C_1$. Therefore at least $\ell - 1 \geq 4$ short jobs must complete before $C_1$. As shown earlier no short jobs are done in parallel by $C_1$. Thus for a *half-B* job $J_1$ we have

$$a_1 + \dots a_{\ell-1} + \ell - 1 < C_1 = \frac{B}{2} + 4 + \ell. \tag{10.5}$$

On the other hand by (10.1) we have

$$a_1 + \dots a_{\ell-1} > \frac{(\ell-1)B}{4}, \tag{10.6}$$

which leads to contradiction since

$$\frac{(\ell-1)B}{4} \geq \frac{B}{2} + 5 \tag{10.7}$$

holds for $\ell - 1 \geq 4$ and $B \geq 33$. This proves that $J_1$ and $J_2$ must be done in parallel and $t_1 > 0$ exists. Also, no job $J_j$, $j \geq 3$ is in $(0, t_1)$.

Then for a *half-B* job $J_1$ we have

$$a_1 + \dots + a_\ell + x = \frac{B}{2} + 4, \tag{10.8}$$

where $0 \leq x \leq 7$ is the total time job $J_1$ is done on $M_1$ in parallel with $J_2$ on $M_2$ in $[0, C_1]$. However by (10.1) we have $\frac{B}{2} < a_1 + a_2$. Hence, since $B$ and $a$ are even and divisible by 11, we have $\ell \leq 1$. Since $t_1 > 0$, we get $\ell = 1$. By (10.8) and (10.1), we get $\frac{B}{2} > a_1 \geq \frac{B}{2} - 3$, which gives contradiction, since $B$ and $a$ are even and divisible by 11. Thus $J_1$ is not a *half-B* job, and it is a *full-B* job. By similar arguments applied to $\mathcal{S}^R$ we show that $J_n$ is a *full-B* job.

We now show that $\mathcal{S}$ looks like the schedule in Fig. 10.2 in the interval $[0, B+11]$ and $[(n-1)(B+11)-8, n(B+11)-8]$. By definition of $t_1$, jobs $J_1$ and $J_j$, for $j \geq 2$, are not processed parallel in the interval $[0, t_1]$. Thus only short jobs can be processed in parallel with $J_1$ on $M_1$ in the interval $[0, t_1]$. Let $A_1 = \{\omega_1^1, \ldots, \omega_\ell^1\}$ be the set of *short* jobs processed in $[0, t_1]$. Because of no-wait requirement, those jobs are not processed outside of $[0, t_1]$ and thus each must complete by $t_1$. Let $0 \leq x \leq 6$ be the total time job $J_1$ is done on $M_1$ in parallel with $J_2$ on $M_2$ in $[0, C_1]$. We have

$$a_1 + \cdots + a_\ell + x = B + 4. \tag{10.9}$$

Since $a_1, \ldots, a_\ell, B$ are divisible by 11, we get $a_1 + \cdots + a_\ell = B$ and $\ell = 3$. Thus $x = 4$. A similar argument can be repeated for the schedule $\mathcal{S}^R$. Thus $t_1 = B + 3$, $C_1 = B + 11$, and the jobs $J_1$ and $J_2$ are processed in parallel in the interval $[B + 3, B + 11]$. Since $x = 4$, we may assume without loss of generality that $J_1$ is done on $M_1$ in parallel with $J_2$ on $M_2$ in $[B + 3, B + 7]$ and $J_1$ is done on $M_2$ in parallel with $J_2$ on $M_1$ in $[B + 7, B + 11]$. Similar arguments apply to $\mathcal{S}^R$. Therefore $\mathcal{S}$ is the same as the schedule in Fig. 10.2 in the intervals $[0, B + 11]$ and $[(n-1)(B + 11) - 8, n(B + 11) - 8]$.

By now we have that all jobs $J_2, \ldots, J_{2n-3}$ are *half-B*. It remains to prove that the remainder of $\mathcal{S}$ looks like the schedule in Fig. 10.2 for the *half-B* jobs and the remaining short jobs.

We first show that jobs $J_2$ and $J_3$ are scheduled in the interval $[B+11, 2B+18]$ in $\mathcal{S}$ as shown in Fig. 10.2. We begin by proving that $J_2$ and $J_3$ are not done in parallel in $\mathcal{S}$. Suppose for contradiction that the two jobs are done in parallel, and that $t_2$ is the earliest moment $J_2$ and $J_3$ are done in parallel in $\mathcal{S}$. Then $J_2$ and $J_3$ are done in parallel in $[t_2, C_2]$ since both $J_2$ and $J_3$ are no-wait. We have $\ell \geq 1$ short jobs completed in a no-wait fashion in $[C_1, t_2]$ if $C_1 < t_2$. Let $z$ be the total time $J_2$ is done on $M_1$ in parallel with $J_3$ on $M_2$ in $[C_1, C_2]$. We have

$$a_1 + \cdots + a_\ell + z = \frac{B}{2}, \tag{10.10}$$

where $a_1, \ldots, a_\ell$ are the processing times of $M_2$ operations of short jobs in $[C_1, t_2]$. By (10.1), $\ell = 1$. Thus $a_1 + z = \frac{B}{2}$, which leads to contradiction, since $z \leq 6$, and both $a_1$ and $B$ are even and divisible by 11. If $C_1 = t_2$, then $J_2$ and $J_3$ would be done in parallel from time $t_2$ for at least $\frac{B}{2}$ units of time and thus one of them needs its operation on $M_2$ to be at least that long which leads to contradiction since long *half-B* jobs have their operations on $M_2$ not longer than 6. Therefore, $J_2$ and $J_3$ are not done in parallel, and a similar argument applied to $\mathcal{S}^R$ shows that $J_{n-1}$ and $J_{n-2}$ are not done in parallel.

To complete this part of the proof consider job $J_4$. Let $t_3$ be the *earliest* moment when $J_3$ and $J_4$ are processed in parallel in $\mathcal{S}$. We now prove that $t_3$ exists. Suppose for contradiction that $J_3$ and $J_4$ are never done in parallel, i.e., $C_3 \leq S_4$. Therefore, the $\ell \geq 7$ remaining time units of $M_2$ operations of jobs $J_2$ and $J_3$ (both are *half- B* jobs) need to be processed in parallel with some operations on $M_1$ to avoid idle time in $\mathcal{S}$. Those operations may only come from *short* jobs, but each of them can deliver only *one* unit-time operation on $M_1$. Thus at least $\ell \geq 7$ short jobs must start after $C_1$ but before $C_3$. Therefore at least $\ell - 1 \geq 6$ short jobs must complete in $[C_1, C_3]$. However, since no two short jobs can be done in parallel in $[C_1, C_3]$, we have

$$a_1 + \ldots a_{\ell-1} + \ell - 1 < C_3 = B + 4 + \ell, \tag{10.11}$$

where $a_1, \ldots, a_{\ell-1}$ are the processing times of $M_2$ operations of short jobs that complete in $[C_1, C_3]$. On the other hand by (10.1) we have

$$a_1 + \ldots a_{\ell-1} > \frac{(\ell - 1)B}{4}, \tag{10.12}$$

which leads to contradiction since

$$\frac{(\ell - 1)B}{4} > B + 5 \tag{10.13}$$

holds for $\ell - 1 \geq 6$ and $B \geq 33$. This proves that $J_3$ and $J_4$ must be done in parallel and $t_3$ exists. In the interval $[C_1, t_3]$, job $J_2$ is not done in parallel with any job $J_j$ where $j < 2$ or $j \geq 3$. Thus jobs $J_2$ and $J_3$ can only be processed with short jobs in parallel in $[C_1, t_3]$. Let $A_2 = \{\omega_1^2, \ldots, \omega_\ell^2\}$ be the set of *short* jobs processed in $[C_1, t_3]$. Because of no-wait requirement, those jobs are not processed outside of $[C_1, t_3]$. We get

$$a_1 + \cdots + a_\ell + y = B + 4, \tag{10.14}$$

where $y \leq 6$ is the total time $J_3$ on $M_1$ is processed in parallel with $J_4$ on $M_2$ in $(t_3, C_3)$. Thus $a_1 + \cdots + a_\ell = B$, $\ell = 3$, and $y = 4$. Also since $\ell = 3$, we have $\{J_2, J_3\} \cap \mathcal{E} \neq \emptyset$ and $\{J_2, J_3\} \cap \mathcal{F} \neq \emptyset$. Therefore, without loss of generality, $\mathcal{S}$ jobs $J_2$ and $J_3$ are scheduled $[B + 11, 2B + 18]$ as shown in Fig. 10.2. Similarly jobs $J_{n-1}$ and $J_{n-2}$ and $[(n - 2)(B + 11), (n - 1)(B + 11) - 4]$.

We can repeat this argument for the remaining pairs of *half-B* jobs and thus by induction we obtain the sets $A_1, \ldots, A_n$ that make up the required partition. We leave details to the reader. □

Hall and Sriskandarajah [11] point out that an $NP$-hardness proof of the problem $O2|\text{pmtn}, no - wait|C_{\max}$ is given in Strusevich [20]. However, to the author's knowledge that proof has not been published.

## 10.2 Short No-Wait Schedules: Test for $C_{\max} \leq 3$

We develop a polynomial-time test that determines if a bipartite multigraph $G = (\mathcal{J}, \mathcal{M}, E)$ with $\Delta(G) = 3$ has a no-wait schedule with makespan not exceeding 3.

*Claim 19* A no-wait schedule with $C_{\max} = 3$ for $G$ exists if and only if there is 3-edge coloring of $G$ with colors 1, 2, or 3 where each job-vertex with degree 2 has one of the edges incident with the vertex colored with color 2.

***Proof*** We have the following correspondence between schedule $\mathcal{S}$ with $C_{\max} = 3$ for $G$ and a 3-edge coloring $C$ of $G$ with colors 1, 2, or 3: job $J_j \in \mathcal{J}$ is processed on machine $M_h \in \mathcal{M}$ in the interval $[i-1, i]$, $i = 1, 2, 3$, in $\mathcal{S}$ if and only if the edge $(J_j, M_h) \in E$ is colored with color $i = 1, 2, 3$ in $C$.

If schedule $\mathcal{S}$ is no-wait, then each job $J_j$ of length 2 is processed in the interval $[0, 2]$ or $[1, 3]$. Thus one of the edges incident with $J_j$ is colored with color 2 in $C$.

On the other hand, if each job-vertex $J_j$ with degree 2 has one of the edges incident with $J_j$ colored with color 2 in $C$, then $J_j$ is processed in the interval $[0, 2]$ or $[1, 3]$ in $\mathcal{S}$ depending on whether the two edges incident with $J_j$ in $G$ are colored with 1 and 2, or 2 and 3 in $C$, respectively. Thus each job is no-wait in the schedule $\mathcal{S}$. □

It remains to test whether a 3-edge coloring of $G$ with colors 1, 2, or 3 where each job-vertex with degree 2 has one of the edges incident with the vertex colored with color 2 exists or not. To develop the test we first turn the multigraph $G$ into a simple graph $H$ with weights on edges. Let $X$ be the set of job-vertices of degree 2 in $G$, and let $Y$ be the set of all vertices of degree 3 in $G$. Clearly, these sets are disjoint. We turn the multigraph $G$ into a simple graph $H$ with weighted edges by replacing each multiedge by a single edge. Each edge in $H$ incident with exactly one vertex in $X \cup Y$ gets weight 1, each edge in $H$ incident with exactly two vertices in $X \cup Y$ gets weight 2, and each edge in $H$ incident with two vertices that are *not* in $X \cup Y$ gets weight 0. We have the following claim.

*Claim 20* The weight of a maximum weight matching in $H$ equals $|X| + |Y|$ if and only if there is 3-edge coloring of $G$ with colors 1, 2, or 3 where each job-vertex with degree 2 has one of the edges incident with the vertex colored with color 2.

***Proof*** Let $M$ be a matching in $H$. For $e = \{u, v\} \in M$, define $A_e = e \cap (X \cup Y)$. We have $|A_e| = w(e)$ by definition of $H$. Also, since $M$ is a matching, $A_e \cap A_{e'} = \emptyset$ for different edges $e, e' \in M$. Thus $|\bigcup_{e \in M} A_e| = \sum_{e \in M} w(e)$.

Suppose the maximum weight matching $M$ in $H$ has weight $|X| + |Y|$. Thus $|\bigcup_{e \in M} A_e| = \sum_{e \in M} w(e) = |X| + |Y|$. On the other hand, $\bigcup_{e \in M} A_e \subseteq X \cup Y$. Thus $|\bigcup_{e \in M} A_e| \leq |X| + |Y|$. Therefore, $\bigcup_{e \in M} A_e = X \cup Y$, and $X \cup Y \subseteq V(M)$, where $V(M)$ is the set of vertices incident with the edges in $M$. The set $X \cup Y$ includes all job-vertices of degree 2 or 3 in $G$. Color the edges in $M$ with color 2 in $G$, and remove them from $G$. The resulting multigraph has $\Delta \leq 2$, and thus by König's edge-coloring theorem, its edges can be colored with two colors 1 and 3.

This gives the required 3-edge coloring of $G$ since each job-vertex with degree 2 has one of the edges incident with the vertex colored with color 2.

Now suppose that there is a 3-edge coloring of $G$ with colors 1, 2, or 3 where each job-vertex with degree 2 has one of the edges incident with the vertex colored with color 2. Let $M$ be the set of edges colored with color 2 in $G$. The $M$ is a matching, and $X \cup Y \subseteq V(M)$ by definition of the coloring. Thus $\bigcup_{e \in M} A_e = X \cup Y$. Therefore, $|\bigcup_{e \in M} A_e| = \sum_{e \in M} w(e) = |X| + |Y|$, and thus the matching has the total weight $|X| + |Y|$ in $H$ as required. □

Therefore we need to check if a maximum weight matching in bipartite $H$ has weight $|X| + |Y|$. This can be done in polynomial time by the Hungarian method for instance, see Lawler [13].

## 10.3 Short No-Wait Schedules: Test for $C_{\max} \leq 4$

We now develop a polynomial-time test to determine if a bipartite multigraph $G = (\mathcal{J}, \mathcal{M}, E)$ has a no-wait schedule with makespan not exceeding 4. We already developed a polynomial-time test for $C_{\max} \leq 3$ in the previous section; thus it remains to develop a polynomial-time test assuming that a no-wait schedule with $C_{\max} \leq 3$ does not exist for $G$. To that end for each machine-vertex $M_h$ in $G$ with degree $\deg(M_h) < 4$, we add $4 - \deg(M_h)$ unique jobs with a single unit-time operation on $M_h$. By doing so for each machine-vertex with degree less than 4 in $G$, we obtain a bipartite multigraph $H$ with each machine workload equal 4, and $\Delta(H) = 4$. We claim the following.

*Claim 21* A no-wait schedule for $G$ with $C_{\max} = 4$ exists if and only if a compact schedule for $H$ with $C_{\max} = 4$ exists.

***Proof*** If a no-wait schedule $\mathcal{S}$ for $G$ with $C_{\max} = 4$ exists, then by adding pendant job-vertices to $G$ the schedule $\mathcal{S}$ can be readily extended to a schedule $\mathcal{S}'$ for $H$ where each machine is occupied in $[0, 4]$. Thus, since $\mathcal{S}$ is no-wait for $G$, $\mathcal{S}'$ is compact for $H$ with $C_{\max} = 4$.

On the other hand, a compact schedule $\mathcal{S}$ for $H$ is no-wait for $G$ once all pendant job-vertices that have been added to $G$ are removed from $\mathcal{S}$. If $\mathcal{S}$ has makespan $C_{\max} = 4$, then the resulting no-wait schedule for $G$ has makespan $C_{\max} \leq 4$. However, a no-wait schedule for $G$ with $C_{\max} \leq 3$ does not exist. Therefore, the resulting schedule has makespan $C_{\max} = 4$. □

By this claim we can limit the test to bipartite multigraphs $G = (\mathcal{J}, \mathcal{M}, E)$ with $\Delta(G) = 4$, and with each machine-vertex of degree $\Delta(G)$, i.e., each machine workload equals $\Delta(G)$. Therefore, the polynomial-time test from Sect. 9.2.3 can be used to test whether or not $G$ has a no-wait schedule with makespan $C_{\max} = 4$. The test works for multigraphs as well, see Giaro [9].

## 10.4 Short No-Wait Schedules: $C_{\max} \leq 5$

In this section we consider the problem $O|\text{pmtn}, \text{no-wait}|C_{\max} \leq 5$ to decide if there is a no-wait schedule with makespan $C_{\max} \leq 5$ for an open shop $G = (\mathcal{J}, \mathcal{M}, E)$. This problem remains open.

**Problem 10.2** What is the complexity status of the $O|\text{pmtn}, \text{no-wait}|C_{\max} \leq 5$ problem?

We now show that instances of the problem with $n > m$ and jobs of length 3, 4, or 5 cannot have no-wait schedules with makespan 5. This follows from the following lemma.

**Lemma 10.2** *For instances with more jobs than machines, i.e., $n > m$, if the shortest job is of length at least* 3*, then a no-wait schedule with makespan $C_{\max} = 5$ does not exist.*

***Proof*** Casselgren and Toft [5] prove the following lower bound for $C_{\max}$ of no-wait schedules

$$\left\lceil \frac{n}{|M|} \right\rceil \min_j \{P_j\} \leq C_{\max}, \tag{10.15}$$

where $M$ is maximum matching in $G = (\mathcal{J}, \mathcal{M}, E)$. Suppose for contradiction that $C_{\max} = 5$ for some no-wait schedule. Since $\min_j\{P_j\} \geq 3$, we get

$$\left\lceil \frac{n}{|M|} \right\rceil \leq 1.$$

Hence, $n \leq |M|$. However, $|M| \leq \min\{n, m\} = m < n$, which gives contradiction and proves that no-wait schedule with makespan $C_{\max} = 5$ does not exist. □

Despite this negative result, optimal no-wait schedules, though longer than 5, can be readily obtained for some classes of open shops. For instance, consider the $(4, 5)$-biregular instances $G = (\mathcal{J}, \mathcal{M}, E)$, where each job is of length 4 and each machine has workload 5. These instances meet the conditions of Lemma 10.3. Thus the minimum makespan $C_{\max} > 5$ for $G$. Casselgren and Toft [5] show a polynomial-time algorithm for obtaining optimal no-wait schedules for this class of instances. We now briefly introduce their idea yet we recast it in open shop scheduling context that seems particularly appealing because of its simplicity. We illustrate this approach with an example. By König's edge-coloring theorem the edges of $G = (\mathcal{J}, \mathcal{M}, E)$ can be colored with colors 1, 2, 3, 4, and 5. The coloring can easily be turned into a schedule $\mathcal{S}$ with $C_{\max} = 5$. Please see Fig. 10.3 for an example of such a schedule. By Lemma 10.3 not all jobs in this schedule are no-wait, for instance job $J_3$ in Fig. 10.3. How can one turn this schedule into a no-wait schedule at the cost of increasing the makespan by as little as possible? The idea is to make a copy of $\mathcal{S}$ and start it at 5. The resulting schedule is a concatenation $\mathcal{S}\mathcal{S}$

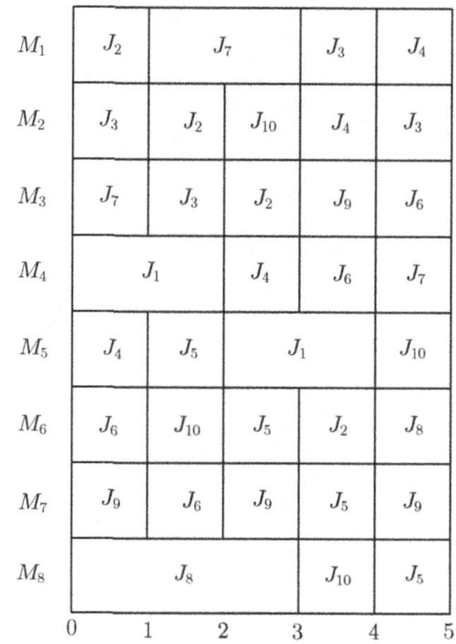


**Fig. 10.3** An optimal schedule $\mathcal{S}$ with makespan $C_{\max} = 5$ for an instance with all jobs of length 4 and all machines with workload 5

of two copies of $\mathcal{S}$, one starts at 0 and the other at 5. Hence, each job occurs twice in $\mathcal{SS}$. We refer to the copy in the interval $[0, 5]$ as an *earlier* copy and to the copy in the interval $[5, 10]$ as a *later* copy. Now delete all jobs that are no-wait in $\mathcal{S}$ from the later copy. These are jobs $J_1$, $J_2$, $J_5$, and $J_{10}$ in Fig. 10.3. Any other job $J_j$ is processed in two intervals $[0, i]$ and $[i + 1, 5]$ for some $i = 1, 2, 3$ in $\mathcal{S}$. This holds since the edge coloring ensures that the edges incident with $J_j$ in $G$ miss exactly one color among 1, 2, 3, 4, or 5, and this cannot be 1 or 5 because otherwise $J_i$ would be no-wait. Delete the operations of $J_j$ processed in $[0, i]$ from the earlier copy of $\mathcal{S}$ and the operations of $J_j$ processed in $[5 + (i + 1), 10]$ from the later copy of $\mathcal{S}$. Therefore, the resulting schedule has job $J_j$ processed in the interval $[i + 1, 5 + i]$, which is of length 4, and thus $J_i$ is no-wait in the resulting schedule. Since $i \le 3$, the resulting schedule has makespan $C_{\max} \le 8$. The resulting schedule for $\mathcal{S}$ shown in Fig. 10.3 is shown in Fig. 10.4. Somewhat surprisingly, the schedules thus obtained are optimal since by the lower bound in (10.15) we have

$$4\left\lceil \frac{n}{|M|} \right\rceil \le C_{\max}.$$

However, $\lceil \frac{n}{|M|} \rceil \ge 2$ since $|M| \le m < n$ for the instances with all jobs of length 4 and all machines with workload 5. Hence, $C_{\max} \ge 8$, which proves that the schedules with makespan 8 are optimal for the instances. The general result of Casselgren and Toft [5] can be formulated as follows.

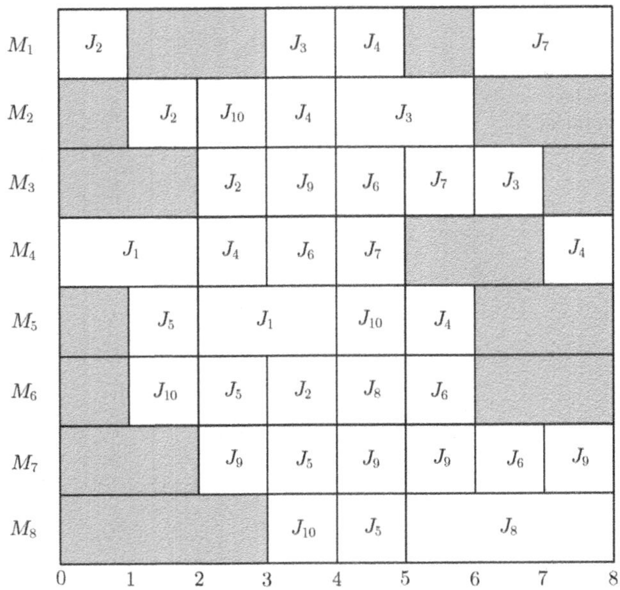


**Fig. 10.4** An optimal no-wait schedule with makespan $C_{\max} = 8$ for an instance in Fig. 10.3

**Lemma 10.3** *If* $\min_j\{P_j\} = \Delta(G) - 1$, *then there is a no-wait schedule with* $C_{\max} \leq 2\Delta(G)-2$. *Moreover, if* $G$ *is* $(\Delta(G)-1, \Delta(G))$*-biregular, then the schedule is optimal.*

Asratian and Kamalian [2] give the following sufficient condition for no-wait schedules to exist for the instances with $n \leq m$.

**Lemma 10.4** *If* $\deg(J_j) \geq \deg(M_h)$ *for each* $(J_j, M_h) \in E$, *then there is a no-wait schedule with* $C_{\max} = \Delta(G)$ *such that each job* $J_j$ *is scheduled in the interval* $[0, \deg(J_j)]$.

The condition is a corollary from a theorem of Geller and Hilton [7]. The idea is that by König's edge-coloring theorem each vertex $v$ of degree $\deg(v) = \Delta(G)$ in $G = (\mathcal{J}, \mathcal{M}, E)$ has the $\Delta(G)$ edges incident with $v$ colored with colors $1, \ldots, \Delta(G)$. Let $\mathcal{M}_{\Delta(G)}$ be the set of all edges colored with $\Delta(G)$ and incident with some vertex of degree $\Delta(G)$. The matching $\mathcal{M}_{\Delta(G)}$ covers all job-vertices with degree $\Delta(G)$. By the lemma assumptions the number of those vertices does not exceed the number of machines $m$. Moreover, if $w$ is a machine-vertex of degree $\Delta(G)$, then $(v, w) \in \mathcal{M}_{\Delta(G)}$ for some job-vertex $v$ of degree $\Delta(G)$. Thus, $|\mathcal{M}_{\Delta(G)}| \leq m$. Therefore, deleting the edges in $\mathcal{M}_{\Delta(G)}$ from $G$ results in a graph $G'$ of degree $\Delta(G) - 1$ that meets the assumption of the lemma. Moreover, scheduling job $J_j$ on machine $M_h$ for $(J_i, M_h) \in \mathcal{M}_{\Delta(G)}$ in the interval $[\Delta(G) - 1, \Delta(G)]$ leaves the interval $[0, \Delta(G) - 1]$ for $G'$. The induction completes the argument.

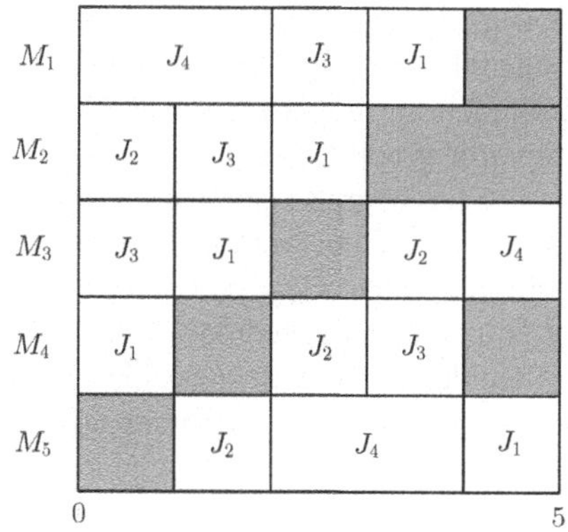


**Fig. 10.5** An optimal no-wait schedule with makespan $C_{\max} = 5$ for the open shop $\mathbb{P}$

To illustrate let us consider the following open shop with $n = 4$ jobs and $m = 5$ machines

$$\mathbb{P} = \begin{bmatrix} 1 & 1 & 1 & 1 & 1 \\ 0 & 1 & 1 & 1 & 1 \\ 1 & 1 & 1 & 1 & 0 \\ 2 & 0 & 1 & 0 & 2 \end{bmatrix}.$$

The optimal no-wait schedule with makespan $C_{\max} = 5$ is shown in Fig. 10.5. The vertices $J_1$ and $J_4$ are of degree 5, and $\mathcal{M}_5 = \{(J_1, M_5), (J_4, M_3)\}$, which fixes the schedule in $[4, 5]$ and reduces the open shop to

$$\mathbb{P}' = \begin{bmatrix} 1 & 1 & 1 & 1 & 0 \\ 0 & 1 & 1 & 1 & 1 \\ 1 & 1 & 1 & 1 & 0 \\ 2 & 0 & 0 & 0 & 2 \end{bmatrix},$$

with degree $\Delta = 4$, and all job-vertices $J_1$, $J_2$, $J_3$, and $J_4$ of degree 4. The matching $\mathcal{M}_4 = \{(J_1, M_1), (J_2, M_3), (J_3, M_4), (J_4, M_5)\}$ fixes the schedule in $[3, 4]$. The reader is encouraged to complete the example.

## 10.5 Short No-Wait Schedules: $C_{\max} \leq 6$

In this section we consider a problem $O|\text{pmtn, no-wait}|C_{\max} \leq 6$ to decide if there is a no-wait schedule with makespan $C_{\max} \leq 6$ for an open shop $G = (\mathcal{J}, \mathcal{M}, E)$. We have the following complexity result.

**Theorem 10.1** *The problems $O|pmtn, no\text{-}wait|C_{\max} \leq 6$ and $O|p_{ij} \in \{0, 1\}$, no-wait$|C_{\max} \leq 6$ are NP-complete in the strong sense.*

***Proof*** By Theorem 9.7 the problem to decide whether or not there is a compact schedule for (3, 6)-biregular bipartite graphs with makespan $C_{\max} = 6$ is NP-complete in the strong sense. This holds for simple graphs. Each (3, 6)-biregular bipartite graph $G$ is a job–machine bipartite graph $G = (\mathcal{J}, \mathcal{M}, E)$ where each job-vertex has degree 3 (three unit-time operations) and each machine-vertex has degree 6 (each machine workload equals 6). Thus $\Delta(G) = 6$. A no-wait schedule for $G$ with makespan $C_{\max} = 6$ keeps each machine busy in the interval $[0, 6]$, and thus it is a compact schedule for $G$ with the same makespan. By definition a compact schedule for $G$ with makespan $C_{\max} = 6$ is a no-wait schedule with the same makespan. Thus the problem $O|p_{ij} \in \{0, 1\}, \text{no-wait}|C_{\max} \leq 6$ is NP-complete in the strong sense. □

## 10.6 No-Wait and Cyclic Compact Open Shop Scheduling

The cyclic compact schedules can be turned into no-wait schedules as follows. Let $\mathcal{S}$ be a cyclic compact schedule for $G = (\mathcal{J}, \mathcal{M}, E)$ with makespan $C_{\max}$. By definition of cyclic schedules each job $J_j$ is either processed in a single interval $[S_j, S_j + P_j]$, where $0 \leq S_j < S_j + P_j \leq C_{\max}$ or *not* processed in a single interval $[A_j, C_{\max} - P_j + A_j]$, where $0 < A_j < P_j$ in $\mathcal{S}$. In the former case $J_j$ is no-wait. In the latter $J_j$ is processed in $[0, A_j]$ and $[C_{\max} - P_j + A_j, C_{\max}]$. Consider the concatenation $\mathcal{S}' = \mathcal{S}\mathcal{S}$ of two copies of a cyclic compact $\mathcal{S}$. The schedule $\mathcal{S}'$ has a copy of job $J_j$, which is *not* no-wait in $\mathcal{S}$, processed in the interval $[C_{\max} - P_j + A_j, C_{\max} + A_j]$. Thus the copy, if any, is no-wait in $\mathcal{S}'$. Clearly each job $J_j$ that is no-wait in $\mathcal{S}$ remains so in $\mathcal{S}'$. Therefore deleting all other copies from $\mathcal{S}'$ results in a no-wait schedule for $G$. The makespan of the resulting schedule equals $C_{\max} + \max_j\{A_j\} \leq C_{\max} + \Delta(G) - 2$. This construction allows us to use the results of Sect. 9.6 to obtain short no-wait schedules. In particular by Theorem 9.9, open shops $G = (\mathcal{J}, \mathcal{M}, E)$ with job lengths limited to 1, 2, 4, or 5 and machine workloads not exceeding 5 have cyclic compact schedules with $C_{\max} = 5$ or $C_{\max} = 6$. Thus there exist no-wait schedules with makespan $C_{\max} = 8$ or $C_{\max} = 9$ for those open shops.

## 10.7 Exact Algorithms and Approximations

A PTAS does not exist for the problem $O|no - wait|C_{\max}$. This follows from the following theorem.

**Theorem 10.2** *If $P \neq NP$, then no polynomial-time algorithm exists for the problem $O|no - wait|C_{\max}$ with the worst-case ratio less than $\frac{7}{6}$.*

***Proof*** Consider the set $\mathcal{I}$ of open shop instances defined in the proof of Theorem 10.1. The problem $\Pi$ defined by $\mathcal{I}$ and by the question whether $I \in \mathcal{I}$ has a no-wait schedule with makespan not exceeding 6 or not is $NP$-complete that follows immediately from the proof of Theorem 10.1. Suppose for contradiction that there is a polynomial-time algorithm $A$ such that $C^A_{\max}/C^*_{\max} < 7/6$ for any instance of $O|no-wait|C_{\max}$. Thus, in particular, $C^A_{\max}/C^*_{\max} < 7/6$ for any instance of $\Pi$. The algorithm $A$ can be used to solve $\Pi$ as follows. If $C^A_{\max} \leq 6$ for an instance $I$, then the answer for $I$ is affirmative. Otherwise, if $C^A_{\max} > 6$ for $I$, then, since all processing times in $I$ are integer, we have $C^A_{\max} \geq 7$ and integer. Thus, since $C^*_{\max} > 6C^A_{\max}/7$, we get $C^*_{\max} > 6$ and the answer for $I$ is negative. Since $C^A_{\max}$ can be computed in polynomial time for each $I \in \mathcal{I}$, we have $\Pi$ in $P$. This implies $P = NP$ since $\Pi$ is $NP$-complete and gives contradiction. □

However, a PTAS exists for $O2|no-wait|C_{\max}$, which was shown by Bansal et al. [3]. Sidney and Sriskandarajah [19] show a $\frac{3}{2}$-approximation algorithm for the two-machine problem. Their algorithm relies on the Gilmore and Gomory algorithm [10] for a special case of the traveling salesman problem. Panwalkar and Koulamas [16] present a further analysis of $O2|\text{no-wait}|C_{\max}$ to improve the running time of the $\frac{3}{2}$-approximation algorithm. They also propose a polynomial-time algorithm for the two-machine, no-wait job-proportionate open shop. The algorithm runs in $O(n \log n)$ time. An example of the schedule obtained by the algorithm is given in Fig. 10.1a. All these results assume tacitly or explicitly non-preemptive schedules. We have seen however in Fig. 10.1b that no-wait preemptive schedules cannot only reduce optimal makespan but they can also be quite different from the optimal non-preemptive schedules. The optimization and approximation algorithms for no-wait preemptive schedules are largely unexplored yet interesting areas worthy further research.

Liaw et al. [14] show a branch and bound algorithm for two-machine no-wait open shop makespan minimization. Naderi and Zandieh [15] propose mixed integer linear programs and metaheuristics for the problem $O|\text{no-wait}|C_{\max}$. Again both papers assume non-preemptive schedules. Brucker et al. [4] and Kubiak et al. [12] provide further complexity results for the no-wait open shop scheduling.

## Problems

**10.1** Show that each instance $G = (\mathcal{J}, \mathcal{M}, E)$ with $\Delta(G) \leq 2$ has a no-wait schedule with $C_{\max} \leq 2$.

**10.2** Complete the induction in the proof of Lemma 10.1.

**10.3** Prove Lemma 10.3.

**10.4** Prove that the problem $O|\text{pmtn, no-wait}|C_{\max} \leq 6$ is $NP$-hard in the strong sense.

## References

1. A. Acampora, S. Shah, Multihop lightwave networks: A comparison of store-and-forward and hot-potato routing, in *Proc. INFOCOM, Bal Harbour, Fl* (IEEE, Piscataway, NJ, 1991), pp. 10–19
2. A.S. Asratian, R.R. Kamalian, Investigation on interval edge-colorings of graphs. J. Comb. Theory B **62**, 34–43 (1994)
3. N. Bansal, M. Mahdian, M. Sviridenko, Minimizing makespan in no-wait job shops. Math. O. R **30**, 817–831 (2005)
4. P. Brucker, B. Jurisch, M. Jurisch, Open-shop problems with unit processing times. Zeitschrift fur Opns. Res. **37**, 59–73 (1993)
5. C.J. Casselgren, B. Toft, One-sided interval edge-colorings of bipartite graphs. Discrete Mathematics **339**, 2628–2639 (2016)
6. M.R. Garey, D.S. Johnson, *Computers and Intractability: A Guide to the Theory of NP-Completeness* (W. H. Freeman, 1979)
7. D.P. Geller, A.J.W. Hilton, How to color the lines of a bigraph. Networks **4**, 281–282 (1974)
8. K. Giaro, NP-hardness of compact scheduling in simplified open and flow shops. Eur. J. Oper. Res. **130**, 90–98 (2001)
9. K. Giaro, Private communication (2021)
10. P.C. Gilmore, R.E. Gomory, Sequencing a one state-variable machine: A solvable case of the traveling salesman problem. Operations Research **12**, 655–679 (1964)
11. N.G. Hall, C. Sriskandarajah, A survey of machine scheduling problems with blocking and no-wait in process. Operations Research **44**, 510–524 (1996)
12. W. Kubiak, C. Sriskandarajah, K. Zaras, A note on the complexity of openshop scheduling problems. INFOR, 29 (1991)
13. E.L. Lawler, *Combinatorial Optimization. Networks and Matroids* (Dover Publications, 2001)
14. C-F. Liaw, C-Y. Cheng, M. Chen, Scheduling two-machine no-wait open shops to minimize makespan. Comput. Oper. Res. **32**, 901–917 (2005)
15. B. Naderi, M. Zandieh, Modeling and scheduling no-wait open shop problems. Int. J. Prod. Econ. **158**, 256–266 (2014)
16. S.S. Panwalkar, C. Koulamas, The two-machine no-wait general and proportionate open shop makespan problem. Eur. J. Oper. Res. **238**, 471–475 (2014)
17. C.H. Papadimitriou, P. Kanellakis, Flow-shop scheduling with limited temporary storage. J. ACM **27**, 533–549 (1980)
18. S. Sahni, Y. Cho, Complexity of scheduling shops with no wait in process. Math. O. R **4**, 448–457 (1979)
19. J.B. Sidney, C. Sriskandarajah, A heuristic for the two-machine no-wait openshop scheduling problem. Naval Res. Logist. **46**, 129–145 (1999)
20. V.A. Strusevich, *Complexity aspects of shop scheduling problems*. PhD thesis, Erasmus Universiteit Rotterdam, 1991
21. T. Szymanski, An analysis of "hot-potato" routing in a fiber optic packet switched hypercube, in *Proc. of INFOCOM, San Francisco, CA* (ACM Press, New York, 1990), pp. 918–925

# Chapter 11
# Applications of Preemptive Open Shop Scheduling

## 11.1 Introduction

Before we present more applications of open shop scheduling to real-life problems and scheduling theory let us briefly summarize the applications discussed in previous chapters of this book. Chapter 4 presents applications to University timetabling, Chap. 6 presents application to scheduling wireless networks with primary interference, Chap. 5 presents applications to product design and scheduling customer orders among other applications of open shop scheduling where operations of a job can be processed concurrently, Chaps. 7 and 8 present application to scheduling large oncology centers, Chap. 9 presents application to timetabling and just-in-time scheduling, and Chap. 10 presents applications to scheduling without intermediate storage and optical networks. Other applications are also listed in Ahmadian et al. [1].

This chapter introduces further applications of open shop scheduling. Those include satellite-switched time-division multiple access method used to allocate the communication bandwidth provided by a satellite link to carry traffic between earth stations; scheduling reconfigurable data centers; file transfer; scheduling crossbar switches to guarantee 100% throughput for traffic with given rates; multimessage unicasting and multicasting; scheduling and bandwidth allocation problem; and scheduling theory. Majority of those applications permit preemptions which allow for polynomial-time algorithms when it comes to makespan minimization. Details will be described in the following sections.



W. Kubiak, *A Book of Open Shop Scheduling*, International Series in Operations Research & Management Science 325,
https://doi.org/10.1007/978-3-030-91025-9_11

## 11.2 Satellite Communication

Satellite communication systems have been built to connect a large number of earth stations. The *satellite-switched time-division multiple access* (SS/TDMA) method has been used to allocate the communication bandwidth provided by a satellite link to carry traffic between earth stations, see Inukai [30], and Lewandowski and Liu [35] for further references to the SS/TDMA method. In a nutshell, a satellite equipped with spot-beam antennas, and a switch periodically switches the connections between uplink beams and downlink beams to connect beam zones. A TDMA frame is a sequence of *time slots* of different durations. In each of the time slots, a different *switching mode* determines a specific set of interconnections so that the traffic between zones is routed without conflicts. The SS/TDMA time slot assignment problem instance is defined by an $n \times n$ non-negative *traffic matrix* $\mathcal{T}$ where $t_{ij}$ is the amount of traffic which is to be routed from the uplink beam $i$ to the downlink beam $j$, and $n$ is the number of spot-beams. The quantities $t_{ij}$ may be expressed in terms of some basic traffic units such as a number of $T1$ channels, Inukai [30], or they can be the amount of time, Inukai [30], and Dell'Amico and Martello [18]. We let $t_{ij}$ to be the amount of time in our discussion. The problem is to find a *switching mode matrix* $S_i$, and a time duration $t_i$ for each time slot so as to meet the traffic demand defined by $\mathcal{T}$ within the shortest possible TDMA frame. A switching mode matrix is a square $n \times n$ matrix where *at most* one entry in each row and column is equal to 1, and all other entries are equal to 0. To find a solution to the time slot assignment problem let us have a closer look at the traffic matrix $\mathcal{T}$. The total demand for its row $i$ is equal to $R_i = \sum_j t_{ij}$ which is the amount of time required by the uplink beam $i$ to transmit information to all downlink beams, and the total demand for its column $j$ is equal to $C_j = \sum_i t_{ij}$ which is the amount of time required by the downlink beam $j$ to receive information from all uplink beams. Clearly, the TDMA frame cannot be shorter than

$$\alpha = \max\{\max_i\{R_i\}, \max_j\{C_j\}\}.$$

Due to Birkhoff–von Neumann theorem we can get the frame of length $\alpha$ as follows. We take the following $(2n) \times (2n)$ matrix:

$$\mathcal{D} = \frac{1}{\alpha}\begin{bmatrix} \mathcal{T} & A \\ B & \mathcal{T}^T \end{bmatrix},$$

where the diagonal matrix

$$
A = \begin{bmatrix} \alpha - R_1 & 0 & \dots & 0 \\ 0 & \alpha - R_2 & \dots & 0 \\ 0 & & \dots & 0 \\ . & . & . & . \\ . & . & . & . \\ . & . & . & . \\ 0 & 0 & \dots & \alpha - R_n \end{bmatrix}
$$

complements each row sum of $\mathcal{D}$ to $\alpha$, and the diagonal matrix

$$
B = \begin{bmatrix} \alpha - C_1 & 0 & \dots & 0 \\ 0 & \alpha - C_2 & \dots & 0 \\ 0 & & \dots & 0 \\ . & . & . & . \\ . & . & . & . \\ . & . & . & . \\ 0 & 0 & \dots & \alpha - C_n \end{bmatrix}
$$

complements each column sum of $\mathcal{D}$ to $\alpha$. By the theorem there are $(2n) \times (2n)$ permutations matrices $P_1, \dots, P_q$ and positive numbers $t_1, \dots, t_q$ such that $t_1 + \dots t_q = 1$ and

$$
\mathcal{D} = t_1 P_1 + \dots + t_q P_q. \tag{11.1}
$$

Let $S_i$ be the submatrix of the permutation matrix $P_i$ lying in the rows $1, \dots, n$ and the columns $1, \dots, n$ of $P_i$, $i = 1, \dots, q$. Clearly, the matrices $S_i$ are switching modes and

$$
\mathcal{T} = \alpha(t_1 S_1 + \dots + t_q S_q).
$$

Thus there are $q$ time slots with durations $\alpha t_1, \dots, \alpha t_q$. Since $t_1 + \dots t_q = 1$ the total duration is equal to $\alpha$. The number of permutation matrices $q \leq n^2 - n + 1$, see Horn and Johnson [29]. The decomposition in (11.1) is not unique. Gonzalez and Sahni [26] would view the matrix $\mathcal{T}$ as an instance of an open shop problem with preemptions and makespan minimization and solve the problem to optimality by their algorithm running in time $O(r(r + n \log n))$ where $r$ is the number of positive entries in $\mathcal{T}$. In their solution a matching of jobs with machines would correspond to a switching mode, and a TDMA frame would correspond to an open shop preemptive schedule. Interestingly enough the algorithm of Gonzalez and Sahni preceded this of Inukai [30]. The latter seems to have been designed exclusively with the SS/TDMA time slot assignment in mind. The reader is referred to Dell'Amico and Martello [18] for detailed comparison of the algorithms by Gonzales and Sahni [26], and Inukai [30] in a broader historical context.

Inukai [30] emphasizes reduction of the number $q$ as an important secondary criterion for the SS/TDMA time slot assignment problem. Clearly, the secondary criterion naturally carries over to the open shop with preemptions problem, $O|\text{pmtn}|C_{\max}$. A well known upper bound on $q$ is $n^2 - n + 1$, see Horn and Johnson [29]. The algorithm of Inukai reduces that bound to $n^2 - 2n + 2$. Farahat and Mirsky [23] prove that this bound is the best possible. Brualdi [4], and Brualdi and Gibson [5] provide further analysis of the Birkhoff algorithm for doubly stochastic matrices. The minimization of $q$ is NP-hard in the strong sense which was recently proved by Dufossé and Uçar [22]. Kulkarni et al. [33] further prove that the minimization of $q$ is not fixed parameter tractable, and show polynomial-time algorithm for $q = 2, 3$. Dufossé et al. [21] show that a family of heuristics based on the original proof of Birkhoff can miss optimal decompositions which was earlier conjectured by Brualdi [4].

Lewandowski and Liu [35] extend the TDMA time slot assignment problem to take into account the ability to demultiplex each of the uplink beams into at most $k$ signals, transmit each of these signals to different downlink beams, and then multiplex up to $k$ of these signals into each of the downlink beams. For the open shop this means that at most $k$ operations of each job can be processed at a time, and at most $k$ jobs can be processed by each machine at a time. Observe that the single-operation machine model studied in Sect. 8.1 permits each stage to processed several jobs at a time by allowing parallel identical machines in each stage, yet it still requires at most one operation of each job to be processed at a time. The concurrent open shops in Chap. 4 on the other hand allow all operations of each job to be processed in parallel. We have the following lower bound for the makespan of the frame:

$$\alpha = \max\left\{\max_{i,j} t_{ij}, \max_i \frac{1}{k} R_i, \max_j \frac{1}{k} C_j\right\},$$

see Lewandowski and Liu [35]. de Werra [13] gives the following decomposition of the traffic matrix $\mathcal{T}$:

$$\mathcal{T} = t_1 Z_1 + \ldots t_q Z_q,$$

where each matrix $Z_1, \ldots, Z_q$ is a 0, 1 matrix with at most $k$ 1's in each row, and at most $k$ 1's in each column, and importantly

$$t_1 + \cdots + t_q = \alpha.$$

The decomposition can be computed in polynomial time using network flow algorithms, see de Werra [10, 11, 13].

## 11.3 Reconfigurable Data Centers

Reconfigurable data centers and software defined networks connect servers within a data center by optical connections, Chen et al. [8]. An advantage of such network technology is that as traffic between servers changes over time, the network topology can be reconfigured to better match the change and to prevent localized bottlenecks. A traffic matrix in a reconfigurable data center represents traffic that needs to be routed among a set of $n$ servers. A route is a matching between senders and receivers. The change between routes however requires moving laser pointers and receivers that comes at a cost, Kulkarni et al. [33]. Therefore, finding a shortest schedule for a given traffic matrix with as few as possible preemptions improves performance of the routing algorithms.

## 11.4 Crossbar Switches

An input-buffered crossbar switch with $n$ input ports and $n$ output ports has a buffer for each input port. In such a switch, time is slotted and synchronized so that packets from different input buffers can be read out simultaneously within a time slot. In a time slot, a crossbar switch sets up a connection pattern corresponding to a permutation matrix. As a permutation matrix is a one-to-one mapping from input ports to output ports, packets destined to the same output ports cannot be transmitted at the same time, Chang et al. [7]. A crossbar switch scheduling algorithm needs to guarantee 100% throughput for traffic with given rates $r_{i,j}$ from input $i$ to output $j$. The rates are specified by an $n \times n$ traffic rate matrix

$$\mathbb{R} = \begin{bmatrix} r_{1,1} & \cdots & r_{1,n} \\ r_{2,1} & \cdots & r_{2,n} \\ \cdot & \cdots & \cdot \\ \cdot & r_{i,j} & \cdot \\ \cdot & \cdots & \cdot \\ r_{n,1} & \cdots & r_{n,n} \end{bmatrix}.$$

The traffic rate matrix needs to be substochastic, total of each row and column must not exceed 1, Horn and Johnson [29], in order for 100% throughput to be achieved at all. Chang et al. [7] propose a scheduling algorithm that guarantees 100% throughput for substochastic matrices. The algorithm first finds a convex combination of permutation matrices for a substochastic $\mathbb{R}$. This stage is based on Birkhoff–von Neumann theorem; however, algorithms for makespan minimization for open shop with preemptions like the one by Gonzales and Sahni [26] can also be used for this stage. The convex combination of permutation matrices is then used by fair queueing algorithms, see Demers et al. [19] and Chapter 10 in Kubiak [32], to switch between permutation matrices to guarantee 100% throughput for traffic defined by the traffic rate matrix $\mathbb{R}$.

## 11.5 Multimessage Unicasting and Multicasting

Gonzalez [27] considers multimessage unicasting and multicasting problem where processors communicate by sending and receiving messages over a fully connected network. Each processor can send or receive a message; however, no processor may send more than one message at a time, and no processor may receive more than one message at a time. There are $t_{i,j}$ messages to be sent from processor $i$ to processor $j$. The problem is to find a shortest possible communication schedule for all messages to be transmitted. This problem reduces to a preemptive open shop where each processor $i$ in the communication network represents a job $J_i$ and a machine $M_i$, and the processing time of operation $O_{i,h}$ equals $t_{i,h}$, Gonzalez [27] and [25]. The multimessage unicasting is a special case of multimessage multicasting where the same message needs to be sent to many processors. The algorithms for the latter transform the multicasting problem into a unicasting problem and apply the open shop algorithms to solve the unicasting problem, see Gonzalez [28].

## 11.6 Scheduling and Wavelength Assignment problem

Bampis and Rouskas [2] study the following scheduling and wavelength assignment problem. Consider sources $s = 1, \ldots, n$ and destinations $d = 1, \ldots, n$ and an $n \times n$ matrix $P$ where $p_{s,d}$ is the number of packets to be transmitted from $s$ to $d$. The transmission happens at a certain wavelength assigned to the destination node $d$. Typically there are fewer wavelengths, $m$, than destination nodes, thus several destination nodes may be assigned to the same wavelength. Hence a wavelength is a machine. For a given assignment of destination nodes to machines, we obtain an open shop where a source node $s$ is a job made up of $m$ operations corresponding to wavelengths. The operation of job $s$ on machine $h$ is made up of all destinations $d$ assigned to the same wavelength $h$. Thus $s$ transmits at wavelength $h$ to all destinations $d$ with that wavelength. If it does so, it cannot use any other wavelength machine at the same time (at most one operation of a job can be processed at a time). To illustrate the problem consider the following source to destination matrix:

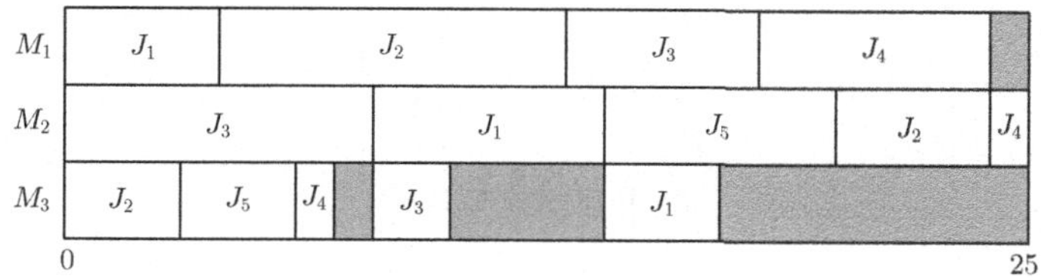


**Fig. 11.1** A schedule and wavelength assignment

$$\mathrm{P} = \begin{bmatrix} 2 & 1 & 3 & 2 & 5 \\ 3 & 3 & 3 & 6 & 1 \\ 1 & 5 & 2 & 4 & 3 \\ 4 & 0 & 1 & 2 & 1 \\ 0 & 2 & 3 & 0 & 4 \end{bmatrix}.$$

Assuming three wavelengths $M_1$, $M_2$, and $M_3$ and the assignment of destinations 1 and 4 to $M_1$, 2 and 5 to $M_2$, and 3 to $M_3$ we obtain a three-machine open shop

$$\mathbb{P} = \begin{bmatrix} 4 & 6 & 3 \\ 9 & 4 & 3 \\ 5 & 8 & 2 \\ 6 & 1 & 1 \\ 0 & 6 & 3 \end{bmatrix}.$$

The source 2 uses the wavelength $M_1$ to transmit 3 packets to $d = 1$ and 6 packets to $d = 4$, it uses the wavelength $M_2$ to transmit 3 packets to $d = 2$ and 1 packet to $d = 5$, and it uses the wavelength $M_3$ to transmit 3 packets to $d = 3$. This corresponds to job $J_2$ with operations having processing time 9 on $M_1$, processing time 4 on $M_2$, and processing time 3 on $M_3$. A schedule for job $J_2$ and the remaining four jobs is given in Fig. 11.1.

## 11.7 Scheduling Theory

The application to scheduling theory have been observed in the study of the two-phase method for preemptive scheduling in de Werra [12, 14]. We illustrate this method using the scheduling on unrelated processors with preemptions to minimize makespan problem, $R|\text{pmtn}|C_{\max}$. An instance of the problem is specified, like in the open shop problem, by an $n \times m$ non-negative real matrix

$$\mathbb{P} = \begin{bmatrix} p_{1,1} & \cdots & p_{1,m} \\ p_{2,1} & \cdots & p_{2,m} \\ \cdot & \cdots & \cdot \\ \cdot & p_{i,h} & \cdot \\ \cdot & \cdots & \cdot \\ p_{n,1} & \cdots & p_{n,m} \end{bmatrix}.$$

However its entries have a different meaning than in the open shop scheduling problem. To explain the difference suppose for the time being that preemptions are not allowed. Then each job $J_i$ is completely processed by one of the $m$ machines. If that machine is machine $M_h$, then the job $J_i$ processing time equals $p_{i,h}$. The remaining processing times of job $J_i$, or the remaining entries in row $i$, are then irrelevant for the solution. Now returning to the preemptive case let us take a convex combination $p_{i,1}x_{i,1}+\cdots+p_{i,m}x_{i,m}$ of the processing times of job $J_i$ , $i = 1, \ldots, n$. For given $n$ convex combinations, one for each job, we obtain the following instance of the open shop:

$$\mathbb{P}_x = \begin{bmatrix} p_{1,1}x_{1,1} & \cdots & p_{1,m}x_{1,m} \\ p_{2,1}x_{2,1} & \cdots & p_{2,m}x_{2,m} \\ \cdot & \cdots & \cdot \\ \cdot & p_{i,h}x_{i,h} & \cdot \\ \cdot & \cdots & \cdot \\ p_{n,1}x_{n,1} & \cdots & p_{n,m}x_{n,m} \end{bmatrix},$$

where

$$\Delta(\mathbb{P}_x) = \left\{ \max_i \left\{ \sum_{h=1}^{m} p_{i,h}x_{i,h} \right\}, \max_h \left\{ \sum_{i=1}^{n} p_{i,h}x_{i,h} \right\} \right\}. \tag{11.2}$$

The solution to the instance $\mathbb{P}_x$ of the open shop problem $O|\text{pmtn}|C_{\max}$ with makespan $C_{\max} = \Delta(\mathbb{P}_x)$ can be found in polynomial time, see Sect. 3.7. The solution solves $R|\text{pmtn}|C_{\max}$ by scheduling fraction $x_{i,h}$ of processing time $p_{i,h}$ of $J_i$ on machine $M_h$. In other words job $J_i$ has now $m$ operations $O_{i,1}, \ldots, O_{i,m}$ with processing times $p_{i,1}x_{i,1}, \ldots, p_{i,m}x_{i,m}$, respectively. Thus to find an optimal solution to $R|\text{pmtn}|C_{\max}$ one needs to find the coefficients $x_{i,h}$ that minimize $\Delta(\mathbb{P}_x)$. This can be done by solving a linear program, see Lawler and Labetoulle [34], Brucker [6] and Błażewicz et al. [3]. A similar two-phase approach based on LP and reduction to open shop works for

- The problem $R|\text{pmtn}, r_i|L_{\max}$, see Lawler and Labetoulle [34], Brucker [6] and Błażewicz et al. [3].
- The problems in Sects. 4.3 and 4.10.
- The problem $O|\text{pmtn}, r_j|L_{\max}$ in Sect. 3.7.2.
- The multiprocessor open shop scheduling in Chap. 8.

## 11.8 Data Migration and File Transfer Scheduling

Data migration problem arises whenever large amount of data need to be rapidly transferred to new locations within storage area networks in order to better respond to changing demand for data. During the migration the network still needs to provide efficient though unlikely best service. Therefore it is important to compute as short as possible data migration schedule that produces the target data layout. Coffman et al. [9], see also Gandhi et al. [24] and Khuller et al. [31], introduce file transfer multigraph $G = (V, E)$ to model the problem. The vertices in $V$ represent storage devices (computers or computer centers). The labeled edges correspond to the files to be transferred between the vertices. The label $p_{i,j}$ of an edge $(i, j)$ represents transfer time required to transfer the file between $i$ and $j$. Each vertex $i$ completes transfers only after all edges incident with $v$ in $G$ complete transfers. The key constraint is that no two edges incident with the same vertex can be transferred at the same time. This constrain can be relaxed by increasing the maximum number of simultaneous file transfers to more than one by increasing the number of communication ports at vertices, see Coffman et al. [9]. The problem can be recast as open shop scheduling with simultaneity constraints, see de Werra [15] de Werra and Erschler [16] de Werra et al. [17], as it was shown in Chap. 6 for wireless networking with primary interference.

From a different angle, each vertex in $G$ corresponds to machine in open shop. Each edge $(i, j)$ labeled with $p_{i,j}$ is a bi-processor job with processing time $p_{i,j}$. The job needs to be processed on machines $i$ and $j$ simultaneously for exactly $p_{i,j}$ units of time. The problem becomes a subproblem of the problem $P|\text{fix}_j|C_{\max}$ with $|\text{fix}_j| = 2$, see Drozdowski [20], where each job is assigned a fixed subset of parallel machines to be processed on simultaneously.

# Problems

**11.1** Find more applications of preemptive open shop scheduling in scheduling theory, see Sect. 11.7.

# References


1. M.M. Ahmadian, M. Khatami, A. Salehipour, T.C.E. Cheng, Four decates of research on the open-shop scheduling problem to minimize makespan. Eur. J. Oper. Res. (2021). https://doi.org/10.1016/j.ejor.2021.03.026
2. E. Bampis, G.N. Rouskas, The scheduling and wavelength assignment problem in optical wdm networks. J. Lightwave Tech. **20**, 782–789 (2002)
3. J. Błażewicz, K. Ecker, E. Pesch, G. Schmidt, J. Węglarz, *Handbook on Scheduling: From Theory to Applications*. International Handbooks on Information Systems (Springer, 2007)

4. R.A. Brualdi, Notes on the Birkhoff algorithm for doubly stochastic matrices. Can. Math. Bull. **25**, 127–147 (1982)
5. R.A. Brualdi, P.M. Gibson, Convex polyhedra of doubly stochastic matrices: I. applications of permanent function. J. Comb. Theory A **22**, 194–230 (1977)
6. P. Brucker, *Scheduling Algorithms* (Springer, 1995)
7. C.-S. Chang, W.-J. Chen, H.-Y. Huang, On Service Guarantees for Input Buffered Crossbar Switches: A Capacity Decomposition Approach by Birkhoff and von Neumann, in *Seventh International Workshop on Quality of Service* (IEEE, 1999), pp. 79–86
8. K. Chen, A. Singla, A. Singh, K. Ramachandran, L. Xu, Y. Zhang, X. Wen, Y. Chen, Osa: An optical switching architecture for data center networks with unprecedented flexibility. IEEE/ACM Trans. Netw. **22**, 498–511 (2014)
9. E.G. Coffman Jr., M.R. Garey, D.S. Johnson, A.S. LaPaugh, Scheduling file transfers. SIAM J. Comput. **14**, 744–780 (1985)
10. D. de Werra, A decomposition property of polyhedra. Mathematical Programming **30**, 261–266 (1984)
11. D. de Werra, Preemptive scheduling, linear programming and network flows. SIAM J. Alg. Disc. Meth. **5**, 11–20 (1984)
12. D. de Werra, On the two-phase method for preemptive scheduling. Eur. J. Oper. Res. **37**, 227–235 (1988)
13. D. de Werra, A note on SS/TDMA satellite communication. Linear Algebra Appl. **135**, 69–77 (1990)
14. D. de Werra, Almost nonpreemptive schedules. Ann. Oper. Res. **26**, 243–256 (1990)
15. D. de Werra, Extensions of coloring models for scheduling purposes. Eur. J. Oper. Res. **92**, 474–492 (1996)
16. D. de Werra, J. Erschler, Open shop with some additional constraints. Graphs Comb. **12**, 81–93 (1996)
17. D. de Werra, N.V.R. Mahadev, U. Peled, Edge chromatic scheduling with simultaneity constraints. SIAM J. Disc. Math., 631–641 (1993)
18. M. Dell'Amico, S. Martello, Open shop, satellite communication and a theorem by Egerváry (1931). Oper. Res. Lett. **18**, 207–211 (1996)
19. A. Demers, S. Keshav, S. Shenkar, Analysis and symulation of a fair queueing algorithm. Internetworking Res. Exp. **1**, 3–26 (1990)
20. M. Drozdowski, Scheduling multiprocessor tasks - an overview. Eur. J. Oper. Res. **94**, 215–230 (1996)
21. F. Dufossé, K. Kaya, I. Panagiotas, B. Uçar, Further notes on Birkhoff-von Neumann decompsition od doubly stochastic matrices. Linear Algebra Appl. **554**, 68–78 (2018)
22. F. Dufossé, B. Uçar, Notes on Birkhoff-von Neumann decomposition of doubly stochastic matrices. Linear Algebra Appl. **479**, 108–115 (2016)
23. H.K. Farahat, L. Mirsky, Permutation endomorphisis and refinment of a theorem of Birkhoff. Proc. Camb. Philos. Soc. **56**, 322–328 (1960)
24. R. Gandhi, M.M. Halldórsson, G. Kortsarz, H. Shachnai, Improved results for data migration and open shop scheduling. ACM Trans. Algorithms **2**, 116–129 (2006)
25. T. Gonzalez, Open shop scheduling, in eds. by Joseph Y-T. Leung, *Handbook on Scheduling: Algorithms, Models, and Performance Analysis* (Chapman and Hall/CRC, 2004), pp. 6–1 to 6–14
26. T. Gonzalez, S. Sahni, Open shop scheduling to minimize finish time. J. ACM **23**, 665–679 (1976)
27. T.F. Gonzalez, Complexity and approximations for multimessage multicasting. J. Par. Dist. Comput. **55**, 215–235 (1998)
28. T.F. Gonzalez, Simple algorithms for multimessage multicasting with forwarding. Algorithmica **29**, 511–533 (2001)
29. R.A. Horn, C.R. Johnson, *Matrix Analysis*, 2nd edn. (Cambridge University Press, 2013)
30. T. Inukai, An efficient SS/TDMA time slot assignment algorithm. IEEE Trans. Commun. **27**, 1449–1455 (1979)

31. S. Khuller, Y. Kim, Y.C. Wan, Algorithms for data migrantion with cloning, in *Proc. of the 22nd ACM Symposium on Principles od Database Systems*, pp. 27–36 (2003)
32. W. Kubiak, *Proportional Optimization and Fairness* (Springer, 2009)
33. J. Kulkarni, E. Lee, M. Singh, Minimum Birkhoff-von Neumann decomposition, in *Proc. 19th International Conference on Integer Programming and Combinatorial Optimization (IPCO)*, pp. 343–354, Waterloo, ON, Canada (2017)
34. E.L. Lawler, J. Labetoulle, On preemptive scheduling on unrelated parallel processors by linear programming. J. ACM **25**, 612–619 (1978)
35. J.L. Lewandowski, C.L. Liu, SS/TDMA stellite communications with k-permutation switching modes. SIAM J. Alg. Disc. Meth. **8**, 519–534 (1987)

# Chapter 12
# Two-Machine Open Shop Scheduling with Time Lags

## 12.1 Makespan Minimization

The open shop with job-dependent time lags has been studied for quite sometime in the literature. The time lags model delays required between job's operations due to necessary transportation needed to move a job from one machine to another for instance, or intermediate processes, or operations that are not constrained by resources. Zhang [18] provides an interesting discussion of the time lag models and their applications in scheduling. Most research on the open shops with time lags has focused on two-machine open shops where each job $J_j$, $j = 1, \ldots, n$, consists of two operations $O_{j,1}$ and $O_{j,2}$ to be processed on two machines $M_1$ and $M_2$, respectively, in any order. The operations $O_{j,1}$ and $O_{j,2}$ processing times equal $p_{j,1} \geq 0$ and $p_{j,2} \geq 0$, respectively, and the time lag is $l_j \geq 0$. In a feasible schedule either machine can process at most one job at a time, each job can be processed by at most one machine at a time, and the *later* operation of job $J_j$ in the schedule needs to wait at least $l_j$ time units to start following the completion of the *earlier* operation of job $J_j$ in the schedule. Yu [16], and Yu et al. [17] prove a series of strong complexity results for the makespan minimization. They prove that the problem is NP-hard in the strong sense even if all operations are unit-time operations. This problem is denoted by $O2|p_{ij} = 1, l_j|C_{\max}$ in the well-known notation of Graham et al. [7]. We now give a different proof of that result.

**Theorem 12.1** *The problem $O2|p_{ij} = 1, l_j|C_{\max}$ is NP-hard in the strong sense.*

***Proof*** Yu [16] shows that the numerical three dimensional matching (N3DM) problem (see Garey and Johnson [6] for the definition of the N3DM problem) is NP-complete in the strong sense even for the instances with $X = \{1, \ldots, n\}$, $Y = \{1, \ldots, n\}$, a multiset of non-negative integers $Z = \{l_1, \ldots, l_n\}$, and an integer $n < e < 2n$ such that



W. Kubiak, *A Book of Open Shop Scheduling*, International Series in Operations Research & Management Science 325,
https://doi.org/10.1007/978-3-030-91025-9_12

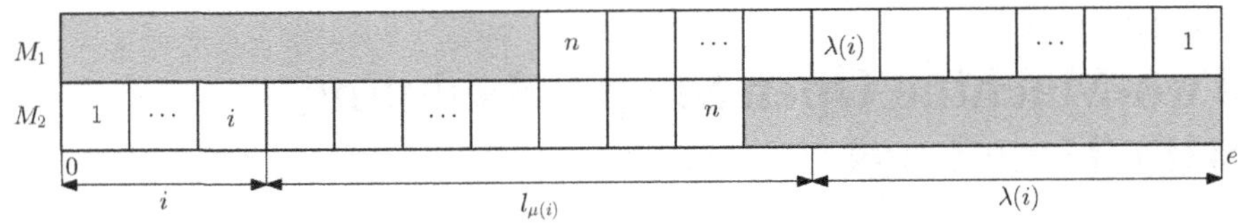


**Fig. 12.1** A schedule for the numerical three dimensional matching

$$\sum_{i=1}^{n} l_i + n(n+1) = ne. \tag{12.1}$$

The problem is to decide whether there are permutations $\lambda$ and $\mu$ of the set $\{1, \ldots, n\}$ such that

$$i + \lambda(i) + l_{\mu(i)} = e \tag{12.2}$$

for $i = 1, \ldots, n$. The problem is referred to as the restricted numerical three dimensional matching (RN3DM) in Yu [16]. For an instance $X = \{1, \ldots, n\}$, $Y = \{1, \ldots, n\}$, $Z = \{l_1, \ldots, l_n\}$, and $n < e < 2n$ of the RN3DM we construct an instance of an open shop with $n$ jobs $J_1, \ldots, J_n$ scheduled on two machines, $M_1$ and $M_2$. Each job has two unit-time operations, one on $M_1$ and the other on $M_2$. For job $J_j$ the delay between its two operations is required to be at least $l_j$, $j = 1, \ldots, n$. The threshold for maximum makespan equals $C_{\max} = e$.

Suppose there are permutations $\lambda$ and $\mu$ that meet the condition (12.2). Schedule job $J_{\mu(i)}$ in $[i-1, i]$ on $M_2$ and in $[e-\lambda(i), e-\lambda(i)+1]$ on $M_1$. We have $i \le e-\lambda(i)$ by (12.2). Hence the two operations of $J_{\mu(i)}$ are not scheduled in parallel. Thus the jobs are scheduled in $[0, n]$ on $M_2$ so that the machine processes exactly one job at a time, and since $\lambda$ is a permutation, the jobs are scheduled in $[e-n, e]$ on $M_1$ so that the machine processes exactly one job at a time. The delay for job $J_{\mu(i)}$ equals $e - \lambda(i) - i = l_{\mu(i)}$ by (12.2), $i = 1, \ldots, n$, see Fig. 12.1. Therefore, we obtain a feasible schedule with $C_{\max} = e$.

Now suppose there is a feasible schedule $\mathcal{S}$ with $C_{\max} \le e$. Consider job $J_i$ in $\mathcal{S}$, its earlier operations is processed in $[x_i - 1, x_i]$ on one machine and its later operation in $[C_i, C_i + 1]$ on the other machine. Moreover $C_i - x_i = \ell_i \ge l_i$. Let $C_{\max} - C_i = y_i$. We have

$$x_i + y_i + \ell_i = C_{\max}, \tag{12.3}$$

for $i = 1, \ldots, n$. Thus by (12.1) we get

$$\sum_{i=1}^{n}(x_i + y_i) \le ne - \sum_{i=1}^{n} l_i \le n(n+1). \tag{12.4}$$

To complete the proof consider the multiset $\{x_1, y_1, \ldots, x_i, y_i, \ldots, x_n, y_n\}$. For each pair $x_i$ and $y_i$, the $x_i$ corresponds to the completion time of one of the two operations of $J_i$, and $y_i$ corresponds to the start of the other of the two operations of $J_i$. The operations must be done on different machines $M_1$ and $M_2$ in $\mathcal{S}$. Thus $x_i$ corresponds to one machine and $y_i$ corresponds to the other. We can split the pair $x_i$ and $y_i$ between two sets $\mathcal{M}_1$ and $\mathcal{M}_2$ depending on the machines $x_i$ and $y_i$ correspond to. We do this for each $i = 1, \ldots, n$ to obtain $\mathcal{M}_1 = \{\alpha_1, \ldots, \alpha_n\}$ and $\mathcal{M}_2 = \{\beta_1, \ldots, \beta_n\}$. Since $\mathcal{S}$ is a feasible schedule both $\mathcal{M}_1$ and $\mathcal{M}_2$ must be sets, i.e., either must include $n$ *different* positive integer numbers. Thus

$$\sum_{i=1}^{n} \alpha_i \geq \frac{n(n+1)}{2}, \tag{12.5}$$

and

$$\sum_{i=1}^{n} \beta_i \geq \frac{n(n+1)}{2}. \tag{12.6}$$

However, by the construction of $\mathcal{M}_1$ and $\mathcal{M}_2$ we have

$$\sum_{i=1}^{n} (x_i + y_i) = \sum_{i=1}^{n} (\alpha_i + \beta_i), \tag{12.7}$$

which implies by (12.4), (12.5), and (12.6) that

$$\sum_{i=1}^{n} \alpha_i = \frac{n(n+1)}{2}$$

and

$$\sum_{i=1}^{n} \beta_i = \frac{n(n+1)}{2}.$$

This is only possible if $\mathcal{M}_1 = \{1, \ldots, n\}$ and $\mathcal{M}_2 = \{1, \ldots, n\}$. Thus there are permutations $\lambda$ and $\mu$ of the set $\{1, \ldots, n\}$ such that $\lambda(i) = x_i$ and $\mu(i) = y_i$ and by (12.3) we have

$$\lambda(i) + \mu(i) + \ell_i = C_{\max} \tag{12.8}$$

for $i = 1, \ldots, n$. Therefore by (12.1)

$$n(n+1) = nC_{\max} - \sum_{i=1}^{n} \ell_i = ne - \sum_{i=1}^{n} l_i$$

which implies

$$n(e - C_{\max}) = \sum_{i=1}^{n} (l_i - \ell_i).$$

Suppose for contradiction that $e > C_{\max}$ or $\ell_i > l_i$ for some $i = 1, \ldots, n$. Then the left hand side is positive or the right hand side is negative since $\ell_i \geq l_i$ for all $i = 1, \ldots, n$ in $\mathcal{S}$. Either case leads to contradiction, thus $C_{\max} = e$ and $\ell_i = l_i$ for all $i = 1, \ldots, n$. Therefore by (12.8)

$$\lambda(i) + \mu(i) + l_i = e \tag{12.9}$$

for $i = 1, \ldots, n$. This proves that there are permutations $\lambda$ and $\mu$ that provide an affirmative answer to the RN3DM problem. This completes the proof. □

Yu [16] then goes on to prove that the problem is NP-hard in the strong sense even if there are only two possible values $l$ and $l'$ of time lags in a job-proportionate open shop, i.e., the problem $O2|p_{i1} = p_{i2}, l_j \in \{l, l'\}|C_{\max}$, and it is also NP-hard in the ordinary sense when only one value $l$ of time lag is permitted in a job-proportionate open shop, i.e., the problem $O2|p_{i1} = p_{i2}, l_j = l|C_{\max}$.

Munier-Kordon and Rebaine [11] observe that if all time lags $l_1, \ldots, l_n$ are distinct, non-negative, and integral, then the problem $O2|p_{ij} = 1, l_j|C_{\max}$ can be solved in $O(n \log n)$ time. Their algorithm orders the jobs in decreasing order of their time lags, $l_1 > \cdots > l_n$. Next, it schedules the *earlier* operation of job $J_i$ in $\left[\lceil \frac{i}{2} \rceil - 1, \lceil \frac{i}{2} \rceil\right]$ on $M_1$ if $i$ is odd, or on $M_2$ if $i$ is even, $i = 1, \ldots, n$. The *later* operation of $J_i$ is scheduled in $\left[\lceil \frac{i}{2} \rceil + l_i, \lceil \frac{i}{2} \rceil + l_i + 1\right]$ on $M_2$ if $i$ is odd, or on $M_1$ if $i$ is even, $i = 1, \ldots, n$. Hence, the later operation of $J_i$ waits exactly $l_i$ time units to start after the completion time of the $J_i$'s earlier operation. In order to prove that the schedule is feasible we observe that for $1 \leq i < j \leq n$, we have $l_i - l_j \geq j - i$. Thus $\lceil \frac{i}{2} \rceil + l_i > \lceil \frac{j}{2} \rceil + l_j$ for $l_i - l_j > 1$ or $j$ even, and $\lceil \frac{i}{2} \rceil + l_i = \lceil \frac{j}{2} \rceil + l_j$ for $l_i - l_j = 1$ and $j$ odd. In the latter case $j = i + 1$ and $i$ is even. Therefore, for any two jobs $J_i$ and $J_j$ their later unit-time operations are scheduled either in different time slots or in the same time slot but on different machines. Finally, the *latest* earlier operation is scheduled in time slot $\left[\lceil \frac{n}{2} \rceil - 1, \lceil \frac{n}{2} \rceil\right]$, and the *earliest* later operation is scheduled in time slot $\left[\lceil \frac{n}{2} \rceil + l_n, \lceil \frac{n}{2} \rceil + l_n + 1\right]$. Since $l_n \geq 0$, we obtain a feasible schedule. The schedule makespan equals $1 + l_1 + 1 = 2 + l_1$ and thus it is clearly optimal. We observe that the total completion time of the schedule equals

$$F = \sum_{i=1}^{n} \left( \left\lceil \frac{i}{2} \right\rceil + l_i + 1 \right)$$

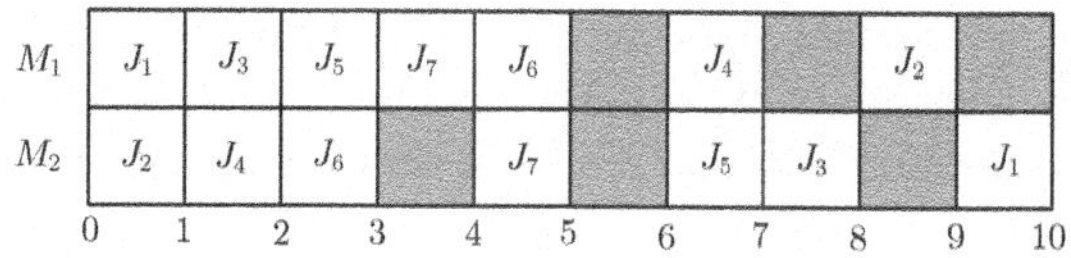


**Fig. 12.2** An optimal schedule with $C_{\max} = 10$ obtained by the Munier-Kordon and Rebaine algorithm [11]

which is optimal as well. Therefore the schedule is ideal, see Coffman et al. [4]. To illustrate the algorithm consider a two-machine open shop with $n = 7$ jobs having the following time lags $l_1 = 8, l_2 = 7, l_3 = 5, l_4 = 4, l_5 = 3, l_6 = 1, l_7 = 0$. The schedule obtained by the algorithm is shown in Fig. 12.2.

Munier-Kordon and Rebaine [11] show an algorithm that produces $\frac{5}{4}$-approximate solutions for the general problem $O2|p_{ij} = 1, l_j|C_{\max}$ which permits jobs with the same time lags. The algorithm relies on the algorithm developed for distinct time lags and described earlier.

The problem $O2|p_{i1} = p_{i2} = 1, l_j \in \{l, l'\}|C_{\max}$ with two distinct time lag values $l$ and $l'$ can be solved in time $O(n \log n)$, Munier-Kordon and Rebaine [11]. Their algorithm however does not run in polynomial time with respect to the succinct input encoding which requires $O(\max\{\log n, \log l, \log l'\})$ bits to specify the problem instance. The question whether there is a polynomial-time algorithm with respect to the succinct input encoding remains open.

For the problems with arbitrary operation processing times, Rebaine and Strusevich [14] give a linear-time algorithm for the instances with short time lags, i.e., time lags that meet the following condition: $\max_j\{l_j\} \leq \min_{ij}\{p_{ij}\}$. Strusevich [15] gives $\frac{3}{2}$-approximation algorithm for the problem $O2|l_j|C_{\max}$. Zhang and van de Velde [19] give a 2-competitive greedy online algorithm for $O2|l_j|C_{\max}$.

Other complexity results and heuristics can be found in Rayward-Smith and Rebaine [12], Dell'Amico and Vaessens [5], and Rebaine [13].

Ageev [1] considers a job-proportionate two-machine open shop with *exact* time lags, i.e., the problem $O2|p_{j1} = p_{j2}, \text{exact } l_j|C_{\max}$, and proves that no $(\frac{3}{2} - \epsilon)$-approximation algorithm, $\epsilon > 0$, running in polynomial time exists for the problem unless $P = NP$.

**Theorem 12.2** *If $P \neq NP$, then no polynomial-time algorithm for $O2|p_{j1} = p_{j2}, \text{exact } l_j|C_{\max}$ exists with the worst case ratio less than $\frac{3}{2}$.*

He further proves that no $(\frac{5}{4} - \epsilon)$-approximation algorithm, $\epsilon > 0$, running in polynomial time exists for the problem with two time lag values 0 an $l$ unless $P = NP$.

**Theorem 12.3** *If $P \neq NP$, then no polynomial-time algorithm for $O2|\text{exact } l_j \in \{0, l\}|C_{\max}$ exists with the worst case ratio less than $\frac{5}{4}$.*

Finally, Ageev [1] points out that the 3-approximate algorithm developed for two-machine flow shop with exact delays by Ageev and Kononov [2] and Leung et al. [10] gives 3-approximate algorithms for $O2|\text{exact}\, l_j|C_{\max}$. A recent literature review is given by Khatami et al. [9]. The review focuses on scheduling with exact time lags. It is worth observing that the open shop scheduling with exact time lags generalizes no-wait open shop scheduling.

## 12.2 Total Completion Time Minimization

Brucker et al. [3] prove that *weighted* total completion time minimization with time lags, the problem $O2|p_{ij} = 1, l_i|\sum w_i C_i$, is NP-hard in the strong sense. They prove that the same holds for the total completion time with jobs being *released* possibly at different times, i.e., the problem $O2|p_{ij} = 1, l_i, r_i|\sum C_i$.

In this section we prove that the problem where all jobs are released at the same time and their weights are all equal, i.e., the problem $O2|p_{ij} = 1, l_i|\sum C_i$ is NP-hard in the strong sense. This result strengthens those earlier complexity results for total completion time, and it answers a question that has been open since the paper by Brucker et al. [3].

**Theorem 12.4** *The problem $O2|p_{ij} = 1, l_j|\sum C_j$ is NP-hard in the strong sense.*

***Proof*** The proof is by reduction from the Restricted Numerical Three Dimensional Matching (RN3DM) problem which is proved NP-complete in the strong sense in Yu [16], see also Yu et al. [17]. An instance of the RN3DM problem consists of sets $X = \{1, \dots, n\}$, $Y = \{1, \dots, n\}$, a multiset of non-negative integers $Z = \{l_1, \dots, l_n\}$, and an integer $n < e < 2n$ such that

$$\sum_{i=1}^{n} l_i + n(n+1) = ne, \tag{12.10}$$

see Yu [16]. The problem is to decide whether there are permutations $\lambda$ and $\mu$ of the set $\{1, \dots, n\}$ such that

$$i + \lambda(i) + l_{\mu(i)} = e \tag{12.11}$$

for $i = 1, \dots, n$. For an instance of the RN3DM problem we construct an instance of an open shop with $2n$ jobs $J_1, \dots, J_n$ and $I_1, \dots, I_n$, scheduled on two machines $M_1$ and $M_2$. Each job has two unit-time operations, one on $M_1$, and the other on $M_2$. For job $J_j$ the delay between its two operations is required to be at least $t_j := l_j + 2n - e > 0$, $j = 1, \dots, n$, similarly for job $I_j$ the delay between its two operations is required to be at least $t_j := l_j + 2n - e > 0$, $j = 1, \dots, n$. The threshold for total completion time equals

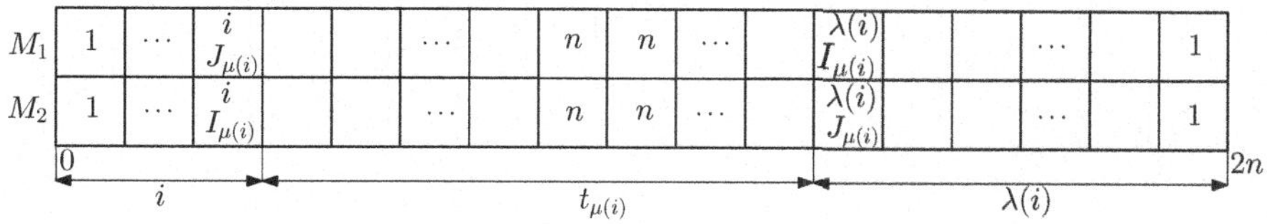


**Fig. 12.3** A schedule for the permutations $\lambda$ and $\mu$ that give an affirmative answer to the RN3DM

$$F = n(n+1) + 2\sum_{j=1}^{n}(t_j + 1) = 3n^2 + n.$$

For permutations $\lambda$ and $\mu$ such that

$$i + \lambda(i) + l_{\mu(i)} = e \tag{12.12}$$

for $i = 1, \ldots, n$: schedule $J_{\mu(i)}$ in $[i-1, i]$ on $M_1$ and in $[2n-\lambda(i), 2n-\lambda(i)+1]$ on $M_2$, and $I_{\mu(i)}$ in $[i-1, i]$ on $M_2$ and in $[2n-\lambda(i), 2n-\lambda(i)+1]$ on $M_1$ for $i = 1, \ldots, n$. Since $2n - \lambda(i) - i = t_{\mu(i)} > 0$ and $\lambda$ is a permutation, the resulting schedule is feasible. Furthermore, the jobs $J_1, \ldots, J_n$ complete at $n+1, \ldots, 2n$ on $M_2$, and the jobs $I_1, \ldots, I_n$ complete at $n+1, \ldots, 2n$ on $M_1$, see Fig. 12.3. Hence the total completion time of the schedule equals $3n^2 + n$ which equals the threshold $F$.

Now, let $\mathcal{S}$ be a feasible schedule with total completion time not exceeding $F$. We first show that $C_{\max} = 2n$ in $\mathcal{S}$. To that end let $x_{\sigma(1)} \le x_{\sigma(2)} \le \cdots \le x_{\sigma(2n-1)} \le x_{\sigma(2n)}$ be the times when the *earlier* operations of the jobs $J_1, \ldots, J_n$, and $I_1, \ldots, I_n$ complete in $\mathcal{S}$. Because of the delay due to time lags the total completion time of $\mathcal{S}$ is at least

$$\sum_{i=1}^{n}(x_{\sigma(2i-1)} + x_{\sigma(2i)}) + 2\sum_{j=1}^{n}(t_j + 1), \tag{12.13}$$

which does not exceed the threshold $F$ for $\mathcal{S}$. Hence

$$\sum_{i=1}^{n}(x_{\sigma(2i-1)} + x_{\sigma(2i)}) \le n(n+1). \tag{12.14}$$

For two machines we have $i \le x_{\sigma(2i-1)} \le x_{\sigma(2i)}$, $i = 1, \ldots, n$. Thus by (12.14) we get $x_{\sigma(2i-1)} = x_{\sigma(2i)} = i$ for $i = 1, \ldots, n$. Therefore each job $J_1, \ldots, J_n$, and $I_1, \ldots, I_n$ completes after time $n$ in $\mathcal{S}$. Let $C_{\pi(1)} \le C_{\pi(2)} \le \cdots \le C_{\pi(2n-1)} \le C_{\pi(2n)}$ be the completion times of the jobs $J_1, \ldots, J_n$, and $I_1, \ldots, I_n$ in $\mathcal{S}$. Clearly $C_{\pi(i)} = n + c_{\pi(i)}$, for some $c_{\pi(i)} \ge 1$, thus

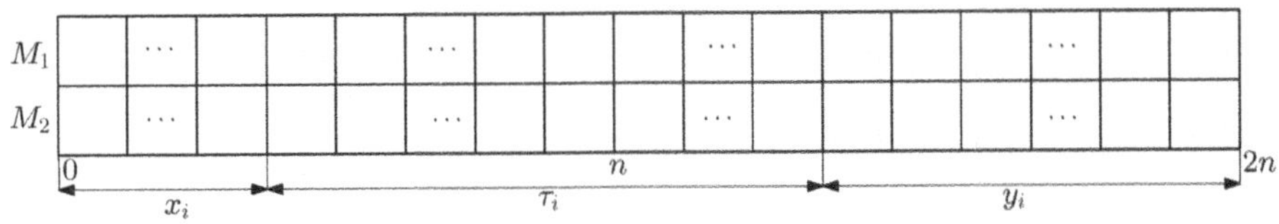


**Fig. 12.4** The relationship between $x_i$ and $y_i$ for job $J_i$

$$\sum_{i=1}^{n}(c_{\pi(2i-1)} + c_{\pi(2i)}) \le n(n+1) \tag{12.15}$$

in $\mathcal{S}$. Again, for two machines we have $i \le c_{\pi(2i-1)} \le c_{\pi(2i)}$, $i = 1, \ldots, n$. Thus by (12.15) we get $c_{\pi(2i-1)} = c_{\pi(2i)} = i$ for $i = 1, \ldots, n$. Therefore all jobs complete by $C_{\max} = 2n$ in $\mathcal{S}$ which is what we set out to show first. We now use that fact in the second part of the proof. Consider job $J_i$ in $\mathcal{S}$, its earlier operation is processed in $[x_i - 1, x_i]$ on one machine and its later operation in $[C_i - 1, C_i]$ on the other machine. Moreover $C_i - 1 - x_i = \tau_i \ge t_i$ in $\mathcal{S}$. Let $2n - (C_i - 1) = y_i$. We have

$$x_i + y_i + \tau_i = 2n, \tag{12.16}$$

for $i = 1, \ldots, n$. Thus by (12.10) and definition of $t_i$ we get

$$\sum_{i=1}^{n}(x_i + y_i) \le ne - \sum_{i=1}^{n} l_i \le n(n+1). \tag{12.17}$$

To complete the proof consider the multiset $\{x_1, y_1, \ldots, x_i, y_i, \ldots, x_n, y_n\}$. For each pair $x_i$ and $y_i$, the $x_i$ *is* the completion time of one of the two operations of $J_i$, and $y_i$ *corresponds* to the start of the other two operations of $J_i$, please see Fig. 12.4 for details.

The operations must be done on different machines $M_1$ and $M_2$ in $\mathcal{S}$. Thus $x_i$ corresponds to one machine and $y_i$ corresponds to the other. We can split the pair $x_i$ and $y_i$ between two sets $\mathcal{M}_1$ and $\mathcal{M}_2$ depending on the machines $x_i$ and $y_i$ correspond to. We do this for each $i = 1, \ldots, n$ to obtain $\mathcal{M}_1 = \{\alpha_1, \ldots, \alpha_n\}$ and $\mathcal{M}_2 = \{\beta_1, \ldots, \beta_n\}$. Since $\mathcal{S}$ is a feasible schedule both $\mathcal{M}_1$ and $\mathcal{M}_2$ must be sets, i.e., either must include $n$ *distinct* positive integer numbers. Thus

$$\sum_{i=1}^{n} \alpha_i \ge \frac{n(n+1)}{2}, \tag{12.18}$$

and

$$\sum_{i=1}^{n} \beta_i \geq \frac{n(n+1)}{2}. \tag{12.19}$$

However, by the construction of $\mathcal{M}_1$ and $\mathcal{M}_2$ we have

$$\sum_{i=1}^{n}(x_i + y_i) = \sum_{i=1}^{n}(\alpha_i + \beta_i), \tag{12.20}$$

which implies by (12.17), (12.18), and (12.19) that

$$\sum_{i=1}^{n} \alpha_i = \frac{n(n+1)}{2}$$

and

$$\sum_{i=1}^{n} \beta_i = \frac{n(n+1)}{2}.$$

This is only possible if $\mathcal{M}_1 = \{1, \ldots, n\}$ and $\mathcal{M}_2 = \{1, \ldots, n\}$. Thus there are permutations $\lambda$ and $\mu$ of the set $\{1, \ldots, n\}$ such that $\lambda(i) = x_i$ and $\mu(i) = y_i$ and thus by (12.16) we have

$$\lambda(i) + \mu(i) + \tau_i = 2n \tag{12.21}$$

for $i = 1, \ldots, n$. Therefore by (12.10)

$$n(n+1) = 2n^2 - \sum_{i=1}^{n} \tau_i = ne - \sum_{i=1}^{n} l_i = 2n^2 - \sum_{i=1}^{n} t_i.$$

Hence $\tau_i = t_i$ for all $i = 1, \ldots, n$. Therefore by (12.21)

$$\lambda(i) + \mu(i) + t_i = 2n \tag{12.22}$$

for $i = 1, \ldots, n$, and by definition of $t_i$

$$\lambda(i) + \mu(i) + l_i = e$$

for $i = 1, \ldots, n$. The permutations $\lambda$ and $\mu$ give an affirmative answer to the RN3DM problem which proves the theorem. □

We observed in Sect. 12.1 that Munier-Kordon and Rebaine algorithm [11] solves the problem $O2|p_{ij} = 1, l_j|\sum C_j$ with distinct time lags in $O(n \log n)$ time.

For the schedule obtained for $O2|p_{ij} = 1, l_j| \sum C_j$ with distinct time lags total weighted completion time equals

$$F_w = \sum_{i=1}^{n} \left( \left\lceil \frac{i}{2} \right\rceil + l_i + 1 \right) w_i,$$

where $w_i$ is the weight of job $J_i$. For the instances of $O2|p_{ij} = 1, l_i| \sum w_i C_i$, where the order $l_1 > \cdots > l_n$ implies $w_1 \geq \cdots \geq w_n$, the schedule obtained for $O2|p_{ij} = 1, l_j| \sum C_j$ with distinct time lags is optimal for $O2|p_{ij} = 1, l_i| \sum w_i C_i$ since by the rearrangement inequality of Hardy, Littlewood, and Polya [8] the weighted sum

$$\sum_{i=1}^{n} \left\lceil \frac{i}{2} \right\rceil w_i,$$

is minimized by matching the positions $1, 1, 2, 2, \ldots, \left\lceil \frac{n}{2} \right\rceil$ in non-decreasing order with the weights $w_1 \geq \cdots \geq w_n$ in non-increasing order.

However the complexity status of $O2|p_{ij} = 1, l_i| \sum w_i C_i$ and $O2|p_{ij} = 1, l_i, r_i| \sum C_i$ remains open for distinct time lags.

## References


1. A. Ageev, Inapproximability lower bounds for open shop problems with exact delays, in eds. by A. Eremeev et al., *OPTA 2018, CCIS 871* (Springer International Publishing AG, 2018), pp. 45–55
2. A. Ageev, A.V. Kononov, Approximation algorithms for scheduling problems with exact delays, in eds. by T. Erlebach, C. Kaklamanis, *WAOA 2006*, volume LNCS vol. 4368 (Springer, 2007), pp. 1–14
3. P. Brucker, S. Knust, T.C.E. Cheng, N.V. Shakhlevich, Complexity results for flow-shop and open-shop scheduling problems with transportation delays. Ann. Oper. Res. **129**, 81–106 (2004)
4. E.G. Coffman Jr., D. Dereniowski, W. Kubiak, An efficient algorithm for finding ideal schedules. Acta Informatica **49**, 1–14 (2012)
5. M. Dell'Amico, R. Vaessens, Flow and open shop scheduling on two machines with transportation times and machine-independent processing times is NP-hard. Materiali di discussione 141, Dipartimento di Economia Politica, Università di Modena, 1995
6. M.R. Garey, D.S. Johnson, *Computers and Intractability: A Guide to the Theory of NP-Completeness* (W. H. Freeman, 1979)
7. R.L. Graham, E.L. Lawler, J.K. Lenstra, A.H.G. Rinnooy Kan, Optimization and approximation in deterministic sequencing and scheduling: A survey. Anns. Discr. Math. **5**, 287–326 (1979)
8. G. Hardy, J. Littlewood, G. Polya, *Inequalities* (Cambridge University Press, 1952)
9. M. Khatami, A. Salehipour, T.C.E. Cheng, Coupled task scheduling with exact delays: Literature review and models. Eur. J. Oper. Res. **282**, 19–39 (2020)
10. J.Y.-T. Leung, H. Li, H. Zhao, Scheduling two-machine flow shops with exact delays. Int. J. Found. Comput. Sci. **18**, 341–359 (2007)

11. A. Munier-Kordon, D. Rebaine, The two-machine open-shop problem with unit-time operations and time delays to minimize the makespan. Eur. J. Oper. Res. **203**, 42–49 (2010)
12. V.J. Rayward-Smith, D. Rebaine, Open shop scheduling with delays. Inf. théorique et Appl. **26**, 439–447 (1992)
13. D. Rebaine, Scheduling the two-machine open shop problem with non-symmetric time delays (Congrès ASAC, 2004)
14. D. Rebaine, V.A. Strusevich, Two-machine open shop scheduling with special transportation times. J. Oper. Res. Soc. **50**, 756–764 (1999)
15. V.A. Strusevich, A heuristic for the two-machine open-shop scheduling problem with transportation times. Discrete Appl. Math. **93**, 287–304 (1999)
16. W. Yu, *The Two-machine Flow Shop Problem with Delays and the One- machine Total Tardiness Problem*. PhD thesis, Eindhoven University of Technology, 1996
17. W. Yu, H. Hoogeveen, J.K. Lenstra, Minimizing makespan in a two-machine flow shop with delays and unit-time operations is NP-hard. J. Scheduling **7**, 333–348 (2004)
18. X. Zhang, *Scheduling with Time Lags*. PhD thesis, Erasmus Universiteit Rotterdam, 2010
19. X. Zhang, S. van de Velde, On-line two-machine open shop scheduling with time lags. Eur. J. Oper. Res. **204**, 14–15 (2010)

# Index





W. Kubiak, *A Book of Open Shop Scheduling*, International Series in Operations Research & Management Science 325,
https://doi.org/10.1007/978-3-030-91025-9

The manufacturer's authorised representative in the EU is Springer Nature Customer Service Centre GmbH, Europaplatz 3, 69115 Heidelberg, Germany. If you have any concerns regarding our products, please contact ProductSafety@springernature.com

Printed and bound by CPI Group (UK) Ltd, Croydon, CR0 4YY
30/09/2026
02225293-0008